ARCHITECTURES OF EARTH SYSTEM

International institutions are prevalent in world politics. More than 1,000 multilateral treaties are in place just to protect the environment alone. And yet, it is also clear that these institutions do not operate in a void but are enmeshed in larger, highly complex webs of governance arrangements. This compelling book conceptualizes these broader structures as the 'architectures' of global governance. Here, over 40 international relations scholars offer an authoritative synthesis of a decade of research on global governance architectures with an empirical focus on protecting the environment and vital earth systems. They investigate the structural intricacies of earth system governance and explain how global architectures enable or hinder individual institutions and their overall effectiveness. The book offers much-needed conceptual clarity about key building blocks and structures of complex governance architectures, charts detailed directions for new research and provides analytical groundwork for policy reform.

FRANK BIERMANN is Professor of Global Sustainability Governance at the Copernicus Institute of Sustainable Development, Utrecht University in the Netherlands. He is the founding chair of the Earth System Governance Project and Editor-in-Chief of *Earth System Governance*.

RAKHYUN E. KIM is Assistant Professor of Global Environmental Governance at the Copernicus Institute of Sustainable Development, Utrecht University in the Netherlands. He is a senior research fellow with the Earth System Governance Project, and co-chair of its Taskforce on Earth System Law.

The **Earth System Governance Project** was established in 2009 as a core project of the International Human Dimensions Programme on Global Environmental Change. Since then, the Project has evolved into the largest social science research network in the area of sustainability and governance. The Earth System Governance Project explores political solutions and novel, more effective governance mechanisms to cope with the current transitions in the socio-ecological systems of our planet. The normative context of this research is sustainable development; earth system governance is not only a question of institutional effectiveness, but also of political legitimacy and social justice.

The **Earth System Governance series** with Cambridge University Press publishes the main research findings and synthesis volumes from the Project's first ten years of operation.

Series Editor

Frank Biermann, Utrecht University, the Netherlands

Titles in print in this series

Biermann and Lövbrand (eds.), *Anthropocene Encounters: New Directions in Green Political Thinking*

van der Heijden, Bulkeley and Certomà (eds.), *Urban Climate Politics: Agency and Empowerment*

Linnér and Wibeck, *Sustainability Transformations: Agents and Drivers across Societies*

Betsill, Benney and Gerlak (eds.), *Agency in Earth System Governance*

Biermann and Kim (eds.), *Architectures of Earth System Governance: Institutional Complexity and Structural Transformation*

ARCHITECTURES OF EARTH SYSTEM GOVERNANCE

Institutional Complexity and Structural Transformation

Edited by

FRANK BIERMANN
Utrecht University

RAKHYUN E. KIM
Utrecht University

CAMBRIDGE
UNIVERSITY PRESS

University Printing House, Cambridge CB2 8BS, United Kingdom

One Liberty Plaza, 20th Floor, New York, NY 10006, USA

477 Williamstown Road, Port Melbourne, VIC 3207, Australia

314–321, 3rd Floor, Plot 3, Splendor Forum, Jasola District Centre,
New Delhi – 110025, India

79 Anson Road, #06-04/06, Singapore 079906

Cambridge University Press is part of the University of Cambridge.

It furthers the University's mission by disseminating knowledge in the pursuit of
education, learning, and research at the highest international levels of excellence.

www.cambridge.org
Information on this title: www.cambridge.org/9781108489515
DOI: 10.1017/9781108784641

© Cambridge University Press 2020

This publication is in copyright. Subject to statutory exception
and to the provisions of relevant collective licensing agreements,
no reproduction of any part may take place without the written
permission of Cambridge University Press.

First published 2020

Printed in the United Kingdom by TJ International Ltd, Padstow Cornwall

A catalogue record for this publication is available from the British Library.

Library of Congress Cataloging-in-Publication Data
Names: Biermann, Frank, 1967– editor. | Kim, Rakhyun, 1983– editor.
Title: Architectures of earth system governance : institutional complexity and structural transformation /
edited by Frank Biermann, Rakhyun Kim.
Description: Cambridge, United Kindom ; New York, NY : Cambridge University Press, 2020. | Series: Earth
System Governance series | Includes bibliographical references and index.
Identifiers: LCCN 2019034640 (print) | LCCN 2019034641 (ebook) | ISBN 9781108489515 (hardback) |
ISBN 9781108784641 (epub)
Subjects: LCSH: Environmental policy – International cooperation. | Environmental protection –
International cooperation. | Environmental law, International.
Classification: LCC GE170 .A74 2020 (print) | LCC GE170 (ebook) | DDC 363.7/0526–dc23
LC record available at https://lccn.loc.gov/2019034640
LC ebook record available at https://lccn.loc.gov/2019034641

ISBN 978-1-108-48951-5 Hardback
ISBN 978-1-108-74730-1 Paperback

Cambridge University Press has no responsibility for the persistence or accuracy of
URLs for external or third-party internet websites referred to in this publication
and does not guarantee that any content on such websites is, or will remain,
accurate or appropriate.

Contents

List of Contributors	*page* vii
Acknowledgements	xxi
1 Architectures of Earth System Governance: Setting the Stage FRANK BIERMANN AND RAKHYUN E. KIM	1
Part I The Building Blocks	35
2 Intergovernmental Institutions RONALD B. MITCHELL, ARILD UNDERDAL, STEINAR ANDRESEN AND CAREL DIEPERINK	37
3 International Bureaucracies DOMINIQUE DE WIT, ABBY LINDSAY OSTOVAR, STEFFEN BAUER AND SIKINA JINNAH	57
4 Transnational Institutions and Networks AGNI KALFAGIANNI, LENA PARTZSCH AND OSCAR WIDERBERG	75
5 Institutional Architectures for Areas beyond National Jurisdiction ORAN R. YOUNG	97
Part II Core Structural Features	117
6 Institutional Interlinkages THOMAS HICKMANN, HARRO VAN ASSELT, SEBASTIAN OBERTHÜR, LISA SANDERINK, OSCAR WIDERBERG AND FARIBORZ ZELLI	119
7 Regime Complexes LAURA GÓMEZ-MERA, JEAN-FRÉDÉRIC MORIN AND THIJS VAN DE GRAAF	137

8 Governance Fragmentation   158
   FRANK BIERMANN, MELANIE VAN DRIEL, MARJANNEKE J. VIJGE
   AND TOM PEEK

Part III   Policy Responses   181

9 Policy Integration   183
   HENS RUNHAAR, BETTINA WILK, PETER DRIESSEN, NIALL DUNPHY, ÅSA
   PERSSON, JAMES MEADOWCROFT AND GERARD MULLALLY

10 Interplay Management   207
    OLAV SCHRAM STOKKE

11 Orchestration   233
    KENNETH W. ABBOTT, STEVEN BERNSTEIN AND AMY JANZWOOD

12 Governance through Global Goals   254
    MARJANNEKE J. VIJGE, FRANK BIERMANN, RAKHYUN E. KIM, MAYA
    BOGERS, MELANIE VAN DRIEL, FRANCESCO S. MONTESANO, ABBIE YUNITA
    AND NORICHIKA KANIE

13 Hierarchization   275
    RAKHYUN E. KIM, HARRO VAN ASSELT, LOUIS J. KOTZÉ, MARJANNEKE
    J. VIJGE AND FRANK BIERMANN

Part IV   Future Directions   297

14 Taking Stock and Moving Forward   299
    FRANK BIERMANN, RAKHYUN E. KIM, KENNETH W. ABBOTT, JAMES
    HOLLWAY, RONALD B. MITCHELL AND MICHELLE SCOBIE

Glossary   322
Index   325

Contributors

Kenneth W. Abbott is Jack E. Brown Chair in Law Emeritus, Professor of Global Studies Emeritus and Senior Sustainability Scholar in the Global Institute of Sustainability, all at Arizona State University, United States. His research focuses on the interdisciplinary study of international institutions, international law and international relations. He studies a wide range of public and private institutions, in fields including environmental protection and sustainability, global health, corruption, emerging technologies and international trade. Before joining Arizona State University, Abbott taught for over twenty five years at Northwestern University, where he held the Elizabeth Froehling Horner Chair and served as director of the university-wide Center for International and Comparative Studies. Abbott is a Lead Faculty member of the Earth System Governance Project, and a member of the editorial boards of *International Theory, Regulation and Governance* and *Journal of International Economic Law*.

Steinar Andresen is a research professor at the Fridtjof Nansen Institute, Oslo, Norway. He has also been associated with the University of Oslo, Department of Political Science as well as the PluriCourts Centre of Excellence, also at the University of Oslo. He has been visiting scholar at the University of Washington in Seattle, Princeton University and The Brookings Institution in Washington, DC, as well as the International Institute for Applied Systems Analysis in Austria, and is a member of the Lead Faculty of the Earth System Governance Project. He has published extensively internationally, both books and journal articles, mostly on international environmental politics.

Steffen Bauer is a senior researcher in the Environmental Governance and Transition to Sustainability programme of the German Development Institute/ Deutsches Institut für Entwicklungspolitik in Bonn, Germany. Bauer heads the institute's project Research and Dialogue for a Climate-Smart and Just Transformation (KLIMALOG) and is a senior research fellow of the Earth System

Governance Project. His research generally addresses questions of international organization, global environmental governance and sustainable development with a special interest in the role of the United Nations and a regional focus on Africa. Bauer's current work is concerned with geopolitical and development implications of international climate policy, the 2030 Agenda for Sustainable Development and pertinent issues of international equity and justice.

Steven Bernstein is a professor of International Relations in the Department of Political Science and co-director of the Environmental Governance Lab at the Munk School of Global Affairs, University of Toronto, Canada. His academic publications span the areas of global governance, global environmental and sustainability politics, international political economy and international institutions. He also co-edits the journal *Global Environmental Politics* and is a Lead Faculty member of the Earth System Governance Project. Current major research projects include 'Coherence and Incoherence in Global Sustainable Development Governance' and 'Transformative Policy Pathways towards Decarbonization'.

Maya Bogers is a PhD candidate within the European Research Council-funded GlobalGoals project, at the Copernicus Institute of Sustainable Development, Utrecht University, the Netherlands. She has an interdisciplinary background with a BSc in natural sciences and an MSc in Innovation Sciences. Taking a keen interest in data science, she specialized in quantitative models and network analysis to measure the effects of innovation policies at the regional, national and international levels. Prior to returning to her *alma mater* to pursue a PhD, she worked for the Dutch Ministry of Foreign Affairs on several international development projects in Africa, primarily focusing on youth development and innovation. Within the GlobalGoals project, she focuses on quantitative models to assess the steering effects of the Sustainable Development Goals on global governance networks.

Frank Biermann is a professor of Global Sustainability Governance with the Copernicus Institute of Sustainable Development, Utrecht University, the Netherlands. He is the director of the GlobalGoals research programme on the steering effects of the Sustainable Development Goals, supported by a European Research Council 'Advanced Grant'; the founder of the Earth System Governance Project, a global transdisciplinary research network; and the editor-in-chief of *Earth System Governance*, the new peer-reviewed journal. Biermann's current research examines multilateral institutions, options for reform of the United Nations, global adaptation governance, Sustainable Development Goals, the political role of science, global justice, non-state climate actions and conceptual innovations such as the notion of the Anthropocene. His recent books are *Earth System Governance: World Politics in the Anthropocene* (2014); *Governing through Goals: Sustainable*

Development Goals as Governance Innovation (co-edited, 2017); and *Anthropocene Encounters: New Directions in Green Political Thinking* (co-edited, 2019).

Dominique De Wit is a PhD candidate at the University of California, Santa Cruz in the United States. She holds a BA in Political Science and an MA in Politics, studying international relations and environmental governance. Her dissertation research centres on the politics of climate change and energy transitions. Drawing from previous research on individual practices and energy behaviour to optimize the integration of decentralized energy technologies, her dissertation explores localities' desire for greater energy control, accountability and a more sustainable world. De Wit engages with concepts of ethics and imaginations of better futures to trace the emergence, development and aggregation of energy transitions and climate change initiatives.

Carel Dieperink is an assistant professor at the Copernicus Institute of Sustainable Development, Utrecht University, the Netherlands. Dieperink has a background in public administration, with an MSc degree from Twente University, and international relations, with a doctoral degree from Utrecht University. He coordinates the Water Governance programme of the Environmental Governance group of the Copernicus Institute. His research and teaching address multi-level water governance and the interplay between international and national institutions dealing with water issues. Recent projects studied the implementation of the European Water Framework and Flood Directives, integrated coastal zone management, international river basin management and knowledge co-production (assessing Dutch and EU water and climate projects). Current projects focus on groundwater overextraction and soil subsidence in the Vietnamese Mekong Delta, the use of applications of information and communication technologies in urban water governance and capacities and policy feedback mechanisms in water governance.

Peter Driessen is a professor of Environmental Governance at the Copernicus Institute of Sustainable Development, Faculty of Geosciences, Utrecht University, and vice-dean of the Faculty of Geosciences. His research contributes to the scholarly debate on sustainability governance by analyzing interventions that have the potential to make governance outcomes more congruent with sustainability goals. Driessen's research covers topics such as water governance, climate adaptation, sustainable urban development, environmental impact assessment, and science–policy interactions. Previously, he was scientific director of the Netherlands national research programme Knowledge for Climate (2007–2015) and principal investigator of a European Union-funded research project on flood risk governance in urban agglomerations across Europe (STARFLOOD Project). Driessen is a member of the Lead Faculty of the Earth System Governance Project.

Niall Dunphy is the director of the Cleaner Production Promotion Unit within the School of Engineering, University College Cork, Ireland. At the University College, he is a principal investigator of the Environmental Research Institute and a founding member of its Carbon Lab initiative. Dunphy is a transdisciplinary researcher and practitioner, leading a diverse team conducting engaged research on the human and societal dimensions of sustainability, with an emphasis on energy and the built environment. Examples of recent projects include 'Human Factor in the Energy System' (ENTRUST, coordinator), 'Co-production of Pathways for a Low-carbon, Climate Resilient Ireland' (Imagining2050, co-investigator) and 'Politico-institutional Framing of Collective Engagements with the Energy System' (EnergyPOLITIES, principal investigator). Dunphy's current research is focused on the theme of society, sustainability and energy, encompassing practices and behaviours, social acceptance and acceptability, participatory co-design and novel business models.

Laura Gómez-Mera is an associate professor of Political Science at the University of Miami, United States. She holds a doctorate from Oxford University and an MSc from the London School of Economics and Political Science, United Kingdom. She has held residential fellowships at the Woodrow Wilson International Center for Scholars in Washington, DC, and at the Kellogg Institute for International Studies at the University of Notre Dame. She has published in *International Studies Quarterly*, *European Journal of International Relations*, *Global Governance*, *Review of International Political Economy*, *Studies in Comparative International Development* and *Latin American Politics and Society*, among others.

Thomas Hickmann is a post-doctoral researcher with the Copernicus Institute of Sustainable Development, Utrecht University, the Netherlands. Until mid-2019, he was a researcher and lecturer at the Faculty of Economics and Social Sciences of the University of Potsdam, Germany, where he co-directed the research project *Carbon Governance Arrangements and the Nation-State: The Reconfiguration of Public Authority in Developing Countries*. His research mainly deals with theories of international relations, global sustainability governance, the role and function of transnational actors in world politics, the effect of new modes of governance on public-administrative systems in the Global South and the interplay of international bureaucracies with subnational and non-state actors. His most recent book is *The Anthropocene Debate and Political Science* (co-edited with Lena Partzsch, Philipp Pattberg and Sabine Weiland, 2019).

James Hollway is an assistant professor of International Relations and Political Science at the Graduate Institute of International and Development Studies, Geneva, Switzerland, where he co-directs the Professional Skills Training programme and is affiliated with the Centre for International Environmental Studies, the Centre for Trade and Economic Integration and the Global Governance Centre. He also leads the fisheries cluster in the Taskforce on Oceans Governance of the Earth System Governance Project. His research focuses on international institutions, especially those governing fisheries or trade, and he develops network theories and methods for studying their interdependencies. He has published in *Global Environmental Politics*, *Journal of Conflict Resolution*, *Journal of International Economic Law*, *Network Science*, *Social Networks* and *Sociological Methodology*, and contributed to the Sixth Global Environment Outlook report (Cambridge University Press 2019).

Amy Janzwood is a PhD candidate studying political science and environmental studies at the University of Toronto, Canada, and a research fellow at the Earth System Governance Project. Her research broadly investigates how civil society actors interact with state and non-state actors on issues like energy, climate change and trade. Her dissertation explores the governance and outcomes of contested energy infrastructure projects in and through Canada. She is also a research associate at the Environmental Governance Lab, where she analyzes how strategies like carbon disclosure and carbon pricing provide pathways to decarbonization.

Sikina Jinnah is an associate professor of Environmental Studies at the University of California at Santa Cruz, United States, and holds an affiliate faculty appointment with the Politics Department. Her first book, *Post-treaty Politics: Secretariat Influence in Global Environmental Governance* (2014), received the 2016 Harold and Margaret Sprout Award for best book in international environmental affairs from the International Studies Association. Her monograph (with Jean-Frédéric Morin), *Greening through Trade: How American Trade Policy is Linked to Environmental Protection Abroad (2020)*, examines the role of preferential trade agreements in securing environmental objectives. She is also the co-author (with Jean-Frédéric Morin and Amandine Orsini) of *Global Environmental Politics: Understanding the Governance of the Earth* (2020), and co-editor (with Simon Nicholson) of *New Earth Politics: Essays from the Anthropocene* (2016). Most recently, Jinnah has been working on governance of climate change engineering, supported by a 2017 Carnegie Fellowship.

Agni Kalfagianni is an associate professor of Transnational Sustainability Governance at the Copernicus Institute of Sustainable Development, Utrecht University, the Netherlands. She specializes in the effectiveness, legitimacy and ethical and justice considerations of private and transnational forms of governance in the sustainability domain. She is (co)editor-in-chief of the Global Environmental Governance book series by Routledge; associate editor of *Earth System*

Governance and member of the editorial board of *International Environmental Agreements* and *Agriculture and Human Values*. Her work has appeared in numerous international peer-reviewed journals including *Global Environmental Change*, *Journal of Business Ethics*, *Globalizations* and edited volumes with major university presses, such as MIT Press.

Norichika Kanie is a professor at the Graduate School of Media Governance, Keio University, Japan, and a senior research fellow at United Nations University Institute for the Advanced Study of Sustainability. He has been a founding member of the scientific steering committee of the Earth System Governance Project. Before joining Keio, he worked at Tokyo Institute of Technology and The University of Kitakyushu, Japan. He serves various committees and steering groups, including the Sustainable Development Goals Promotion Round-Table in the Sustainable Development Goals Promotion Headquarters, the Government of Japan and Promotion of Overcoming Population Decline and Vitalizing Local Economy in Japan, Cabinet Office, Government of Japan. He also led the Project on Sustainability Transformation beyond 2015, with support from the Environment Research and Technology Development Fund, Ministry of the Environment, Japan. He recently published *Governing through Goals: Sustainable Development Goals as Governance Innovation* (2017, co-edited with Frank Biermann). He holds a PhD in Media and Governance from Keio University, Japan.

Rakhyun E. Kim is an assistant professor of Global Environmental Governance at the Copernicus Institute of Sustainable Development, Utrecht University, the Netherlands. He is an interdisciplinary scholar in the field of international environmental law and governance with an interest in the application of complexity and network approaches to studying the architecture of global governance. He is lead author of the United Nations Environment Programme's sixth *Global Environment Outlook*, and he assumes leadership roles in Task Forces on Earth System Law and on Ocean Governance within the Earth System Governance Project. Since 2019, Kim is a member of the editorial board of *Review of European, Comparative & International Environmental Law*, and from 2014 to 2017 he served as book review editor for *Transnational Environmental Law*. He is the winner of the 2013 Oran R. Young Prize, awarded for the best conference paper of an early-career scholar at an Earth System Governance conference. Previously, he held positions at United Nations University Institute for the Advanced Study of Sustainability and the Australian National University College of Law.

Louis J. Kotzé is a research professor of Law at the Faculty of Law, North-West University, South Africa, and visiting professor of environmental law at the University of Lincoln, United Kingdom. His research focuses on the

Anthropocene, environmental constitutionalism, human rights and earth system law, with over 130 publications on these themes. Recent books include *Research Handbook on Human Rights and the Environment* (with Anna Grear, Edward Elgar 2015), *Global Environmental Constitutionalism in the Anthropocene* (Hart 2016), *Environmental Law and Governance for the Anthropocene* (Hart 2017) and *Sustainable Development Goals: Law, Theory and Implementation* (with Duncan French, Elgar 2018). He is co-editor of the *Journal of Human Rights and the Environment* and associate editor of *Earth System Governance*. In 2016, he obtained a second PhD at Tilburg University, the Netherlands, and he has been awarded a European Commission Horizon 2020 Marie Curie Fellowship to lead a research project during 2018–2019 at the University of Lincoln entitled *Global Ecological Custodianship: Innovative International Environmental Law for the Anthropocene*.

James Meadowcroft holds the Canada Research Chair in Governance for Sustainable Development at Carleton University, Canada. He is a professor in both the Department of Political Science and the School of Public Policy and Administration. His research is focused on the ways governments are adjusting their practices and policies to cope with problems of the environment and sustainable development. Meadowcroft has written widely on environmental politics and policy, democratic participation and deliberative democracy, sustainable development strategies, environmental limits and socio-technical transitions. Recent work focuses on energy and the transition to a low-carbon society, and includes publications on carbon capture and storage, smart grids, the development of Ontario's electricity system, green technology and negative carbon emissions. He has recently co-edited (with Daniel J. Fiorino) *Conceptual Innovation in Environmental Policy* (2017).

Ronald B. Mitchell is a professor of Political Science and Environmental Studies at the University of Oregon, United States, and a Lead Faculty member of the Earth System Governance Project. Mitchell is an expert on international environmental politics and law. He founded and directs the International Environmental Agreements Database Project, a catalogue of all environmental treaties. His research seeks to identify and explain patterns in the design and effectiveness of international environmental treaties. He has published four books and over 40 articles and chapters in edited volumes, including in *International Organization*, *Global Environmental Politics*, *Annual Review of Environment and Resources* and *International Environmental Agreements*. He has received three US National Science Foundation grants and was the recipient of the American Political Science Association's 2018 Elinor Ostrom Career Achievement Award. He co-directed the Dissertation Initiative for the Advancement of Climate Change

Research from 2003 to 2014. He teaches courses on international relations, international environmental politics and international organization.

Francesco Saverio Montesano is a PhD candidate within the GlobalGoals project at the Copernicus Institute of Sustainable Development, Utrecht University, the Netherlands. He holds an MSc in Global Politics from the London School of Economics, and throughout his academic and professional career he has specialized in international relations, global governance and foreign policy analysis. Prior to joining the Copernicus Institute of Sustainable Development, he worked as a researcher at the Clingendael Institute in The Hague and at the College of Europe in Bruges, where he focused on the foreign and security policy of key global actors, with a special emphasis on China and the European Union. Within the GlobalGoals project, he studies the steering effects of the Sustainable Development Goals at the international level.

Jean-Frédéric Morin is a professor at the Political Science Department of Laval University, Québec City, Canada, where he holds the Canada Research Chair in International Political Economy. Before accepting this research chair, he was a faculty member of Université libre de Bruxelles, Belgium and researcher at McGill University, Canada. His most recent research projects look at treaty design, institutional interactions and policy diffusion in the fields of trade, intellectual property, investment and environment. His recent publications have appeared in *International Studies Quarterly*, *European Journal of International Relations*, *Review of International Political Economy* and *Global Environmental Politics*.

Gerard Mullally is a lecturer in the Department of Sociology, University College Cork, Ireland. At the University College Cork, he is also a research associate with the Cleaner Production Promotion Unit, the Environmental Research Institute, the Institute for Social Sciences in the 21st Century and the Centre for Marine and Renewable Energy. Mullally's research focuses on transdisciplinary approaches to sustainability in research, teaching and learning, and societal engagement with an emphasis on localizing the Sustainable Development Goals. He is lead scientist on Imagining 2050: Engaging, Envisioning and Enabling Dialogue on Pathways towards a Low Carbon, Climate Resilient Ireland, funded by the Irish Environmental Protection Agency. He was a member of the Earth System Governance Project's Taskforce on Environmental Policy Integration coordinated by the Copernicus Institute of Sustainable Development, Utrecht University. His most recent book is *Transdisciplinary Perspectives on Transitions to Sustainability* (co-edited, 2017).

List of Contributors

Sebastian Oberthür is a professor of Environment and Sustainable Development and Director of Doctoral Studies at the Institute for European Studies at the Vrije Universiteit Brussel, Belgium. Trained as a political scientist with a strong background in international law, he focuses on issues of international and European environmental, climate and energy governance, with an emphasis on institutional issues and perspectives. From 2005 to 2015, Oberthür served as Academic Director at the Institute for European Studies. From 2006 to 2013, he was a member of the Compliance Committee of the Kyoto Protocol to the United Nations Framework Convention on Climate Change. Oberthür has dealt with issues of international and European environmental and climate policy since the early 1990s. He is a member of the Lead Faculty of the Earth System Governance Project.

Abby Lindsay Ostovar is a Water Policy Specialist for Montgomery & Associates, focusing on groundwater sustainability in California. She is also an adjunct faculty member in the School of International Services at American University and California State University Monterey Bay, United States. Ostovar's current research examines collaborative water governance in Peru, integrating the role of socio-spatial relations and knowledge and belief systems to better understand the linkages between participation, justice and adaptive capacity. She has published in *Global Environmental Politics*, *Environmental Science and Policy*, and *Journal of Environmental Studies and Sciences*. Ostovar is a research fellow of the Earth System Governance Project and a senior fellow in the Environmental Leadership Programme. Previously, she coordinated the Partnership on Technology Innovation and the Environment, and managed the United States State Department's trade-related environmental cooperation programmes, advancing environmental protection in over 15 countries. Ostovar holds a PhD from American University, MA degrees from the Fletcher School of Law and Diplomacy and Department of Urban and Environmental Policy and Planning at Tufts University, and a BSc from the University of Mary Washington.

Lena Partzsch is a professor of Environmental and Development Policy at the University of Freiburg, Germany. Her research interests lie in the fields of sustainability governance and international relations. She received the 2016 Best Paper Award of *GAIA: Ecological Perspectives for Science and Society*, and *Global Governance* chose her article, 'Take action now: The legitimacy of celebrity power in international relations', as the best paper of 2018 for translation into Spanish and publication in *Foro Internacional*. Partzsch's most recent book is *The Anthropocene Debate and Political Science* (co-edited with Sabine Weiland, Philipp Pattberg and Thomas Hickmann, 2018).

Tom Peek is a junior lecturer in Earth System Governance at Utrecht University, the Netherlands. Within the Earth System Governance Project, he has coordinated and contributed to the Earth System Governance Harvesting Initiative and served as assistant manager for the 2018 Utrecht Conference on Earth System Governance. Furthermore, Peek is part of the GlobalGoals research programme. His MSc thesis has studied the integration and coherence of the implementation of the Sustainable Development Goals in the United Nations system.

Åsa Persson is research director and deputy director of the Stockholm Environment Institute and adjunct lecturer at Linköping University, Department for Thematic Studies, Sweden, and a member of the scientific steering committee of the Earth System Governance Project. Persson leads research development and investment across all regional centres of the Stockholm Environment Institute and conducts research on global sustainable development governance. She has published widely on environmental policy integration, climate adaptation governance and planetary boundaries. Her research on the Sustainable Development Goals and adaptation finance has fed into policy processes. Her current research focuses on comparative politics of implementation of Nationally Determined Contributions under the climate convention and the Sustainable Development Goals, funded by the Swedish Research Council Formas. Persson is a member of several advisory boards and of the editorial board of the journal *Earth System Governance*.

Hens Runhaar is an associate professor of Governance of Nature and Biodiversity at the Copernicus Institute of Sustainable Development, Utrecht University, and professor of Governance of Biodiversity in Agricultural Landscapes at Wageningen University and Research, the Netherlands. His research focuses on steering towards more nature-inclusive agriculture by public and private actors and by reforming the incumbent agricultural regime, and on the uptake and upscaling of nature-based solutions in urban areas. Runhaar has published over 75 scientific papers and (co)edited four special journal issues, including the special issue 'Environmental Policy Integration: Taking Stock of Policy Practice in Different Contexts' in *Environmental Science and Policy*. From 2015 to 2018 he coordinated the international working group on Environmental Policy Integration within the Earth System Governance Taskforce on Conceptual Foundations.

Lisa Sanderink is a researcher and PhD candidate in the Department of Environmental Policy Analysis at the Institute for Environmental Studies, Vrije Universiteit Amsterdam, the Netherlands. Her research focuses on institutional complexity of global climate change and energy governance and scrutinizes coherence and effectiveness across renewable energy institutions. She is a member of the CLIMENGO project that aims to map the institutional complexity of global climate and energy governance, to evaluate its effectiveness and legitimacy and to develop a knowledge-base for decision-makers. She is also a research fellow with the Earth System Governance Project. Sanderink holds a BA in European Studies with a specialization in International and European Law from the University of Amsterdam, and an MSc in Environment and Resource Management from Vrije Universiteit Amsterdam.

Olav Schram Stokke is a professor of Political Science at the University of Oslo, Norway, the director of the university's cross-disciplinary bachelor programme on International Relations, and a research professor at the Fridtjof Nansen Institute, where he also served as research director for many years. Previous affiliations include the Centre for Advanced Study at the Norwegian Academy of Science and Letters and the International Institute of Applied Systems Analysis. His area of expertise is international relations with special emphasis on institutional analysis, resource and environmental management and regional cooperation in Polar regions. Among his recent books are *Disaggregating International Regimes: A New Approach to Evaluation and Comparison* (2012), *Managing Institutional Complexity: Regime Interplay and Global Environmental Change* (2011), and *International Cooperation and Arctic Governance* (2007, paperback 2010, Chinese version 2014).

Michelle Scobie is a lecturer and researcher at the Institute of International Relations at The University of the West Indies, St. Augustine, Trinidad and Tobago, and a Lead Faculty member of the Earth System Governance Project. She is also co-editor of the *Caribbean Journal of International Relations and Diplomacy*. She has practiced as an attorney-at-law in Trinidad and Tobago and Venezuela. She was the first corporate secretary of the Trinidad and Tobago Heritage and Stabilisation Fund, and is a member of the Caribbean Studies Association, the International Studies Association, the University of the West Indies Oceans Governance Network, the Earth System Governance Project, the Future Earth Knowledge Action Network and the International Studies Association Long Range Planning Committee. Her research areas include international law, international environmental law and developing states' perspectives on global and regional environmental governance. Her most recent book is *Global Environmental Governance and Small States: Architectures and Agency in the Caribbean*.

Arild Underdal is a professor emeritus of Political Science, more specifically International Relations, at the University of Oslo, Norway. He is also associated with the Centre for International Climate Research (CICERO) in Oslo. His main research activities have focused on international environmental politics, ranging from management of marine fisheries (the topic of his PhD dissertation) to energy institutions and policies. Publications include *Environmental Regime Effectiveness* (2002, ed. Miles et al.). He has served in various leadership positions, including those of vice-rector (1993–95) and rector of the University of Oslo (2002–2006). In addition, he has served as chair of several external boards (among others the Stockholm Resilience Centre) and committees (among others the Science Advisory Committee of the International Institute for Applied Systems Analysis, IIASA). He was recently elected Foreign Associate of the United States National Academy of Sciences.

Harro van Asselt is a professor of Climate Law and Policy with the University of Eastern Finland Law School. He has over 15 years of research experience and is an expert on interactions between international climate change governance and other fields of international governance. Before joining the University of Eastern Finland, van Asselt worked at the Stockholm Environment Institute, where he remains an associate. He also worked at the Environmental Change Institute at the University of Oxford, United Kingdom, and at the Institute for Environmental Studies at the Vrije Universiteit Amsterdam, the Netherlands. He is the author of *The Fragmentation of Global Climate Governance* (2014), co-editor of *Governing Climate Change* and *The Politics of Fossil Fuel Subsidies and Their Reform* (both Cambridge University Press 2018), and he has more than 80 publications in peer-reviewed academic journals and books. He is editor of *Review of European, Comparative & International Environmental Law*, associate editor of *Carbon and Climate Law Review* and sits on the editorial board of *Global Environmental Politics* and *International Environmental Agreements*. He holds a PhD with distinction from Vrije Universiteit Amsterdam.

Thijs Van de Graaf is an associate professor of International Politics at Ghent University, Belgium. He is also a non-resident fellow with the Payne Institute, Colorado School of Mines, and with the Initiative for Sustainable Energy Policy at Johns Hopkins University, United States. His research covers the intersection of energy security, climate policy and international relations. His recent books include *The Palgrave Handbook of the International Political Economy of Energy* (2016) and *The Politics and Institutions of Global Energy Governance* (2013). Van de Graaf leads the international Research Network on Fragmentation and Complexity in Global Governance and recently served as the lead writer of the report of the IRENA Global Commission on the Geopolitics of Energy Transformation. He is a member of the editorial board of *Energy Research and Social Science*, *International Environmental Agreements* and *Palgrave Communications*.

Melanie van Driel is a PhD candidate within the GlobalGoals project at the Copernicus Institute of Sustainable Development, Utrecht University, the

Netherlands. Before, she was a lecturer in public administration at the Faculty of Governance and Global Affairs of Leiden University, the Netherlands. Van Driel studied political science at the University of Amsterdam, specializing in both policy and governance and in international relations, with a focus on China, Central Eurasia, the Middle East and North Africa and the post-Soviet countries of Eastern Europe. Within the GlobalGoals project, she studies the steering effect of the Sustainable Development Goals at the international level. She is also book review editor with *Journal of Comparative Sociology*.

Marjanneke J. Vijge is an assistant professor of Sustainability Governance in the Developing World at the Copernicus Institute of Sustainable Development, Utrecht University, the Netherlands. Her current research focuses on policy coherence around climate and food security governance in developing countries, in particular regarding the implementation of the Sustainable Development Goals and climate goals. Vijge has a broad range of expertise in multilevel sustainability governance with a focus on developing countries. She has written over 20 publications on the governance of climate, forests, REDD+, extractive industries and environmental governance more generally, ranging from international negotiations to on-the-ground projects. She has done research in India, Kenya, Indonesia and Myanmar, and worked with different United Nations agencies in Europe, Africa and Asia.

Oscar Widerberg is an assistant professor at the Institute for Environmental Studies at the Vrije Universiteit Amsterdam, the Netherlands, working on governance for sustainability. His research focuses on how cities, regions, companies and other non-state and subnational actors influence global governance. He has published academic articles, policy reports, opinion pieces and book chapters on a variety of earth system governance issues, primarily on climate change. Recent publications include a chapter in the UN Environment Gap Report on non-state and subnational climate action. Before joining Vrije Universiteit Amsterdam, Widerberg worked in consulting, advising international public authorities, primarily European institutions, on energy, environment and climate policies.

Bettina Wilk is a project officer in green infrastructure and nature-based solutions for ICLEI – Local Governments for Sustainability, Freiburg, Germany. Prior to that she was a researcher in urban resilience, climate adaptation and policy integration with the Copernicus Institute of Sustainable Development, Utrecht University, the Netherlands. Wilk holds an MSc in Environmental Governance from Utrecht University and an MSc in Social and Cultural Anthropology from the University of Vienna, Austria.

Oran R. Young is a professor emeritus in the Bren School of Environmental Science and Management at the University of California, Santa Barbara, the United States, and a Lead Faculty member of the Earth System Governance Project. His research focuses on the role of social institutions as the principal elements of governance systems, with particular reference to addressing needs for governance associated with large-scale environmental problems like climate change. Young combines theoretical analysis dealing with the formation, evolution and effectiveness of governance systems with applied work relating to the polar regions, the atmosphere and the oceans. He has played several leadership roles in the global change research community and in the community concerned with the international relations of the Arctic. His most recent book is *Governing Complex Systems: Social Capital for the Anthropocene* (2017). Young is a frequent contributor to projects dealing with the rise of human-dominated systems and an active participant in efforts to strengthen the engagement between practitioners and analysts as a basis for responding effectively to needs for governance in the Anthropocene.

Abbie Yunita is a PhD candidate at the Copernicus Institute of Sustainable Development, Utrecht University, the Netherlands. She works within the European Research Council-funded GlobalGoals project, where she focuses on the steering effects of the Sustainable Development Goals at the country level. She is particularly interested in the ways in which the Sustainable Development Goals partake in and reshape the process of governance across diverse contexts, and how they create new grounds for innovation, collaboration and contestation. She has an interdisciplinary background in international studies and geography and holds an MSc in Development Studies from the Graduate Institute of International and Development Studies in Geneva. Prior to commencing her PhD, she worked with the Sustainable India Finance Facility, a partnership between UN Environment, World Agroforestry Centre and BNP Paribas, primarily in research and stakeholder engagement in Andhra Pradesh, India.

Fariborz Zelli is an associate professor at the Department of Political Science at Lund University, Sweden. Prior to joining Lund, he worked at the German Development Institute, Germany, and the Tyndall Centre for Climate Change Research in the United Kingdom. His major research interests include global climate governance, deforestation, environmental peacebuilding and institutional complexity. His books include *Governing the Climate-Energy Nexus* (co-ed., forthcoming with Cambridge University Press), *Environmental Politics and Governance in the Anthropocene* (co-ed., 2016), *Encyclopaedia of Global Environmental Governance and Politics* (co-ed., 2015), a special issue with *Global Environmental Politics* on institutional fragmentation (guest-ed., 2013) and *Global Climate Governance beyond 2012* (co-ed., Cambridge University Press 2010).

Acknowledgements

The 2009 Science and Implementation Plan of the Earth System Governance Project prioritized five analytical themes, one of which was research on governance 'architectures'. Debates on governance architectures have pervaded all project activities since then. Architecture was a core topic at all annual conferences of the Earth System Governance Project, from Amsterdam in 2007 to the later conferences in Colorado, Lund, Tokyo, Norwich, Canberra, Nairobi, Utrecht and Oaxaca, where special panel streams were created to discuss the architecture research programme. The analysis of architectures of earth system governance was central also in a variety of other project activities, and it has found entry in the new 2018–2028 Science and Implementation Plan of the Earth System Governance Project as well.

This volume harvests the key findings of the first ten years of research on the architectures of earth system governance. In putting this book together, however, we had to make one fundamental choice: while the Earth System Governance Project conceptualizes governance architectures as a structural phenomenon that is evident *at all levels of governance* – from local through to global governance – this book is limited to harvesting the insights regarding the *global level* of governance and politics. It is here where we believe that the architecture concept has been most fruitful and still holds most promise for future research.

The first outline of this synthesis book was discussed in December 2016 at the Nairobi Conference on Earth System Governance. Over the following two years, different chapters were presented at a variety of conferences, including annual conferences of the Earth System Governance Project, meetings of the International Studies Association, and a few other venues. All 14 chapters have been intensely discussed and reviewed, and while space constraints forced us to keep all chapters to a maximum length, we are confident that the book, as a whole, presents an insightful overview and analysis of where the debate on earth system governance architectures is located today.

We wish to thank, first, all the 41 authors for their contributions, and for their patience with the synthesis project that took, in the end, longer than anticipated. Many thanks also to the anonymous reviewers from Cambridge University Press who helped to improve the manuscript significantly with their many insightful comments and suggestions. We are also grateful for the support of the International Project Office of the Earth System Governance Project with collecting data on the wealth and breadth of research articles that have been published on governance architectures over the last decade. In addition, we wish to warmly thank our research assistant Julia Fritzsche for her relentless efforts in harmonizing the 14 chapters that we received. The copyeditor Harold Johnson deserves a special mention as well for his tireless proofreading of the entire manuscript. Last but not least, we are grateful to the staff of Cambridge University Press for shepherding this volume through the production process, and especially to Emma Kiddle for overseeing this all along from the first proposal stage through to review and production.

We see this book as a snapshot of where the architecture debate stands today: at a juncture that starts off with institutional complexity and persistent fragmentation but also allows us to look ahead at various types of policy interventions and options for structural reform. The state of our global environment is deteriorating with rapid speed – and we have no doubt that more effective and more equitable global architectures of earth system governance are an important part of the solution. We hope that this book has helped, from the perspective of social science analysis, to drive this debate further.

1

Architectures of Earth System Governance
Setting the Stage

FRANK BIERMANN AND RAKHYUN E. KIM

Since the emergence of the modern state system, governments have sought to regulate their affairs through international treaties and other types of intergovernmental agreement. For example, when it became known that emissions of certain chemicals destroyed the stratospheric ozone layer, governments agreed on a global treaty to ban such emissions. When *Titanic* sank in 1912, governments negotiated the International Convention for the Safety of Life at Sea. More than 1,300 international treaties have been concluded just to address environmental concerns, from the Agreement on the Conservation of Populations of European Bats to the United Nations Framework Convention on Climate Change. There is even an international Agreement Governing the Activities of States on the Moon and Other Celestial Bodies and an Agreement on the Rescue of Astronauts.

In political science, the last 40 years have seen a tremendous number of studies on the emergence, maintenance and effectiveness of such international institutions, including fully fledged international organizations, specialized bodies and programmes, as well as intergovernmental treaty secretariats. And yet, it has become increasingly evident that such international institutions do not operate in a void. All institutions operate instead within complex webs of larger governance settings. Many regulations and policies under the United Nations Framework Convention on Climate Change, for instance, also affect the protection of biological diversity, along with policies on energy, trade, patent rights, civil aviation, shipping and migration. Climate change is addressed not only in meetings of the parties to the climate convention but also by the International Maritime Organization, the United Nations Security Council, the United Nations Development Programme, the International Bank for Reconstruction and Development and many other institutions. In short, international institutions in all their forms are important building blocks of global governance. And yet, they are only one part of the story. They are all part of a larger whole that shapes, enables and at times hinders the functioning of single institutions. We thus must look at this larger picture. Without a better understanding of the entire system of global governance, we cannot grasp the functioning of its constituent parts.

In recent years, scholars began to refer to such complex institutional settings as governance architectures, using the powerful metaphor of buildings with copious rooms, lavish apartments, winding staircases and meandering corridors, all part of one interrelated system while keeping independent roles and spaces.[1] The concept of global governance architectures has filled a major conceptual void in the scholarship on international relations, which earlier, in the heydays of regime analysis from the 1970s onwards, rather focused on single institutions and their dyadic interactions. The concept of governance architecture now shifts the debate to situations in which a governance area is regulated by multiple institutions broadly understood as international organizations, regimes and norms in complex settings.

A key advantage of the architecture concept is its ability to allow for comparisons between and across issue areas, regions and time periods, with the possibility to study the variant effects of governance architectures. The policy relevance of such studies is clear: if a certain type of governance architecture is shown to be more effective, important policy conclusions can be drawn. The focus shifts then from the more limited examination of treaties and other agreements and their political impacts to the broader debate on the structures that shape the overall interplay, dynamics and effectiveness of global governance. We thus move, as it were, upwards from looking at single parts to assessing the larger whole. It is not about having better cake: it is about reinventing and restructuring the entire bakery.

Politically, such an analytical move allows for a new vision on the fundamental restructuring of world politics and global governance. While earlier research focused on the effectiveness of singular treaties or distinct international organizations, an 'architecture lens' invites a much deeper debate about the overall structure of global governance – and its overhaul. Research on the architectures of global governance, therefore, might assist in charting ways for a transformation of contemporary global policymaking.

In describing, comparing and assessing governance architectures in world politics, various classifications are possible. The analytical focus can lie on material structures, such as legal provisions, institutional designs, bureaucracies or financial resources. Characteristics of architectures might then include the degrees of institutional fragmentation, integration, polycentricity, multinodality, multilevelness, complexity, dynamics, density, modularity or hierarchy. Beyond these classical study areas, the analytical focus can also lie on ideational and discursive structures as part of broader governance architectures. Ideational structures, for instance, can

---

[1] See, for instance, Biermann et al. (2009a, 2009b); Kanie et al. (2010); Zelli (2011); Hackmann (2012); Zelli and van Asselt (2013); Biermann (2014); Abbott and Bernstein (2015); Holzscheiter, Bahr and Pantzerhielm (2016); Scobie (2019).

refer to 'an overarching structure of values' (Conca 2006: 26) that 'exert a powerful influence on social and political action' (Reus-Smit 2013: 224). Examples are overarching norms, values and belief systems that go beyond single issues, as they are articulated in general principles of international law.

Multiple causal relations can link such broader structural features and governance outcomes. Many analysts have tried to explain the problem-solving effectiveness of governance architectures, for example, whether certain types of governance architectures are better able to steer societies towards the prevention of global warming or the protection of biological diversity. Scholars have also begun to investigate broader questions of equity and of who gets what through the functioning of governance architectures. As structural constraints, governance architectures determine who has access to resources or justice and how burdens and responsibilities are distributed (Kanie et al. 2010). For instance, the principle of 'common but differentiated responsibilities and respective capabilities', a central part of numerous environmental agreements, has had a remarkable impact on the eventual distributional outcomes of these agreements.

Research on the emergence, functioning and impacts of governance architectures has been a central part of the science programme of the Earth System Governance Project, the leading global research network in the field of governance and sustainable development. This network, which has prioritized research on architectures as one of its five main analytical themes (Biermann et al. 2009a),[2] has mobilized a sizeable amount of research in this field, which in turn contributed numerous publications.

This book brings this and other research together. It is designed to harvest the work on governance architectures in the first ten years of operation of the Earth System Governance Project from 2009 to 2018 (including the work published in 2019); to highlight new debates and trends; and to chart new research directions. Importantly, the research covered in this book is not limited to studies by researchers formally affiliated with the Project. Instead, the book throws a wider net and seeks to draw together all the vibrant debates on governance architecture in all their complexity, diversity and dynamics.

Given the special research interest of our group of collaborators, our empirical focus is on architectures of *earth system governance*. Earth system governance is broader than traditional environmental policy and emphasizes the global complexity of integrated socio-ecological systems. Key concerns of earth system governance are, therefore, broad and often interdependent challenges such as ocean

---

[2] The other four main analytical themes are agency (e.g., Betsill, Benney and Gerlak 2019), adaptiveness (e.g., Boyd and Folke 2012), accountability (e.g., Biermann and Gupta 2011; Gupta and Mason 2014; Park and Kramarz 2019), and allocation and access (e.g., Gupta and Lebel 2010; Kanie et al. 2010). See Biermann (2014) for an integrated discussion of all five themes.

acidification, land use change, food system disruptions, climate change, environment-induced migration, species extinction, ocean governance and changing regional water cycles, as well as more traditional environmental concerns such as air pollution. While earth system governance is a broad area of scholarly inquiry, it also has its conceptual boundaries. Questions of international security, global communication, trade regulation, terrorism or human rights are less studied within the earth system governance research community, unless there are clear links to the functioning of socio-ecological systems, for example in the nexus of climate change impacts and local conflict.

This introductory chapter sets the scene for this book. We first define the concept of architectures of earth system governance and demarcate boundaries between the elements and processes of an architecture. We then provide a review of the past decade's research, addressing the five key questions on the analytical problem of architecture that were advanced in the 2009 Science and Implementation Plan of the Earth System Governance Project (Biermann et al. 2009a, 2010). In the concluding Chapter 14, we continue our review with a discussion of recent research trends and debates around complexity, dynamics and transformation, as well as key methodological approaches, challenges and advances in the field.

Conceptualization

Architecture

The term architecture has been used in many ways to refer to the fundamental structural characteristics of an institution (e.g., Aldy, Barrett and Stavins 2003), a regime (e.g., Hare et al. 2010) or a broader governance system (e.g., Young 2008). We define governance architectures in this book as the overarching system of public and private institutions, principles, norms, regulations, decision-making procedures and organizations that are valid or active in a given area of global governance (drawing on the definition by Biermann et al. 2010). Architecture can thus be described as the *macro-level* of governance, which can be regarded as a 'bird's-eye view' on the global governance landscape. This macro-level is not to be understood as a static entity, but rather as a fluid, dynamic architecture that continuously evolves according to external and internal pressures and governance processes.

Three elements are key in this conceptualization.

First, our conceptualization denotes an *overarching system*, which is broader than a single institution but narrower than an all-encompassing global order (Biermann 2014: 81–2). Therefore, architecture specifically refers to structures at the overarching governance level, which in turn consist of numerous building blocks such as state and non-state actors, transnational networks, intergovernmental institutions and regime complexes.

Second, our conceptualization of architecture refers to institutional settings that *shape decisions of actors and institutions* which exist and interact in a given policy domain. Some degree of influence of architectures on agency is therefore assumed in terms of both the effectiveness of individual institutions and the overall governance system (see also Betsill, Benney and Gerlak 2019).

Third, the impact of an architecture extends to *all levels of governance*. While a key unit of analysis is the macrostructure at the global level, the level of analysis is not limited to the global. Examples include multilevel governance architectures in areas such as climate mitigation, ocean governance, biodiversity conservation and so on.

Importantly, the notion of governance architectures does not assume the existence of a single architect. Instead, in many cases, governance architectures emerge from incremental and tedious processes of institutionalization that are decentralized and hardly planned. In other words, instead of a single architect, many architects are involved in shaping an architecture, even though individually, most have little control over the overall design and performance. The resulting structural configuration then influences how institutions interact by limiting the choices and opportunities for actors or by creating and shaping opportunities (Betsill, Benney and Gerlak 2019). A governance architecture is, in other words, in constant flux, evolving through the interaction between individual institutions at the micro level and the dynamic structure at the macro level.

The notion of architectures of global governance has emerged over the last two decades. Twenty years ago, most institutional research focused on single institutions, such as the ozone regime or individual treaties on wildlife or fisheries. Comparative analyses of such institutions then led to a better understanding of the creation, maintenance and effectiveness of single international institutions and their relations with national policies (Brown Weiss and Jacobson 1998; Young 1999). As more and more institutions were established for new, interconnected issues, the global governance landscape became increasingly crowded (Brown Weiss 1993; Young 1996). Institutions started interacting and forming interlinkages, which has become a central analytical challenge (Young 2002; J. Kim 2004; Oberthür and Gehring 2006; UNEP 2007, 2012, 2019; Chambers 2008; Pattberg 2010; Oberthür and Stokke 2011; Pattberg et al. 2018). Eventually, such interlinkages gave rise to larger complexes of interlocking institutions or actor configurations (Kanie, Andresen and Haas 2013). While institutions in these complexes are only loosely coupled, they nonetheless often formed dense clusters that are sparsely connected to other clusters. As such, not every institution is connected to all others uniformly, just as some governance issues relate more closely to one than to others.

Naturally, distinguishable clusters appear, and each of these inherently emergent, macro-scale modular structures is what we refer to as an architecture of global governance.

Research on governance architectures is marked by a remarkable co-evolution of analytical and normative debates, theory and practice, scholarly discourse and calls for political reform. Theoretical debates – for instance on the orchestration effects of intergovernmental organizations – go hand in hand with elaborate calls for institutional reforms, for example for the creation of a world environment organization or the amendment of the Charter of the United Nations. Debates in theory and practice often run in parallel; at times, they are interrelated, for instance when the urge of policy dilemmas and inconsistent governmental actions stimulates further advancement and research in academia.

The analytical research focuses here on studying the structures of global governance architectures and explaining their institutional variation and varying degrees of performance. One example are studies on the fragmentation of global governance architectures (see Chapter 8). These studies investigate the causes and consequences of governance fragmentation that results, for instance, from international agreements being negotiated by specialized government ministries in forums that are detached from negotiations of other agreements. Some scholars have argued that the current architecture of earth system governance is not conducive to the development of coordinated and synergistic approaches to collective problem-solving in the face of increasing global interdependencies (Young 2008 and 2019). Such research findings have had significant policy implications and impact, as evident in the emphasis on policy and institutional coherence in the United Nations agreement in 2015 on seventeen 'Sustainable Development Goals' (see Chapter 12; see also Nilsson, Griggs and Visbeck 2016).

The normative debate focuses on how architectures should be redesigned or reconstructed in a way that improves the performance of institutions in achieving their primary objectives. A key guiding consideration has been the 'problem of fit' of architectures and institutions to underlying problems (Young 2002; Folke et al. 2007; Galaz et al. 2008; Ekstrom and Young 2009; Ekstrom and Crona 2017). The logic behind this approach is that '[m]ore effective architectures of earth system governance may come about by better linking the study of nature with the study of governance' (Biermann et al. 2009a: 82). Politically, the challenge here is how to shape governance architectures in a way that they better fit the structure and dynamics of the regulatory target, including at the level of the earth system (Kim and Mackey 2014). This needs interdisciplinary research that draws, for instance, on the insights of earth system scientists who study the structure of the earth system (Schellnhuber et al. 2004; Steffen et al. 2004; Reid et al. 2010) and how different ecosystems and natural processes are interacting (Rockström et al. 2009; Steffen et al. 2015, 2018). Such research needs to inform the work of governance scholars who explore the institutional implications for the broader architectures of global governance.[3] However, the analytical and normative dimensions of research on architecture are not mutually exclusive. Instead, they cut across the interface between earth system analysis and governance theories (Biermann 2007).

Building Blocks, Structural Features and Policy Responses

Any global governance architecture consists of building blocks that share structural features and that are subject to a variety of policy interventions by governments and other international actors (see Figure 1.1). This book has been organized around these building blocks (Part I), structural features (Part II) and policy responses (Part III).

Despite the rise of non-state actors and transnational institutions, the central actors in global governance are still the governments of sovereign states. The literature on the state as actor in global governance is endless; hence we did not include a separate chapter on state actors but refer instead to the comprehensive review by Compagnon and colleagues (2012) in an earlier synthesis project. The first major building blocks that we are discussing in this book are *international institutions* that governments create to jointly tackle common problems. In world politics, such institutions with shared norms, principles and rules and decision-making procedures are often referred to as regimes (see Chapter 2). International institutions function through *international bureaucracies* that governments set up to administer and support such institutions. Such bureaucracies can be, for instance, treaty secretariats, such as the relatively large secretariat in support of the United Nations Framework Convention on Climate Change. Also the administrative apparatus of the World Bank or the International Maritime Organization fall into this broad category. While international bureaucracies formally function only to serve the interests of national governments, recent research has compellingly shown that international bureaucracies often enjoy large degrees of political freedom and play independent roles in global governance (Hickmann et al. 2019). They are hence important building blocks of global governance architectures with some degree of autonomy (see Chapter 3).

Furthermore, non-state actors have gained much influence in global governance, along with traditional state actors. In some areas of global governance, such as forest governance, non-state actors even seem to play central governing roles.

---

[3] See, for instance, Vidas (2011); Biermann et al. (2012); Kim and Bosselmann (2013); Scott (2013); Bridgewater, Kim and Bosselmann (2014); Biermann (2014); Dryzek (2014); Robinson (2014); Vordermayer (2014); Vidas et al. (2015); Biermann et al. (2016); Kim and van Asselt (2016); Kotzé (2016); Pattberg and Zelli (2016); Young et al. (2017); Craik et al. (2018); Pickering (2018).

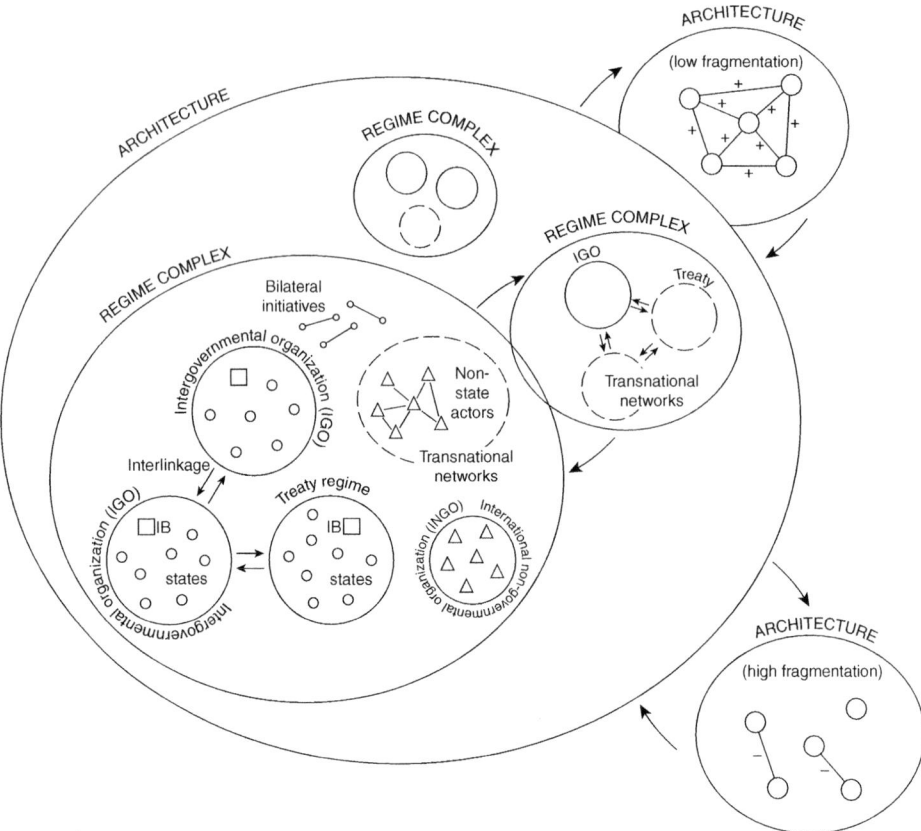

Figure 1.1: Architectures of global governance

Importantly, non-state actors often create their own *transnational institutions*, which we define as institutions set up by private actors with only a marginal role for governments or without any involvement of governments. These transnational institutions and networks are another major building block in any global governance architecture that will be discussed in this book in a separate chapter (see Chapter 4). Finally, vast areas of our planet are *beyond the jurisdiction and effective control* of governments, including the high seas and Antarctica, along with outer space. Other areas are diffuse, virtual and difficult to control by governments, such as the Internet. Thus, this book includes a chapter on institutions operating beyond the jurisdiction or control of national governments, as a final major building block of architectures of earth system governance (see Chapter 5).

These building blocks interact within broader global governance architectures in multiple ways. This has led in the last decade to a prolific research strand that we report on in Part II of this book. The first chapter in that part focuses on the *interlinkages* between two or only a few institutions (see Chapter 6). The following chapter reviews research on the emergence of larger sets of interrelated and interdependent institutions, often described as *regime complexes* (see Chapter 7). In studying an entire governance architecture in which institutions interact, the third chapter of Part II investigates variant degrees of *fragmentation* in global governance architectures (see Chapter 8).

Finally, these institutional interlinkages and the fragmentation of global governance architectures have forced governments, non-state actors, the decision-making bodies of international conventions and treaties as well as leaders of international bureaucracies to develop a variety of policy responses (see Part III of this book). These responses in turn have led to substantial research programmes, which again have often informed policy decisions and pushed forward reforms. Typical responses by state representatives, international bureaucrats and treaty bodies have been, for example, the push for the mutual integration of policies through the advancement of systematic *policy integration* or *mainstreaming* (see Chapter 9).

In situations of institutional interlinkages – especially when norm conflicts and weak governance performance were identified – governments and treaty bodies reacted by purposeful attempts at *interplay management* (see Chapter 10). This research is inevitably linked to the study of institutional interlinkages, which is discussed by Hickmann and colleagues earlier in this book (see Chapter 6). While research on institutional interlinkages analyzes the effects of such interactions – often with a focus on identifying and explaining synergistic or conflicting interactions – the research on interplay management investigates the effectiveness of policy interventions to minimize disruptive effects of conflictive interlinkages.

A new strand of research explores the role of international bureaucracies, but also other actors, as *orchestrators* in global governance. If a governance architecture can be likened to the ensemble of musicians in an orchestra, international bureaucracies – such as the United Nations Environment Programme – might then be the conductors, or the orchestrators, through non-hierarchical, non-compulsory soft ways of global institutional steering (see Chapter 11). A special case, much related to orchestration, is the policy of negotiating *global goals* as joint, widely accepted standards of conduct, with the expectation that public and private actors align their behaviour with such goals. The 17 Sustainable Development Goals, agreed by the United Nations General Assembly in 2015, are the most prominent example of such 'governing through goals' (see Chapter 12). In some cases, however, governments might choose to reform the broader institutional architectures beyond simple interplay management and orchestration by more structural change that establishes *hierarchical orders* among institutions and actors in global governance. We call this policy strategy *hierarchization*, acknowledging that such a strategy might evolve in numerous ways (see Chapter 13).

Figure 1.1 presents a schematic diagram of the architectures of global governance as we conceptualize them in this book. As mentioned above, the concept of governance architecture resembles the macro level structure and offers a bird's-eye view on the global governance landscape. While the overarching global governance architecture comprises the entirety of actors and institutions, individual governance architectures within a given domain may consist of one or more regime complexes, which are meso-level structures. When only one regime complex operates within a global governance domain, it would equate with an architecture. Key elements of a regime complex in turn include, at the micro level, intergovernmental institutions such as international treaties and organizations, as well as transnational networks and non-governmental organizations. These elementary building blocks along with emergent meso- and macrostructures form complicated interlinkages with one another and interact both horizontally and vertically. The increasing diversity and multiplicity of the building blocks and structural features has made global governance more and more complex over time. Such a conceptualization of global governance architectures underscores the fact that any attempt at structural transformation must be informed by a deep understanding of the intricate system of actors, norms, institutions and their networks.

Key Research Questions and Findings

Research on architectures of earth system governance has addressed questions relating to the emergence, design and effectiveness of governance systems as well as the overall integration of global, regional, national and local policymaking. It has essentially been about exploring and explaining causal relationships between architectural features and governance outcomes. We now present key findings to the five core research questions outlined a decade ago in the 2009 Science and Implementation Plan of the Earth System Governance Project (Biermann et al. 2009a).

Performance

The first question that the Earth System Governance Project sought to address over the last decade was, *how is the performance of individual environmental institutions – and of clusters of institutions – affected by the fact that institutions are embedded in larger architectures?*

The performance of architectures can be measured in multiple ways. A central focus in earth system governance research has been the problem-solving effectiveness of architectures (Breitmeier, Underdal and Young 2011; Young 2011a, 2018).

Other studies have also investigated other variables of governance performance, such as legitimacy, accountability and equity (Gupta and Lebel 2010; Biermann and Gupta 2011; Karlsson-Vinkhuyzen and McGee 2013). Research on performance in terms of problem-solving (output legitimacy) and on performance in terms of democratic and transparent procedures (input legitimacy) are deeply related, however (Scharpf 1999). As argued by Oran Young (2011a: 19855), the '[m]aintenance of feelings of fairness and legitimacy is important to effectiveness, especially in cases where success requires active participation on the part of the members of the group over time'.

The performance of a global governance architecture is generally expected to be shaped by certain institutional qualities. One defining characteristic of global governance architectures has been described as fragmentation (see Chapter 8), and many studies have sought to address key questions around the causes and consequences of, and responses to, the fragmentation of governance architectures (e.g., Zelli 2011; Van de Graaf 2013; Zelli and van Asselt 2013). Similar situations have been described from a different perspective as regime complexes (see Chapter 7), or as situations of institutional complexity or polycentricity (Jordan et al. 2018b). As for these characteristics of an architecture, the debate about the consequences of different degrees of fragmentation or the complexity of broader governance architectures has been most controversial. The key question has been whether high degrees of fragmentation or complexity of an architecture affect its performance as well as key institutions operating therein. However, the research community is far from conclusive in the current state of debate. Essentially, there are four broad clusters of assessments.

(1) First, in situations of fragmented architectures or convoluted regime complexes, there is agreement that institutions or actors are not uniformly affected by this broader architecture. For example, effects on the performance of an international institution depend on where the institution is located in an institutional architecture. One may ask, is the institution in the centre or periphery or does it assume a bottleneck position or not? Recent research has shown, for instance, that the texts and decisions of multilateral agreements with central positions attract far more citations in later agreements, hence spreading their rules or expanding legal effects (Kim 2013). Furthermore, the centrality of an international organization's position in a network is positively associated with the output of that organization (Murdie 2013). These institutional networks serve like a public good, and central positions are associated with a higher level of social capital (Young 2017). This also implies that an institution's performance depends not only on its position within an architecture, but also on the characteristics of that architecture. Social network analysis is one useful tool to study such complex arrangements (Kim 2019; see also Chapters 6 and 14).

(2) Second, a cluster of studies expects a strongly fragmented governance architecture to be less effective than a more integrated architecture (M. Young 2012; Biermann 2014; van Asselt 2014). Here, the argument is that disconnects between international institutions operating in silos result in limited cooperation or coordination among them. Most institutional interlinkages are believed to be disruptive or conflictive in character rather than synergistic or cooperative (Biermann et al. 2009b). More specifically, institutional fragmentation is seen as hindering progress in negotiating stringent targets and actions; this limits incentives of subnational actors to take urgently needed action and reduces the overall credibility, stability and coherence of the entire architecture of global governance (Biermann 2014: 83–97). Some researchers observe that fragmented governance architectures are intentionally created by states that benefit from them and maintained by power asymmetry, and in turn unevenly privilege more powerful countries (Benvenisti and Downs 2007; Alter and Meunier 2009). International 'forum-shopping' by astute national governments for pursuing their parochial interests is a case in point (Murphy and Kellow 2013).

The fragmentation of governance architectures is also seen as causing, or as being unable to prevent, negative spillovers and externalities (Ostrom 2010a; Truelove et al. 2014). For example, because global climate governance is fragmented, carbon leakages may occur if businesses transfer production to countries with lower emission restrictions (Carlarne 2008; Eichner and Pethig 2011). An emerging concept here is environmental problem-shifting, referring to instances where policies to solve one environmental problem transfer damage elsewhere or transform one type of pollution into another (Yang et al. 2012; van den Bergh et al. 2015). Problem-shifting is increasingly observed between international environmental regimes, presenting a risk to the overall performance of earth system governance (Kim and Bosselmann 2013; Kim and van Asselt 2016).

(3) Third, an opposing line of studies offers a more positive assessment of the performance of governance architectures that are highly fragmented or complex. For instance, scholars have claimed that fragmentation is ubiquitous and inherent to any governance architecture (Zelli 2011; Kim 2013; Pauwelyn 2014) and that in most systems everything is *not* connected (directly) to everything else, while most actors and institutions can be reached in a small number of steps. There is some evidence that the density of most systems is low, which can make a governance architecture appear as if it is excessively fragmented yet without significantly affecting its functioning.

Furthermore, contrary to the assumption that institutional proliferation leads to further fragmentation, some authors suggest the presence of defragmentation as a self-organized counteracting process that allows for keeping some cohesiveness in governance architectures despite the increasing number of institutions (Kim

2013). This process is also responsible for creating modular structures that are internally densely connected but externally sparsely connected through 'weak ties' (Granovetter 1973). A seemingly sparse and fragmented architecture could hence also be a well-functioning system (Stevens 2018). In this view, institutional fragmentation does not mean anarchy (Galaz et al. 2012) and there could be also 'healthy' fragmentation (Kanie 2007). For example, regime complexes consisting of loosely coupled, partially overlapping institutions are found in some studies as being more adaptable over time and flexible across issues when compared to other more integrated international regimes (Keohane and Victor 2011).

(4) Fourth, a more recent position emerged in the middle ground between these opposing poles of the debate. According to these scholars, the problem is 'not fragmentation per se, but rather the coordination (or lack of it) of fragmented or differentiated institutions' (Zürn and Faude 2013: 120). What matters then is the character or nature of the *institutional interplay* and *interplay management*, which is 'just as likely to produce positive or even synergistic results as it is to lead to interference between or among regimes' (Young 2011a: 19856; see also Oberthür and Pożarowska 2013). The key question is to what extent conflicts between regimes can be resolved through a better division of labour, or through negotiations that lead to mutual accommodation (see Chapter 10). Often, this line of research draws on an overtly optimistic view of governments that carry a general interest in complementarity within global governance architectures (Gehring and Faude 2014), especially when international cooperation in one regime impedes international cooperation in another (Johnson and Urpelainen 2012).

The debate on the implications of institutional fragmentation for global governance is far from resolved. Attempts are being made to test various theories, for example through a comparison across architectures or over time, and explain their performance in relation to the degree of fragmentation or other structural features (see Chapter 8). But we do not expect a broad agreement on how best to improve the performance of governance architectures soon.

Multilevel Governance

The second question emphasized by the Earth System Governance Project in 2009 was, *what is the relative performance of different types of multilevel governance architectures?* This question entails the comparative analysis of governance systems as multilevel structures where many interacting authorities are at work, including local, national, regional and global (Zürn 2012; also Peel, Golden and Keenan 2012). The analytical focus has been on vertical institutional interaction and variability in the performance of institutions within such multilevel governance systems (Biermann 2014: 83–97). Regime complexes of various international institutions could, for example, be conceptualized and analyzed as multilevel governance architectures to the extent that there is a vertical dimension between institutions (Abbott 2014; Hickmann 2017). The same holds of course for the interaction when regional integration organizations – like the European Union – provide a further level of governance between states and international institutions.

Systematic comparisons of multilevel governance architectures require typologies of such architectures. Often the vertical dimension is unpacked along two axes: *levels* of political jurisdiction, from global institutions to local administrations, and spatial *scale* (Van Doren et al. 2018) (with scale generally seen as the spatial, temporal, quantitative or analytical dimensions to measure and study any phenomenon, and levels as units of analysis located at different positions on a scale, see Gibson et al. 2000; Cash et al. 2006: 2). Scale and level determine the frame within which architectures are designed, contested and evaluated. A core question in the study of multilevel governance architectures is, therefore, how problems are framed in terms of scale and level and what implications the framing has for the performance of institutions.

For example, research has sought to identify the appropriate scale at which to address a problem such as climate change or ocean acidification (Kim 2012; Billé et al. 2013). As for climate change, for instance, it has been concluded that 'local thinking must be coupled with global and national scales of action to achieve the levels of carbon dioxide reductions needed to avoid dangerous climate impacts' (Sovacool and Brown 2009). Some analysts have employed the neologism of 'glocal' to refer to characteristics of both local and global considerations, for example in multilevel water governance (Gupta, Pahl-Wostl and Zondervan 2013). Furthermore, scholars focusing on polycentric governance argue that climate change should not be framed and approached exclusively as a global issue. They argue that actors and authorities at all levels should take initiatives for, among others, policy experimentation (Hoffmann 2011; see also Ostrom 2010b; Galaz et al. 2012; Betsill et al. 2015; Dorsch and Flachsland 2017). In fact, such a polycentric arrangement, where top-down and bottom-up processes are simultaneously at play, is becoming widespread. In global climate governance, for instance, we can observe both governance through legally binding, multilateral agreements (Hare et al. 2010) and governance through voluntary commitments and non-state action (Chan et al. 2015).

Successful policy measures can be translated to other levels or contexts through scaling (Hoffmann 2011). But under what circumstances can policies be scaled up and down? Context-specific barriers exist that hamper scaling of successful initiatives, and it is important that these are identified and addressed (Van Doren et al. 2016; Fuhr, Hickmann and Kern 2018). Furthermore, scaling processes change how policy measures and instruments operate and may also render them less effective. Likewise, it is important to understand whether there can be consistent architectures if policies cannot be easily scaled up and down administrative levels (Biermann et al. 2009a). Key conditions identified in the literature include the presence of boundary organizations, such as the Intergovernmental Science-Policy Platform on Biodiversity and Ecosystem Services, which act as brokers connecting institutions across scales or levels (Gupta, Pistorius and Vijge 2016; Holzscheiter 2017; Morin et al. 2017).

In this regard, the role of subnational or transnational *regions* has received new attention in multilevel governance (Balsiger and VanDeveer 2012). Regional agreements such as the Convention for the Protection of the Marine Environment of the North-East Atlantic make up two-thirds of all international environmental agreements. Some of these institutions form regional governance architectures that have been mapped and analyzed (Fidelman and Ekstrom 2012). While the significance of regional governance has often been overshadowed by global counterparts such as the United Nations Convention on the Law of the Sea, researchers have rediscovered the important role that regional agreements play in vertically and horizontally linking governance levels, especially as intermediary institutions bridging the national and the global (Balsiger and Prys 2016).

Fundamental Norms

The third architecture question that the Earth System Governance Project formulated in 2009 was, *what are overarching and crosscutting norms of earth system governance?*

Not all norms are equal in their status, scope or strength. A relatively small number of norms run through most, if not all, institutions and architectures of global governance. One prominent example is the widely accepted norm of the precautionary approach, which requires states to take measures to prevent serious or irreversible environmental harm despite scientific uncertainty. Such overarching or crosscutting norms have formed the normative basis on which many multilateral institutions are negotiated, agreed and implemented. Research in this field has sought: (1) to show such overarching or crosscutting norms that are common to many institutions in the field of earth system governance; (2) to analyze and understand key functions they perform; and (3) to explore ways in which we can bring about progressive development of these norms. These three research foci are discussed below in turn.

(1) First, regarding norm identification, two different approaches have been employed. For identifying *overarching* norms, researchers have assessed the legal status of candidate norms. Here, one would ask whether a norm has reached

a certain status in international law, for example, as customary international law or a general principle of international law. This approach is typically taken up by legal scholars who assume a hierarchy among norms, with some norms being superior or more widely applicable than others (Shelton 2006). For identifying *crosscutting* norms, researchers have looked for norms that appear as central across a range of global governance architectures. Here, the analytical concern is less on the actual legal status of a norm, but rather on the extent to which it appears as a constitutional principle or a basic norm. An example of such a crosscutting norm is the duty to maintain or restore the integrity of the earth's ecosystem (Kim and Bosselmann 2013; Bridgewater, Kim and Bosselmann 2014). Here the argument is that, although there is no explicit recognition among governments that global ecological integrity is a higher order norm than others, it is nonetheless an emergent 'common denominator' that all international soft law instruments such as the 1992 United Nations Declaration on Environment and Development have in their core (Kim and Bosselmann 2013).

It should be noted that this norm identification, sometimes referred to as codification, is done not only by academics but also by policymakers and other practitioners. Probably the most important organization with a codification mandate is the International Law Commission of the United Nations. In earth system governance, the International Union for Conservation of Nature has played an unofficial but significant role through the drafting and updating of an International Covenant on Environment and Development since 1995. Most recently, the United Nations General Assembly has initiated an intergovernmental process for developing a framework agreement for international environmental law, the Global Pact for the Environment. This pact is intended to entrench key fundamental principles that could serve to integrate the fragmented field of earth system governance (see Chapter 13; see also Kotzé and French 2018).

(2) Second, researchers have been interested in determining key functions that these overarching or crosscutting norms perform. One key function is the steering or adjudication of international institutions (Hickmann 2017). There is an emerging consensus in the literature that basic norms are key ingredients for maintaining some degree of polycentric order in global governance architectures (Jordan et al. 2018b). Here, overarching or crosscutting norms may 'provide a means to settle disputes and reduce the level of discord between units to a manageable level' (Jordan et al. 2018a: 19). Therefore, these overarching or crosscutting norms may also contribute to orchestration efforts of international organizations and coordinate the plethora of international institutions (see Chapter 11). In this light, some scholars argue that the absence of a clearly defined goal of international environmental law, for example, is conducive to conflicts and competition between international environmental regimes. In other words, the apparent ineffectiveness of

earth system governance could be attributed in significant part to the weak constitutional character of international environmental law (Kim and Bosselmann 2013).

Although the absence of an overarching norm is likely to be a key barrier for international institutions to function in a mutually beneficial manner, this theoretical proposition has yet to be empirically tested and supported. One area in which empirical evidence can be gathered are the Sustainable Development Goals (see Chapter 12). These goals are an outcome of political processes that codified overarching and crosscutting norms of earth system governance (see Chapter 13; see also Biermann, Kanie and Kim 2017). Will these global goals have any measurable effects on integrating and harmonizing policies that point in different directions into a more coherent whole? If so, what are the key mechanisms of governance through goals? Some see the Sustainable Development Goals as a well-integrated network of targets that will help coordinate competing policy objectives (Le Blanc 2015). Others are more cautious and argue that the goals merely reflect the fragmented set of existing commitments (Kim 2016; see also Kim and Bosselmann 2015; Underdal and Kim 2017; Young et al. 2017). They argue that it is imperative to agree on one overarching goal for the Sustainable Development Goals (Costanza et al. 2015).

(3) Third, earth system governance research has considered the more normative question of how the current set of overarching or crosscutting norms should be reconfigured or reinterpreted in light of rapid global change. For example, researchers have begun to engage cautiously in prescribing which norms should be recognized as most fundamental, especially in the face of challenges of the Anthropocene (Robinson 2014). Here, governance scholars often draw on scientific understandings of the earth system. The concept of planetary boundaries (Rockström et al. 2009; Steffen et al. 2015) has been particularly useful for legal and political theorizing regarding limitations of the sovereignty of nations (Biermann 2012; Kim and Bosselmann 2013; Ebbesson 2014; Kotzé and Kim 2019). For example, certain standards that are fundamental to protecting planetary boundaries could be accepted by the international community as jus cogens norms, from which no derogation is permitted. The restrictions on emissions of ozone-depleting substances, in accordance to the provisions of the Montreal Protocol, is a good example of what could be recognized as such a peremptory norm in the Anthropocene (Biermann 2014: 112–8).

There has been a considerable amount of political and academic effort to facilitate and formalize the recognition of, and respect for, planetary boundaries as a fundamental norm of earth system governance. Examples include the lobbying by scientists in the run-up to the 2012 United Nations Conference on Sustainable Development ('Rio+20') to insert planetary boundaries into the 2030 Agenda for

Sustainable Development as well as a proposal for a framework convention on planetary boundaries (Fernández and Malwé 2018). However, the recognition of planetary boundaries has been perceived as conflicting with some other fundamental norms including global social equity (Steffen and Stafford-Smith 2013). Consequently, the effort did not succeed, and the 2030 Agenda does not include any reference to planetary boundaries. Nonetheless, the Sustainable Development Goals reflect a novel normative development of global governance, and some scholars have put forward the argument that some of the existing but outdated norms and principles of international law should be reinterpreted in light of the new political consensus reached by the international community (Kim 2017).

Interaction between Policy Domains

The fourth question emphasized by the Earth System Governance Project was, *what are the environmental consequences of non-environmental governance systems?* Earth system governance is not just about protecting the environment but more broadly about achieving sustainable development. Therefore, institutional interactions between environmental institutions and those in other policy domains such as trade, energy, transport and agriculture become critical. Mapping of the interaction between the Sustainable Development Goals and a corresponding call for institutional 'championing of one or a subset of the goals is a recent example (Nilsson, Griggs and Visbeck 2016). Research has been organized around analytical and normative dimensions: first, to understand the nature of interaction between institutions in different policy domains (Gehring and Oberthür 2009) and second, to strike a balance between environmental and non-environmental policy objectives and to address trade-offs (Boyle 2007; Pauwelyn 2007).

One of the most researched cross-domain interactions is between trade and environment. Traditionally, trade liberalization was viewed as undermining global environmental governance (Conca 2000). These scholars condemn free trade agreements that favour intensive production and long-distance transportation for causing additional pollution. They also fear that trade agreements can limit the regulatory capacity of political leaders to enact environmental regulations that frustrate the interest of exporters and investors (Morin and Bialais 2018). In particular, researchers have found potential conflicts between climate measures and the law of the World Trade Organization and examined to what extent the principle of integration might address trade-offs (Voigt 2009; Kim 2016).

More recently, however, the nexus between trade and environment has changed through a new generation of preferential trade agreements from the 2000s, such as the 2009 United States–Peru Trade Promotion Agreement (Morin, Pauwelyn and

Hollway 2017). These agreements have created strategic linkages between trade and environment (Jinnah and Morgera 2013) through including environmental provisions (Morin, Dür and Lechner 2018). Contrary to traditional wisdom, research on the new trade–environment nexus suggests the possibility of trade agreements having positive implications for consolidating the effectiveness of environmental agreements (Morin and Bialais 2018). The dynamism of the trade regime, in particular, has been identified as a leverage point for strengthening environmental governance (Morin and Jinnah 2018).

Other non-environmental governance systems studied for their impact on the environment include food and energy (Boas, Biermann and Kanie 2016). As for food, the research community has investigated, for example, how threats to food security and associated norms of the right to food influence negotiations on the architectures of earth system governance (Ziervogel and Ericksen 2010). The governance of the nitrogen and phosphorus cycles has also emerged as an important topic for addressing the food and environment nexus (Ahlström and Cornell 2018; Morseletto 2019). Furthermore, policies around controversial energy sources such as biofuel and hydropower have come under scrutiny for their negative consequences for environmental concerns (Pittock 2010; Scott et al. 2014).

Although research has so far pointed to some positive trends, the general finding still is that more policy integration or coordination is needed across policy domains. Then, the question arises how to respond to negative consequences of non-environmental governance systems on the environment. One key political response is environmental policy integration (see Chapter 9), which refers to the incorporation of environmental goals into non-environmental policy sectors with the aim of targeting the underlying driving forces rather than merely symptoms of environmental degradation (Persson et al. 2018; see also Jordan and Lenschow 2010). More recently, the concept of mainstreaming has become popular in research as well, for example regarding the mainstreaming of climate and biodiversity policies (e.g., Nunan, Campell and Foster 2012; Runhaar, Driessen and Uitenbroek 2014; Karlsson-Vinkhuyzen et al. 2017; Runhaar et al. 2018).

Ultimately, any response to interactions between policy domains entails a normative decision. Sustainable development is widely seen as the overarching objective of the international society; how to reconcile the three pillars of sustainable development – society, economy and environment – remains far from clear. Some argue that environmental concerns must be prioritized because the earth and its natural resources are finite; others argue that socio-economic concerns are as important. The latest version of this long-standing debate is the tension between the 'ecological ceiling' and the 'social foundation' (Raworth 2017). A group of scholars sought to ease the stalemate by proposing the protection of the earth's life-support systems and poverty reduction as the twin priorities (Griggs et al. 2013). Yet, critics argue that the twin priorities simply reflect the dichotomy between (environmental) sustainability and development, which has always been the core of the problem with the concept of sustainable development (Kim and Bosselmann 2015).

Architectural Voids

Finally, in 2009 the Earth System Governance Project called for more research on *how to explain instances of non-governance*, or the occurrence of 'voids' in the governance architecture.

This line of research started from the observation that in earth system governance we often do not have specialized institutions in areas or on issues that are widely identified as being problematic. For example, deforestation proceeds with tremendous speed, yet no global treaty exists to tackle this issue. While governments agreed in 1992 on a global convention on halting climate change and on protecting biological diversity, no treaty on deforestation was negotiated, even though this issue was also on the agenda at that time. How can we explain such instances of 'non-governance' or 'architectural voids'? While the puzzle of the lacking agreement on global deforestation has attracted some attention by researchers and found compelling explanations (Dimitrov 2005), numerous other issues of non-governance are still insufficiently understood. Therefore, the study of governance architecture must also explore the gaps and empty spaces in or between them (Dimitrov, Sprinz and DiGiusto 2007). In comparative research designs, studies on non-governance will also help better explain the emergence and performance of those institutions that have been agreed. Moreover, research on why there are no global institutions in some areas will further our understanding of the consequences of governance architectures. For example, the absence of a global biofuel regime has created a risky North-South allocation pattern (Lima and Gupta 2013).

Some studies in this direction have improved understanding on why we do not have a world environment organization despite 40 years of political struggles (Vijge 2013), why there is no comprehensive law of the atmosphere (Sand 2017) and why we have no comprehensive treaty that covers the Arctic region (Young 2011b). However, there are still areas where our understanding is lacking. For instance, natural scientists have identified nine areas where they believe that our global system is facing planetary boundaries with fundamental value for human survival. While some of these boundaries are heavily institutionalized – for example by the climate convention of 1992 – others are hardly subject to global policy, notably the planetary boundaries on nitrogen and phosphorus (Ahlström and Cornell 2018; Morseletto 2019) and ocean acidification (Kim 2012). In addition, issues such as the global spread of plastics, especially into the marine environment, have been addressed in earth system governance only recently (Dauvergne 2018;

Nielsen, Holmberg and Stripple 2019). In short, the research programme on global non-governance is by far not concluded.

Two reasons, however, might explain why non-governance is less prominent and possibly also less problematic. First, institutions are becoming more adaptive. New and emerging problems do keep surfacing, potentially creating governance gaps. But in most cases, these gaps do not last for very long, with existing institutions quickly adapting to fill them. For example, ocean acidification has emerged as a new global issue. While there is no specialized treaty on this issue, the problem has nonetheless been taken up by the climate regime (Campbell et al. 2016). Similarly, the issue of deforestation has been taken up by the international climate convention since the introduction of Reducing Emissions from Deforestation and forest Degradation (REDD+) in 2005 (Gupta, Pistorius and Vijge 2016). Another example is the debate on geoengineering, that is, artificially cooling the planet by for instance spraying aerosols into the atmosphere to reflect parts of the sunlight or various other technologies. One such technology – the 'fertilization' of oceans to increase the uptake of carbon dioxide from the atmosphere – quickly came under scrutiny because of the unknown environmental risks and side effects. In 2008, parties to the Convention on Biological Diversity agreed on a moratorium on ocean fertilization, and the issue has remained a priority topic as well for the London Convention on dumping of wastes at sea (Pasztor, Scharf and Schmidt 2017), both halting global non-governance at least in the area of ocean fertilization.

Second, in areas of seeming non-governance by states, we often see informal and unconventional forms of steering. The Sustainable Development Goals, as a novel and broad governing mechanism, are one example that leaves little space unaffected by the goals (see Chapter 12; see also Kanie and Biermann 2017). Some also argue that, contrary to the popular claim that geoengineering is still a largely ungoverned space (Nicholson, Jinnah and Gillespie 2017; Jinnah 2018), certain types of geoengineering are governed through 'unacknowledged sources of steering' such as high-level authoritative assessments (Gupta and Möller 2018: 1). This point also reflects the emergence of new non-governmental actors, mechanisms and institutions, which go beyond traditional forms of state-led treaty-based regimes and that fill in areas marked by governmental non-action (Biermann and Pattberg 2008; Biermann 2010).

Methodological Approaches

The 2009 Science and Implementation Plan of the Earth System Governance Project lists several research methods as useful for the study of earth system governance. They include a wide range of approaches such as case studies, statistical techniques, discourse and content analysis, legal analysis, social network analysis, agent-based modelling, systems analysis and qualitative comparative analysis (Biermann et al. 2009a: 77–85). These methods have been employed by researchers to varying degrees. Yet despite this methodological pluralism in earth system governance research, methodological challenges persist. While some of these challenges are common across the field (O'Neill et al. 2013), others are unique to research on governance architectures. We now discuss key methodological approaches used in research on global governance architectures and point to some challenges for future research directions.

The most widely used methodological approaches are qualitative small-n case studies. Data are usually gathered through semi-structured expert interviews, surveys and document analysis. For example, in a study of the fragmentation of global climate governance, van Asselt (2014) conducted three case studies of regime interplay between the United Nations climate regime on the one hand, and on the other hand minilateral clean technology agreements, the Convention on Biological Diversity and the World Trade Organization on the other. Similarly, Jinnah (2014) conducted four case studies to better understand how treaty secretariats help to manage the dense interplay of issues, rules and norms between international regimes. Most studies on governance architectures still tend to draw on comparable methodological approaches.

Some studies employed medium-n analysis, including a growing number of studies on international regime complexes. The framework presented by Keohane and Victor (2011), for example, has been used widely. Many studies that map transnational regime complexes have also used the 'governance triangle' framework by Abbott (2012). These analyses include in their scope not only intergovernmental institutions but also those with constituent actors that range from state and civil society organizations to firms (Guerra 2018). Such mapping exercises allow for a systematic comparison; one can see the different compositions of governance architectures and compare across them.

There is a growing number of quantitative studies using large-n datasets. Here, researchers often describe, analyze and explain an entire global governance architecture by mapping actor or institutional configurations or connectivity patterns (for a review, see Kim 2019). For example, studies were conducted with 681 international non-governmental organizations working on human rights (Murdie 2013), 747 multilateral environmental agreements (Kim 2013) and 680 trade agreements (Morin et al. 2017). In a similar vein, Green (2010) has examined the question to what extent states decide to delegate authoritative functions to private actors, such as rule-making, adjudication, implementation, monitoring and enforcement activities. Other studies include those that looked at the entire set of intergovernmental organizations (Greenhill and Lupu 2017) as well as international fisheries agreements (Hollway and Koskinen 2016). Furthermore, another innovative large-n study used over

1.5 million tweets to assess the influence of the climate treaty secretariat (Kolleck et al. 2017). In many of these large-*n* analyses, the network approach has been particularly popular and useful.[4] Network analyses have uncovered the structural and dynamic patterns as well as power asymmetries in a range of global governance architectures (Lazer 2011).

Against this backdrop, we identify at least four key methodological and empirical challenges for the future study of global governance architectures.

First, given the complexity of global governance architectures, proving causality between constituent processes remains methodologically challenging. Some advances have been made through techniques such as exponential random graph models that allow for determining which generative processes have resulted in observed structural configurations (Chapter 14). Furthermore, statistical regression techniques have been useful to find correlations between network variables such as centrality and non-network variables such as outputs of individual institutions, and to find causality between the so-called network effects (Murdie 2013; also Hafner-Burton, Kahler and Montgomery 2009). More recently, qualitative comparative analysis has been used by earth system governance researchers to carve out combinations of necessary conditions for explaining the effectiveness of governance. Yet despite all progress, establishing causality remains a key challenge in research on global governance architectures (Young 2013).

Second, we lack methodological tools to capture and analyze an entire governance architecture. It is reasonable to expect that an architecture is complex, and therefore is not a simple sum of its parts. However, most of our methodological tools are designed for a reductionist approach, that is, trying to understand the structure and function of a whole by studying individual institutions and then inferring how the whole would function by aggregating them. We need a method that will link two distinct scales, the micro scale at which individual institutions work and the macro scale at which architectures evolve. The network approach introduced above does capture the whole, but it also has weaknesses. For example, it is limited because it only investigates structures, leaving processes in the network often insufficiently considered. A better combination of network analysis and agent-based modelling might be a promising future research path.

Third, research on how two architectures interact and co-evolve has only started to emerge. Existing research on the interaction between policy domains has focussed on how key institutions interact across architectural boundaries, for example regarding cooperation between the Convention on International Trade in Endangered Species of Wild Fauna and Flora and the World Trade Organization.

---

[4] See, for instance, Fiedelman and Ekstrom 2012; Kim 2013; Green 2013; Murdie 2013; Böhmelt and Spilker 2016; Hollway and Koskinen 2016; Widerberg 2016; Greenhill and Lupu 2017; Kolleck et al. 2017; Morin et al. 2017; Ahlström and Cornell 2018.

However, research that maps the interface between two entire architectures and explains their interaction is still at a nascent stage. A recent example is the study on how two populations interact from an organizational ecology perspective (Morin 2018). While promising advances are being made, methodological challenges increase as we zoom out to a higher level of abstraction while not losing detail in resolution of individual institutions.

Fourth, while more data are not always the answer to challenges of social science research (Watts 2017; see also Lazer et al. 2009), it is still a challenge to access quality longitudinal, relational data for advancing our understanding of the global institutional complexity. Datasets on international institutions have been developed and maintained such as the International Environmental Agreements Database (Mitchell 2003), the Transnational Climate Initiatives Database (Bulkeley et al. 2012), the Design of Trade Agreements Database (Dür, Baccini and Elsig 2014), and the Trade and Environment Database (Morin, Dür and Lechner 2018). Other relevant databases include those maintained by international organizations such as ECOLEX (by International Union for Conservation of Nature, United Nations Environment Programme, and Food and Agriculture Organization), InforMEA (United Nations Environment Programme) and the Yearbook of International Organizations (Union of International Associations). However, some of these datasets are not open access, and more importantly, relational data are not always collected or complete. ECOLEX, for example, has information on references to others that are law-related, and the Yearbook of International Organizations offers self-reported longitudinal data on inter-organizational relationships, but these data can be expensive for individual researchers (Murdie 2013; Wilson, Davis and Murdie 2016).

Organization of the Book

We structured this book in four parts.

Part I presents the *building blocks* of architectures of global governance, namely intergovernmental institutions (including treaties and regimes) (Chapter 2), international bureaucracies (secretariats and intergovernmental organizations) (Chapter 3) and non-state, transnational institutions and networks (Chapter 4), along with a chapter on the special challenges of governance in areas beyond national jurisdiction (Chapter 5).

Part II discusses the core structural features of global governance architectures at micro, meso and macro levels. At the micro level, we look at dyadic interlinkages between institutions (Chapter 6). We then move on to study the meso level of regime complexes of loosely coupled institutions (Chapter 7), and governance fragmentation at the macro level of whole architectures (Chapter 8).

Part III lays out research on policy responses to fragmentation and complexity in global governance architectures. Our author teams look into policy integration, which seeks to incorporate environmental concerns and objectives into non-environmental policy areas (Chapter 9); interplay management, which tries to limit conflicts caused by institutional interlinkages (Chapter 10); orchestration through intermediaries (Chapter 11); governance through goals, such as the Sustainable Development Goals (Chapter 12); and finally, the hierarchization of governance architectures (Chapter 13).

Part IV concludes our book and presents, in Chapter 14, new directions in policy and research.

References

Abbott, K. W. (2012). The transnational regime complex for climate change. *Environment and Planning C: Government and Policy*, 30, 571–90.
Abbott, K. W. (2014). Strengthening the transnational regime complex for climate change. *Transnational Environmental Law*, 3, 57–88.
Abbott, K. W., & Bernstein, S. (2015). The High-level Political Forum on Sustainable Development: Orchestration by default and design. *Global Policy*, 6, 222–33.
Ahlström, H., & Cornell, S. E. (2018). Governance, polycentricity and the global nitrogen and phosphorus cycles. *Environmental Science and Policy*, 79, 54–65.
Aldy, J. E., Barrett, S., & Stavins, R. N. (2003). Thirteen plus one: A comparison of global climate policy architectures. *Climate Policy*, 3, 373–97.
Alter, K. J., & Meunier, S. (2009). The politics of international regime complexity. *Perspectives on Politics*, 7, 13–24.
Balsiger, J., & Prys, M. (2016). Regional agreements in international environmental politics. *International Environmental Agreements*, 16, 239–60.
Balsiger, J., & VanDeveer, S. D. (2012). Navigating regional environmental governance. *Global Environmental Politics*, 12, 1–17.
Betsill, M. M., Benney, T. M., & Gerlak, A. K. (eds.) (2019). *Agency in earth system governance*. Cambridge, UK: Cambridge University Press.
Betsill, M. M., Dubash, N., Paterson, M., van Asselt, H., Vihma, A., & Winkler, H. (2015). Building productive links between the UNFCCC and the broader global climate governance landscape. *Global Environmental Politics*, 15 (2), 1–10.
Benvenisti, E., & Downs, G. W. (2007). The Empire's new clothes: Political economy and the fragmentation of international law. *Stanford Law Review*, 60, 595–631.
Biermann, F. (2007). 'Earth system governance' as a crosscutting theme of global change research. *Global Environmental Change*, 17, 326–37.
Biermann, F. (2010). Beyond the intergovernmental regime: Recent trends in global carbon governance. *Current Opinion in Environmental Sustainability*, 2, 284–8.
Biermann, F. (2012). Planetary boundaries and earth system governance: Exploring the links. *Ecological Economics*, 81, 4–9.
Biermann, F. (2014). *Earth system governance: World politics in the Anthropocene*. Cambridge, MA: The MIT Press.
Biermann, F., Abbott, K. W., Andresen, S. et al. (2012). Navigating the Anthropocene: Improving earth system governance. *Science*, 335, 1306–7.
Biermann, F., Bai, X., Bondre, N. et al. (2016). Down to earth: Contextualizing the Anthropocene. *Global Environmental Change*, 39, 341–50.

Biermann, F., Betsill, M. M., Gupta, J. et al. (2009a). *Earth system governance: People, places and the planet. Science and implementation plan of the Earth System Governance Project*. Bonn: Earth System Governance Project.

Biermann, F., Betsill, M. M., Gupta, J. et al. (2010). Earth system governance: A research framework. *International Environmental Agreements*, 10, 277–98.

Biermann, F., & Gupta, A. (2011). Accountability and legitimacy in earth system governance: A research framework. *Ecological Economics*, 70, 1856–64.

Biermann, F., Kanie, N., & Kim, R. E. (2017). Global governance by goal-setting: The novel approach of the UN Sustainable Development Goals. *Current Opinion in Environmental Sustainability*, 26–27, 26–31.

Biermann, F., & Pattberg, P. (2008). Global environmental governance: Taking stock, moving forward. *Annual Review of Environment and Resources*, 33, 277–94.

Biermann, F., Pattberg, P., van Asselt, H., & Zelli, F. (2009b). The fragmentation of global governance architectures: A framework for analysis. *Global Environmental Politics*, 9, 14–40.

Billé, R., Kelly, R., Biastoch, A. et al. (2013). Taking action against ocean acidification: A review of management and policy options. *Environmental Management*, 52, 761–79.

Boas, I., Biermann, F., & Kanie, N. (2016). Cross-sectoral strategies in global sustainability governance: Towards a nexus approach. *International Environmental Agreements*, 16, 449–64.

Böhmelt, T., & Spilker, G. (2016). The interaction of international institutions from a social network perspective. *International Environmental Agreements*, 16, 67–89.

Boyd, E., & Folke, C. (eds.) (2012). *Adapting institutions: Governance, complexity and social-ecological resilience*. Cambridge, UK: Cambridge University Press.

Boyle, A. (2007). Relationship between international environmental law and other branches of international law. In D. Bodansky, J. Brunnée, & E. Hey (eds.), *The Oxford handbook of international environmental law*. Oxford: Oxford University Press.

Breitmeier, H., Underdal, A., & Young, O. R. (2011). The effectiveness of international environmental regimes: Comparing and contrasting findings from quantitative research. *International Studies Review*, 13, 579–605.

Bridgewater, P., Kim, R. E., & Bosselmann, K. (2014). Ecological integrity: A relevant concept for international environmental law in the Anthropocene? *Yearbook of International Environmental Law*, 25, 61–78.

Brown Weiss, E. (1993). International environmental law: Contemporary issues and the emergence of a new world order. *Georgetown Law Journal*, 81, 675–710.

Brown Weiss, E., & Jacobson, H. K. (eds.) (1998). *Engaging countries: Strengthening compliance with international environmental accords*. Cambridge, MA: The MIT Press.

Bulkeley, H., Andonova, L., Bäckstrand, K. et al. (2012). Governing climate change transnationally: Assessing the evidence from a database of sixty initiatives. *Environment and Planning C: Government and Policy*, 30, 591–612.

Campbell, L. M., Gray, N. J., Fairbanks, L. et al. (2016). Global oceans governance: New and emerging issues. *Annual Review of Environment and Resources*, 41, 517–43.

Carlarne, C. (2008). Good climate governance: Only a fragmented system of international law away? *Law and Policy*, 30, 450–80.

Cash, D. W., Adger, W. N., Berkes, F. et al. (2006). Scale and cross-scale dynamics: Governance and information in a multilevel world. *Ecology and Society*, 11.

Chambers, W. B. (2008). *Interlinkages and the effectiveness of multilateral environmental agreements*. Tokyo: United Nations University Press.

Chan, S., van Asselt, H., Hale, T. et al. (2015). Reinvigorating international climate policy: A comprehensive framework for effective nonstate action. *Global Policy*, 6, 466–73.

Compagnon, D., Chan, S., & Mert, A. (2012). The changing role of the state. In F. Biermann, & P. Pattberg (eds.), *Global environmental governance reconsidered* (pp. 237–63). Cambridge, MA: The MIT Press.

Conca, K. (2000). The WTO and the undermining of global environmental governance. *Review of International Political Economy*, 7, 484–94.

Conca, K. (2006). *Governing water: Contentious transnational politics and global institution building*. Cambridge, MA: The MIT Press.

Costanza, R., McGlade, J., Lovins, H., & Kubiszewski, I. (2015). An overarching goal for the UN Sustainable Development Goals. *Solutions*, 5, 13–16.

Craik, N., Jefferies, C. S. G., Seck, S. L., & Stephens, T. (eds.) (2018). *Global environmental change and innovation in international law*. Cambridge, UK: Cambridge University Press.

Dauvergne, P. (2018). Why is the global governance of plastic failing the oceans? *Global Environmental Change*, 51, 22–31.

Dimitrov, R. S. (2005). Hostage to norms: States, institutions and global forest politics. *Global Environmental Politics*, 5 (4), 1–24.

Dimitrov, R. S., Sprinz, D. F., & DiGiusto, G. M. (2007). International nonregimes: A research agenda. *International Studies Review*, 9, 230–58.

Dorsch, M. J., & Flachsland, C. (2017). A polycentric approach to global climate governance. *Global Environmental Politics*, 17, 45–64.

Dryzek, J. S. (2014). Institutions for the Anthropocene: Governance in a changing earth system. *British Journal of Political Science*, 46, 937–56.

Dür, A., Baccini, L., & Elsig, M. (2014). The design of international trade agreements: Introducing a new dataset. *Review of International Organizations*, 9, 353–75.

Ebbesson, J. (2014). Planetary boundaries and the matching of international treaty regimes. *Scandinavian Studies in Law*, 59, 259–84.

Eichner, T., & Pethig, R. (2011). Carbon leakage, the green paradox, and perfect future markets. *International Economic Review*, 52, 767–805.

Ekstrom, J. A., & Crona, B. I. (2017). Institutional misfit and environmental change: A systems approach to address ocean acidification. *Science of the Total Environment*, 576, 599–608.

Ekstrom, J. A., & Young, O. R. (2009). Evaluating functional fit between a set of institutions and an ecosystem. *Ecology and Society*, 14.

Fernández, E. F., Malwé, C. (2018). The emergence of the 'planetary boundaries' concept in international environmental law: A proposal for a framework convention. *Review of European, Comparative & International Environmental Law*, 28, 48–56

Fidelman, P., & Ekstrom, J. A. (2012). Mapping seascapes of international environmental arrangements in the coral triangle. *Marine Policy*, 36, 993–1004.

Folke, C., Pritchard, Jr., L., Berkes, F., Colding, J., & Svedin, U. (2007). The problem of fit between ecosystems and institutions: Ten years later. *Ecology and Society*, 12.

Fuhr, H., Hickmann, T., & Kern, K. (2018). The role of cities in multi-level climate governance: Local climate policies and the 1.5 °C target. *Current Opinion in Environmental Sustainability*, 30, 1–6.

Galaz, V. (2011). Double complexity: Information technology and reconfigurations in adaptive governance. In E. Boyd, & C. Folke (eds.), *Adapting institutions: Governance, complexity and social-ecological resilience* (pp. 193–215). Cambridge, UK: Cambridge University Press.

Galaz, V., Crona, B., Österblom, H., Olsson, P., & Folke, C. (2012). Polycentric systems and interacting planetary boundaries: Emerging governance of climate change–ocean acidification–marine biodiversity. *Ecological Economics*, 81, 21–32.

Galaz, V., Olsson, P., Hahn, T., Folke, C., & Svedin, U. (2008). The problem of fit between governance systems and environmental regimes. In O. R. Young, L. A. King, & H. Schroeder (eds.), *Institutions and environmental change: Principal findings, applications and research frontiers* (pp. 147–86). Cambridge, MA: The MIT Press.

Gehring, T., & Faude, B. (2014). A theory of emerging order within institutional complexes: How competition among regulatory international institutions leads to institutional adaptation and division of labor. *Review of International Organizations*, 9, 471–98.

Gehring, T., & Oberthür, S. (2006). Comparative empirical analysis and ideal types of institutional interaction. In S. Oberthür, & T. Gehring (eds.), *Institutional interaction in global environmental governance: Synergy and conflict among international and EU policies* (pp. 307–71). Cambridge, MA: The MIT Press.

Gehring, T., & Oberthür, S. (2009). The causal mechanisms of interaction between international institutions. *European Journal of International Relations*, 15, 125–56.

Gibson, C., Ostrom, E., & Ahn, T.-K. (2000). The concept of scale and the human dimensions of global change: A survey. *Ecological Economics*, 32, 217–39.

Granovetter, M. S. (1973). The strength of weak ties. *American Journal of Sociology*, 78, 1–22.

Green, J. F. (2010). Private authority on the rise: A century of delegation in multilateral environmental agreements. In J. Tallberg, & C. Jönsson (eds.), *Transnational actors in global governance* (pp. 155–76). Basingstoke: Palgrave.

Green, J. F. (2013). Order out of chaos: Public and private rules for managing carbon. *Global Environmental Politics*, 13, 1–25.

Greenhill, B., & Lupu, Y. (2017). Clubs of clubs: Fragmentation in the network of intergovernmental organizations. *International Studies Quarterly*, 61, 181–95.

Griggs, D., Stafford-Smith, M., Gaffney, O. et al. (2013). Sustainable development goals for people and planet. *Nature*, 495, 305–7.

Guerra, F. (2018). Mapping offshore renewable energy governance. *Marine Policy*, 89, 21–33.

Gupta, A., & Möller, I. (2018). De facto governance: How authoritative assessments construct climate engineering as an object of governance. *Environmental Politics*, 28, 480–501.

Gupta, A., Pistorius, T., & Vijge, M. J. (2016). Managing fragmentation in global environmental governance: The REDD+ partnership as bridge organization. *International Environmental Agreements*, 16, 355–74.

Gupta, J., & Lebel, L. (2010). Access and allocation in earth system governance: Water and climate change compared. *International Environmental Agreements*, 10, 377–95.

Gupta, A., & Mason, M. (eds.) (2014). *Transparency in global environmental governance: Critical perspectives*. Cambridge, MA: The MIT Press.

Gupta, J., Pahl-Wostl, C., & Zondervan, R. (2013). 'Glocal' water governance: A multi-level challenge in the Anthropocene. *Current Opinion in Environmental Sustainability*, 5, 573–580.

Hackmann, B. (2012). Analysis of the governance architecture to regulate GHG emissions from international shipping. *International Environmental Agreements*, 12, 85–103.

Hafner-Burton, E. M., Kahler, M., & Montgomery, A. H. (2009). Network analysis for international relations. *International Organization*, 63, 559–92.

Hare, W., Stockwell, C., Flachsland, C., & Oberthür, S. (2010). The architecture of the global climate regime: A top-down perspective. *Climate Policy*, 10, 600–14.

Hickmann, T. (2017). The reconfiguration of authority in global climate governance. *International Studies Review*, 19, 430–51.

Hickmann, T., Widerberg, O., Lederer, M., & Pattberg, P. (2019). The United Nations Framework Convention on Climate Change Secretariat as an orchestrator in global climate policymaking. *International Review of Administrative Sciences*, in press.

Hoffmann, M. J. (2011). *Climate governance at the crossroads: Experimenting with a global response after Kyoto*. Oxford: Oxford University Press.

Hollway, J., & Koskinen, J. (2016). Multilevel embeddedness: The case of the global fisheries governance complex. *Social Networks*, 44, 281–94.

Holzscheiter, A. (2017). Coping with institutional fragmentation? Competition and convergence between boundary organizations in the global response to polio. *Review of Policy Research*, 34, 767–89.

Holzscheiter, A., Bahr, T., & Pantzerhielm, L. (2016). Emerging governance architectures in global health: Do metagovernance norms explain inter-organisational convergence? *Politics and Governance*, 4, 5–19.

Jinnah, S. (2014). *Post-treaty politics: Secretariat influence in global environmental governance*. Cambridge, MA: The MIT Press.

Jinnah, S. (2018). Why govern climate engineering? A preliminary framework for demand-based governance. *International Studies Review*, 20, 272–82.

Jinnah, S., & Morgera, E. (2013). Environmental provisions in American and EU free trade agreements: A preliminary comparison and research agenda. *Review of European, Comparative & International Environmental Law*, 22, 324–39.

Johnson, T., & Urpelainen, J. (2012). A strategic theory of regime integration and separation. *International Organization*, 66, 645–77.

Jordan, A., Huitema, D., Schoenefeld, J., van Asselt, H., & Forster, J. (2018a). Governing climate change polycentrically: Setting the scene. In A. Jordan, D. Huitema, H. van Asselt, & J. Forster (eds.), *Governing climate change: Polycentricity in action?* (pp. 3–26). Cambridge, UK: Cambridge University Press.

Jordan, A., Huitema, D., van Asselt, H., & Forster, J. (eds.) (2018b). *Governing climate change: Polycentricity in action?* Cambridge, UK: Cambridge University Press.

Jordan, A., & Lenschow, A. (2010). Environmental policy integration: A state of the art review. *Environmental Policy and Governance*, 20, 147–58.

Kanie, N. (2007). Governance with multilateral environmental agreements: A healthy or ill-equipped fragmentation? In L. Swart, & E. Perry (eds.), *Global environmental governance: Perspectives on the current debate* (pp. 67–86). New York: Center for UN Reform Education.

Kanie, N., Andresen, S., & Haas, P. M. (eds.) (2013). *Improving global environmental governance: Best practices for architecture and agency*. Cambridge, MA: The MIT Press.

Kanie, N., & Biermann, F. (eds.) (2017). *Governing through goals: Sustainable Development Goals as governance innovation*. Cambridge, MA: The MIT Press.

Kanie, N., Nishimoto, H., Hijioka, Y., & Kameyama, Y. (2010). Allocation and architecture in climate governance beyond Kyoto: lessons from interdisciplinary research on target setting. *International Environmental Agreements*, 10, 299–315.

Karlsson-Vinkhuyzen, S. I., & McGee, J. (2013). Legitimacy in an era of fragmentation: The case of global climate governance. *Global Environmental Politics*, 13, 56–78.

Karlsson-Vinkhuyzen, S., Kok, M. T. J., Visseren-Hamakers, I. J., & Termeer, C. J. A. M. (2017). Mainstreaming biodiversity in economic sectors: An analytical framework. *Biological Conservation*, 210, 145–56.

Keohane, R. O., & Victor, D. G. (2011). The regime complex for climate change. *Perspectives on Politics*, 9, 7–23.

Kim, J. A. (2004). Regime interplay: The case of biodiversity and climate change. *Global Environmental Change*, 14, 315–24.

Kim, R. E. (2012). Is a new multilateral environmental agreement on ocean acidification necessary? *Review of European, Comparative and International Environmental Law*, 21, 243–58.

Kim, R. E. (2013). The emergent network structure of the ultilateral environmental agreement system. *Global Environmental Change*, 23, 980–91.

Kim, R. E. (2016). The nexus between international law and the Sustainable Development Goals. *Review of European, Comparative & International Environmental Law*, 25, 15–26.

Kim, R. E. (2017). Should deep seabed mining be allowed? *Marine Policy*, 82, 134–7.

Kim, R. E. (2019). Is global governance fragmented, polycentric, or complex: The state of the art of the network approach. *International Studies Review*.

Kim, R. E., & Bosselmann, K. (2013). International environmental law in the Anthropocene: Towards a purposive system of multilateral environmental agreements. *Transnational Environmental Law*, 2, 285–309.

Kim, R. E., & Bosselmann, K. (2015). Operationalizing sustainable development: Ecological integrity as a *Grundnorm* of international law. *Review of European, Comparative & International Environmental Law*, 24, 194–208.

Kim, R. E., & Mackey, B. (2014). International environmental law as a complex adaptive system. *International Environmental Agreements*, 14, 5–24.

Kim, R. E., & van Asselt, H. (2016). Global governance: Problem shifting in the Anthropocene and the limits of international law. In E. Morgera, & K. Kulovesi (eds.), *Research handbook on international law and natural resources* (pp. 473–95). Cheltenham: Edward Elgar.

Kolleck, N., Well, M., Sperzel, S., & Jörgens, H. (2017). The power of social networks: How the UNFCCC Secretariat creates momentum for climate education. *Global Environmental Politics*, 17, 106–26.

Kotzé, L. J. (2016). *Global environmental constitutionalism in the Anthropocene*. Portland: Bloomsbury Publishing.

Kotzé, L. J., & French, D. (2018). A critique of the global pact for the environment: A stillborn initiative or the foundation for *Lex Anthropocenae*? *International Environmental Agreements*, 18, 811–38.

Kotzé, L. J, & Kim, R. E. (2019). Earth system law: The juridical dimensions of earth system governance. *Earth System Governance*, 1.

Lazer, D. (2011). Networks in political science: Back to the future. *Political Science and Politics*, 44, 61–8.

Lazer, D., Pentland, A., Adamic, L. et al. (2009). Computational social science. *Science*, 323, 721–3.

Le Blanc, D. (2015). Towards integration at last? The Sustainable Development Goals as a network of targets. *Sustainable Development*, 23, 176–87.

Lima, M. G. B., & Gupta, J. (2013). The policy context of biofuels: A case of non-governance at the global level? *Global Environmental Politics*, 13, 46–64.

Mitchell, R. B. (2003). International environmental agreements: A survey of their features, formation, and effects. *Annual Review of Environment and Resources*, 28, 429–61.

Morin, J. F. (2018). Concentration despite competition: The organizational ecology of technical assistance providers. *Review of International Organizations*, 70, 1–33.

Morin, J. F., & Bialais, C. (2018). *Strengthening multilateral environmental governance through bilateral trade deals. Policy Brief No. 123*. Waterloo, ON Canada: Centre for International Governance Innovation.

Morin, J. F., Dür, A., & Lechner, L. (2018). Mapping the trade and environment nexus: Insights from a new data set. *Global Environmental Politics*, 18, 122–39.

Morin, J. F., & Jinnah, S. (2018). The untapped potential of preferential trade agreements for climate governance. *Environmental Politics*, 27, 541–65.

Morin, J. F., Louafi, S., Orsini, A., & Oubenal, M. (2017). Boundary organizations in regime complexes: A social network profile of IPBES. *Journal of International Relations and Development*, 20, 543–77.

Morin, J. F., Pauwelyn, J., & Hollway, J. (2017). The trade regime as a complex adaptive system: Exploration and exploitation of environmental norms in trade agreements. *Journal of International Economic Law*, 20 (2), 365–90.

Morseletto, P. (2019). Confronting the nitrogen challenge: Options for governance and target setting. *Global Environmental Change*, 54, 40–9.

Murdie, A. (2013). The ties that bind: A network analysis of human rights international nongovernmental organizations. *British Journal of Political Science*, 44, 1–27.

Murphy, H., & Kellow, A. (2013). Forum shopping in global governance: Understanding states, business and NGOs in multiple arenas. *Global Policy*, 4, 139–49.

Nicholson, S., Jinnah, S., & Gillespie, A. (2017). Solar radiation management: A proposal for immediate polycentric governance. *Climate Policy*, 18, 322–34.

Nielsen, T. D., Holmberg, K., & Stripple, J. (2019). Need a bag? A review of public policies on plastic carrier bags – Where, how and to what effect? *Waste Management*, 87, 428–40.

Nilsson, M., Griggs, D., & Visbeck, M. (2016). Map the interactions between Sustainable Development Goals. *Nature*, 534, 320–2.

Nunan, F., Campbell, A., & Foster, E. (2012). Environmental mainstreaming: The organisational challenges of policy integration. *Public Administration and Development*, 32, 262–77.

Oberthür, S., & Gehring, T. (eds.) (2006). *Institutional interaction in global environmental governance: Synergy and conflict among international and EU policies*. Cambridge, MA: The MIT Press.

Oberthür, S., & Pożarowska, J. (2013). Managing institutional complexity and fragmentation: The Nagoya Protocol and the global governance of genetic resources. *Global Environmental Politics*, 13, 100–18.

Oberthür, S., & Stokke, O. S. (eds.) (2011). *Managing institutional complexity: Regime interplay and global environmental change*. Cambridge, MA: The MIT Press.

O'Neill, K., Weinthal, E., Marion Suiseeya, K. R., Bernstein, S., Cohn, A., Stone, M. W., & Cashore, B. (2013). Methods and global environmental governance. *Annual Review of Environment and Resources*, 38, 441–71.

Ostrom, E. (2010a). Nested externalities and polycentric institutions: Must we wait for global solutions to climate change before taking actions at other scales? *Economic Theory*, 49, 353–69.

Ostrom, E. (2010b). Polycentric systems for coping with collective action and global environmental change. *Global Environmental Change*, 20, 550–7.

Park, S., & Kramarz, T. (2019). *Global environmental governance and the accountability trap*. Cambridge, MA: The MIT Press.

Pasztor, J., Scharf, C., & Schmidt, K. U. (2017). How to govern geoengineering? *Science*, 357, 231.

Pattberg, P. (2010). Public–private partnerships in global climate governance. *Wiley Interdisciplinary Reviews: Climate Change*, 1, 279–87.

Pattberg, P., Chan, S., Sanderink, L., & Widerberg, O. (2018). Linkages: Understanding their role in polycentric governance. In A. Jordan, D. Huitema, H. van Asselt, & J. Forster (eds.), *Governing climate change: Polycentricity in action?* (pp. 169–87). Cambridge, UK: Cambridge University Press.

Pattberg, P., & Zelli, F. (eds.) (2016). *Environmental politics and governance in the Anthropocene: Institutions and legitimacy in a complex world*. London: Routledge.

Pauwelyn, J. (2007). *Conflict of norms in public international law: How WTO law relates to other rules of international law*. Cambridge, UK: Cambridge University Press.

Pauwelyn, J. (2014). At the edge of chaos? Foreign investment law as a complex adaptive system, how it emerged and how it can be reformed. *ICSID Review*, 29, 372–418.

Peel, J., Godden, L., & Keenan, R. J. (2012). Climate change law in an era of multi-level governance. *Transnational Environmental Law*, 1, 245–80.

Persson, Å., Runhaar, H., Karlsson-Vinkhuyzen, S., Mullally, G., Russel, D., & Widmer, A. (2018). Environmental policy integration: Taking stock of policy practice in different contexts. *Environmental Science and Policy*, 85, 113–15.

Pickering, J. (2018). Ecological reflexivity: Characterising an elusive virtue for governance in the Anthropocene. *Environmental Politics*, in press.

Pittock, J. (2010). Better management of hydropower in an era of climate change. *Water Alternatives*, 3, 444–52.

Raworth, K. (2017). *Doughnut economics: Seven ways to think like a 21st-century economist*. White River Junction: Chelsea Green Publishing.

Reid, W. V., Chen, D., Goldfarb, L. et al. (2010). Earth system science for global sustainability: Grand challenges. *Science*, 12, 916–17.

Reus-Smit, C. (2013). Constructivism. In S. Burchill, & A. Linklater (eds.), *Theories of international relations* (pp. 217–40). New York: Palgrave.

Robinson, N. A. (2014). Fundamental principles of law for the Anthropocene? *Environmental Policy and Law*, 44, 13–27.

Rockström, J., Steffen, W., Noone, K. et al. (2009). A safe operating space for humanity. *Nature*, 461, 472–5.

Runhaar, H., Driessen, P., & Uittenbroek, C. (2014). Towards a systematic framework for the analysis of environmental policy integration. *Environmental Policy and Governance*, 24, 233–46.

Runhaar, H., Wilk, B., Persson, Å., Uittenbroek, C., & Wamsler, C. (2018). Mainstreaming climate adaptation: Taking stock about 'What Works' from empirical research worldwide. *Regional Environmental Change*, 18, 1–10.

Sand, P. H. (2017). The discourse on 'protection of the atmosphere' in the International Law Commission. *Review of European, Comparative & International Environmental Law*, 26, 201–9.

Scharpf, F. W. (1999). *Governing in Europe: Effective and democratic?* Oxford: Oxford University Press.

Schellnhuber, H. J., Crutzen, P. J., Clark, W. C., Claussen, M., & Held, H. (2004). *Earth system analysis for sustainability*. Cambridge, MA: The MIT Press.

Scobie, M. (2019). *Global environmental governance and small states: Architectures and agency in the Caribbean*. Cheltenham: Edward Elgar.

Scott, D., Hitchner, S., Maclin, E. M., & Dammert B, J. L. (2014). Fuel for the fire: Biofuels and the problem of translation at the tenth conference of the parties to the convention on biological diversity. *Global Environmental Politics*, 14, 84–101.

Scott, K. N. (2013). International law in the Anthropocene: Responding to the geoengineering challenge. *Michigan Journal of International Law*, 34, 309–58.

Shelton, D. (2006). Normative hierarchy in international law. *American Journal of International Law*, 100, 291–323.

Sovacool, B. K., & Brown, M. A. (2009). Scaling the policy response to climate change. *Policy and Society*, 27, 317–28.

Steffen, W., Richardson, K., Rockström, J. et al. (2015). Planetary boundaries: Guiding human development on a changing planet. *Science*, 347, 1259855.

Steffen, W., Rockström, J., Richardson, K. et al. (2018). Trajectories of the earth system in the Anthropocene. *Proceedings of the National Academy of Sciences of the United States of America*, 2, 201810141 8.

Steffen, W., Sanderson, A., Tyson, P. et al. (2004). *Global change and the earth system: A planet under pressure*. Berlin: Springer.

Steffen, W., & Stafford-Smith, M. (2013). Planetary boundaries, equity and global sustainability: Why wealthy countries could benefit from more equity. *Current Opinion in Environmental Sustainability*, 5, 1–6.

Stevens, C. (2018). Scales of integration for sustainable development governance. *International Journal of Sustainable Development and World Ecology*, 25, 1–8.

Truelove, H. B., Carrico, A. R., Weber, E. U., Raimi, K. T., & Vandenbergh, M. P. (2014). Positive and negative spillover of pro-environmental behaviour: An integrative review and theoretical framework. *Global Environmental Change*, 29, 127–38.

Underdal, A., & Kim, R. E. (2017). The Sustainable Development Goals and multilateral agreements. In N. Kanie, & F. Biermann (eds.), *Governing through goals: Sustainable Development Goals as governance innovation* (pp. 241–9). Cambridge, MA: The MIT Press.

UNEP (2007). *Global environment outlook 4: Environment for development*. Nairobi: United Nations Environment Programme.

UNEP (2012). *Global environment outlook 5: Environment for the future we want*. Nairobi: United Nations Environment Programme.

UNEP (2019). *Global environment outlook 6: Healthy planet, healthy people*. Nairobi: United Nations Environment Programme.

van Asselt, H. (2014). *The fragmentation of global climate governance: Consequences and management of regime interactions*. Cheltenham: Edward Elgar.

Van de Graaf, T. (2013). Fragmentation in global energy governance: Explaining the creation of IRENA. *Global Environmental Politics*, 13, 14–33.

Van den Bergh, J., Folke, C., Polasky, S., Scheffer, M., & Steffen, W. (2015). What if solar energy becomes really cheap? A thought experiment on environmental problem shifting. *Current Opinion in Environmental Sustainability*, 14, 170–9.

Van Doren, D., Driessen, P. P., Runhaar, H., & Giezen, M. (2018). Scaling-up low-carbon urban initiatives: Towards a better understanding. *Urban Studies*, 55, 175–94.

Van Doren, D., Giezen, M., Driessen, P. P. J., & Runhaar, H. (2016). Scaling-up energy conservation initiatives: Barriers and local strategies. *Sustainable Cities and Society*, 26, 227–39.

Vidas, D. (2011). The Anthropocene and the international law of the sea. *Philosophical Transactions of the Royal Society A*, 369, 909–25.

Vidas, D., Fauchald, O. K., Jensen, Ø., & Tvedt, M. W. (2015). International law for the Anthropocene? Shifting perspectives in regulation of the oceans, environment and genetic resources. *Anthropocene*, 9, 1–13.

Vijge, M. J. (2013). The promise of new institutionalism: Explaining the absence of a world or united nations environment organisation. *International Environmental Agreements*, 13, 153–76.

Voigt, C. (2009). *Sustainable development as a principle of international law: Resolving conflicts between climate measures and WTO law*. Leiden: Martinus Nijhoff Publishers.

Vordermayer, M. (2014). 'Gardening the great transformation': The Anthropocene concept's impact on international environmental law doctrine. *Yearbook of International Environmental Law*, 25 (1), 79–112.

Watts, D. J. (2017). Should social science be more solution-oriented? *Nature Human Behaviour*, 1, 5–15.

Widerberg, O. (2016). Mapping institutional complexity in the Anthropocene: A network approach. In P. Pattberg, & F. Zelli (eds.), *Environmental politics and governance in the Anthropocene: Institutions and legitimacy in a complex world* (pp. 81–102). London: Routledge.

Wilson, M., Davis, D. R., & Murdie, A. (2016). The view from the bottom: Networks of conflict resolution organizations and international peace. *Journal of Peace Research*, 53, 442–58.

Yang, Y., Bae, J., Kim, J., & Suh, S. (2012). Replacing gasoline with corn ethanol results in significant environmental problem-shifting. *Environmental Science and Technology*, 46, 3671–8.

Young, M. A. (2012). *Regime interaction in international law: Facing fragmentation.* Cambridge, UK: Cambridge University Press.

Young, O. R. (1996). Institutional linkages in international society: Polar perspectives. *Global Governance*, 2, 1–23.

Young, O. R. (ed.) (1999). *The effectiveness of international environmental regimes: Causal connections and behavioural mechanisms.* Cambridge, MA: The MIT Press.

Young, O. R. (2002). *The institutional dimensions of environmental change: Fit, interplay, and scale.* Cambridge, MA: The MIT Press.

Young, O. R. (2008). The architecture of global environmental governance: Bringing science to bear on policy. *Global Environmental Politics*, 8, 14–32.

Young, O. R. (2011a). Effectiveness of international environmental regimes: Existing knowledge, cutting-edge themes, and research strategies. *Proceedings of the National Academy of Sciences of the United States of America*, 108, 19853–60.

Young, O. R. (2011b). If an arctic ocean treaty is not the solution, what is the alternative? *Polar Record*, 47, 327–34.

Young, O. R. (2012). Arctic tipping points: Governance in turbulent times. *AMBIO*, 41, 75–84.

Young, O. R. (2013). Sugaring off: Enduring insights from long-term research on environmental governance. *International Environmental Agreements*, 13, 87–105.

Young, O. R. (2017). *Governing complex systems: Social capital for the Anthropocene.* Cambridge, MA: The MIT Press.

Young, O. R. (2018). Research strategies to assess the effectiveness of international environmental regimes. *Nature Sustainability*, 1, 461–5.

Young, O. R., Underdal, A., Kanie, N., & Kim, R. E. (2017). Goal setting in the Anthropocene: The ultimate challenge of Planetary Stewardship. In N. Kanie, & F. Biermann (eds.), *Governing through goals: Sustainable Development Goals as governance innovation* (pp. 53–74). Cambridge, MA: The MIT Press.

Young, O. R. (2019). Constructing diagnostic trees: A stepwise approach to institutional design. *Earth System Governance*, 1.

Zelli, F. (2011). The fragmentation of the global climate governance architecture. *Wiley Interdisciplinary Reviews: Climate Change*, 2, 255–70.

Zelli, F., & van Asselt, H. (2013). The institutional fragmentation of global environmental governance: Causes, consequences, and responses. *Global Environmental Politics*, 13, 1–13.

Ziervogel, G., & Ericksen, P. J. (2010). Adapting to climate change to sustain food security. *Wiley Interdisciplinary Reviews: Climate Change*, 1, 525–40.

Zürn, M. (2012). Global governance as multi-level governance. In D. Levi-Faur (ed.), *The Oxford handbook of governance* (pp. 730–44). Oxford: Oxford University Press.

Zürn, M., & Faude, B. (2013). On fragmentation, differentiation, and coordination. *Global Environmental Politics*, 13, 119–30.

Part I

The Building Blocks

2

Intergovernmental Institutions

RONALD B. MITCHELL, ARILD UNDERDAL, STEINAR ANDRESEN AND CAREL DIEPERINK

Intergovernmental institutions are key building blocks in the architectures of earth system governance. This chapter reviews two research programmes on intergovernmental institutions that have dominated the past decade: one into factors that explain their *formation and design*, and another into factors that explain their *effectiveness*. We organize our review around the structures, agents and processes that scholars use to understand how intergovernmental institutions form and operate. We summarize recent theoretical insights and empirical findings and identify some remaining open questions. Our empirical focus is research on intergovernmental institutions in the field of earth system governance, while we believe that our findings have a broader theoretical value for other realms of world politics.

Conceptualization

Early research on international cooperation focused on intergovernmental institutions as dependent variables, explaining when and why states created them despite the disincentives generated by the anarchy of international society (Krasner 1983). Beginning in the 1990s, scholars took up follow-on research assessing the influence of such institutions as independent variables, seeking to distinguish conditions under which they did or did not affect state behaviour (for reviews, see Mitchell 2010; Young 2011). Despite different terminologies, such as treaties, regimes or institutions, all of these scholars study intentional cooperation among states to solve shared problems. Chapters 3–13 in this volume elaborate in more detail on research on the many alternative strategies of earth system governance (see also Andonova and Mitchell 2010).

We focus this chapter on research into intergovernmental institutions in earth system governance, which we define as *cooperative arrangements among national governments to address transboundary environmental problems*. Such institutions vary considerably in their levels of institutionalization and legal formalization (Böhmelt and Pilster 2010; Vabulas and Snidal 2013). We concentrate on formal intergovernmental cooperation common to treaties and other legal instruments, but our definition is broad enough also to capture informal intergovernmental cooperation and the often-unwritten norms, rules and decision-making procedures that underpin formalized instruments and institutions. Our review notes the many agents that help develop, manage and implement intergovernmental institutions but leaves in-depth discussions of other important aspects to other chapters in this volume, notably regarding international bureaucracies (Chapter 3), institutional interlinkages (Chapter 6), regime complexes (Chapter 7), governance fragmentation (Chapter 8) and interplay management (Chapter 10).

Creating, Designing and Adapting Intergovernmental Institutions

Scholars have asked three key questions about intergovernmental institutions as dependent variables.

(1) First, what explains the *pattern* and *timing* of intergovernmental efforts to address human impacts on the earth system? Most studies focus on explaining the emergence of *formal* intergovernmental institutions, especially those grounded in international law (Brunnée, Bodansky and Hey 2007; Bodansky 2010). Although governments sometimes create 'soft law' institutions, most scholars focus on formal legal instruments such as treaties, conventions or protocols, if only because they help document the 'that', 'when' and 'what' of intergovernmental cooperation (Bodansky 2015).

(2) Second, what explains how states *design* the requirements, organizational structures and processes of intergovernmental institutions (Young 2010; Biermann and Pattberg 2012; Jinnah 2014)? Scholars seek to explain variation in treaty goals, membership rules, obligational precision, ambition and depth, the form of monitoring and implementation strategies.

(3) Third, when and why do intergovernmental institutions *learn*, *develop* and *adapt* (Siebenhüner 2008; Abbott and Snidal 2013; Abbott and Bernstein 2015; Beunen and Patterson 2016)? The profiles of intergovernmental institutions vary considerably: some are signed but never enter into force, others take effect but are never modified, others are designed to make regular decisions in response to changing circumstances and yet others significantly expand their substantive scope over time (Young 1999: 28).

In all three questions, scholars focus on intergovernmental environmental institutions as dependent variables and explain their creation, design and adaptation by reference to structural forces, the interests and power of political actors and the strategies by which the latter pursue their goals in light of the former. We now discuss such structural forces, the interests and power of actors and their strategies in turn.

Structures

Objective environmental and behavioural conditions and subjective social conditions constitute the structural context that influences the 'whether', 'when' and 'what' of intergovernmental environmental institutions (Underdal and Hanf 2000; Bernauer 2002). Examining a range of cases, Miles and colleagues (2002) posited that problem structures (particularly the arrangement of states' interests and power) made resolving transnational environmental problems easier (benign) or harder (malign). Their distinction reflected earlier typologies of problem structure (see the review in Dombrowsky 2007). Government support, opposition or indifference to resolving transboundary environmental problems reflects the material costs of both action and inaction, as evident in many negotiations including those on climate change (Sprinz and Vaahtoranta 1994; Dimitrov 2016; Falkner 2016). Negotiations prove easier for tragedy-of-the-commons problems among mixed-motive states that both cause and are impacted by a problem than for those between upstream states that cause a problem and downstream victim states (Mitchell 2010). Capacities also matter: developing states often have incentives to protect endangered species and habitats as sources of ecotourism but may fail to do so because of financial and administrative incapacities (Mitchell 2010).

The nature of environmentally harmful activities also constrains institutional design. Environmental problems vary in the number of responsible and concerned states (Koremenos, Lipson and Snidal 2001). Problems among few states (e.g., river basins or polar bears) are easier to resolve than global problems (e.g., marine pollution or climate change), and choices of membership rules matter in both. Concern among powerful states promotes institutional creation, especially if they have interdependencies or resources that are valued by the states causing the problem. States will accept more ambitious institutional rules if they share high levels of concern but will require weaker rules and escape clauses if concern is weak or contested. States' power, interests and capacities influence whether states base treaties on reciprocal obligations or on asymmetric bargains of side-payments or coercion (Mitchell 2009). The ability to readily observe problematic behaviours and their perpetrators (e.g., marine oil versus chemical pollution or terrestrial point versus non-point pollution) can foster agreement about the existence of an environmental problem, its causes and appropriate regulatory responses. Problems in which member states have strong incentives to cheat on any solution are likely to include strong monitoring, enforcement and withdrawal provisions. Existing intergovernmental institutions, environmental or otherwise, can provide venues that, depending on their mandates, foster or inhibit the creation of new institutions to address environmental problems. All of these structural factors help determine the emergence and form of intergovernmental environmental institutions.

Yet, environmental problems are not objective and unproblematically identifiable. Knowledge and social construction play important roles in the creation of intergovernmental institutions. Material aspects of environmental problems influence the impacts, knowledge and concern states have about a problem. The causal links between human behaviours and environmental harm are clear and immediate for some problems but more complex and temporally and spatially attenuated for others. Debates on climate science clarify that scientific uncertainty depends as much on politics and the incidence of costs and benefits as it does on the science itself (Oreskes and Conway 2010). Problems differ in whether economic interests align with or oppose environmental interests. Movement towards intergovernmental environmental cooperation begins only when concern in one or more states prompts them to put 'the problem' on the international agenda (Carpenter 2007). Such concern, in turn, depends on governments having compelling knowledge of impacts and causes, accepting framings of the problem as warranting international action and coming to see the material and normative implications of action as outweighing the forces of political, social, economic and normative inertia (Jinnah 2011; Vanhala and Hestbaek 2016; Allan and Hadden 2017).

Finally, structural factors can change over time, promoting or inhibiting institutional formation and change (Young 2010). Dynamic human–environment interactions require institutional designs that promote adaptive management and institutional modification to deal with incomplete contracting, as when setting annual fishing quotas or modifying endangered species lists. For other problems, states recognize that levels of scientific knowledge or political concern are likely to change rapidly and, therefore, they include provisions that make it easier to adopt protocols and amendments that strengthen obligations or expand institutional scope (Shi 1999; Urpelainen 2013). In yet other problems, technological and economic changes may generate reductions in environmental impacts, reducing the need for institutional modification. Research on such institutional dynamics and life cycles remains underdeveloped (see Young 2011).

Actors, Agency and Strategies

Structural factors generate patterns in institutional formation and design, shaping choices in ways that make certain institutions more likely than others. However, those patterns emerge only because specific actors use specific social and institutional strategies to pursue their material and non-material goals. Actors' strategic choices (and non-choices) determine whether institutions incorporate designs that neatly match structural predictions or reflect innovations that contradict them. We contend that the 'structure–agency debate' poses an overstated dichotomy: structures influence outcomes by making certain institutional choices less available or attractive or by creating incentives to create innovative alternatives (see Betsill, Benney and Gerlak 2019). Institutional choices at one point in time, in turn, can transform the structural context, either freeing up or locking in constraints that favour certain outcomes over others (Seto et al. 2016).

A paradigmatic story of the roles of different actors in institutional formation and design might go as follows. Scientists identify a previously unrecognized, apparently harmful human impact on the environment; non-governmental organizations and the media disseminate and frame that information and domestic polities express growing concern, as exemplified in climate change negotiations (Schroeder and Lovell 2012; Betzold, Bernauer and Koubi 2016). In response, some governments call for collective international action because unilateral solutions seem inadequate. Highly concerned states and environmental non-governmental organizations push to establish an intergovernmental institution, facing opposition from states that are less vulnerable or less concerned and economic actors whose interests are threatened. Opposing forces may prompt advocates to accept broad institutions that stop at fostering scientific research or delineating hortatory goals or, alternatively, to design institutions with more selective memberships of powerful countries willing and able to implement meaningful actions (Victor 2011: 6). State and non-state advocates can propose designs that make it costly to oppose action or can design institutional obligations and incentives that attract participation by reluctant states. Over time, advocates can use institutional inertia to promote increasingly stringent obligations, greater participation and more effective implementation.

The developmental trajectories of intergovernmental institutions protecting the stratospheric ozone layer and mitigating climate change align reasonably well with this paradigm, as do those of many less-visible institutions. Many institutional origin stories reflect the influence of epistemic communities and global environmental assessments, pressures from environmental non-governmental organizations and civil society, opposition by multinational corporations and proactive leadership by countries and individuals (Mitchell et al. 2006; Ivanova 2010; Böhmelt and Betzold 2013; Brun 2016). Narratives of particular institutions focus on within-case variation, highlighting the choices, actors and processes that explain the timing and design of an institution. Structural forces fade into the background, if only because they are better suited to cross-institutional explanations.

At their most convincing, such narratives either identify how the indeterminacy of structure allows actors leeway in institutional design or clarify how actors' strategies at critical junctures promoted outcomes contrary to those favoured by structural forces. Thus, the role of UNEP's executive director, Mostafa Tolba, in the ozone depletion negotiations is compelling precisely because those negotiations succeeded *before* scientists had fully resolved scientific uncertainty on the problem (Parson 2003). Many scholars have documented how structural conditions favour creation of a new institution but that the institution emerges only because of the leadership of individuals, states or non-state actors (Ivanova 2010; Corneloup and Mol 2014; Hale 2016).

Negotiators have choices and those choices matter for the shape an institution takes. States must make choices about membership. For example, although few states were harvesting whales, fur seals or polar bears, the 1911 fur seal and 1973 polar bear treaties limited membership to states engaged in those harvests, but the 1946 whaling treaty allowed any state to join. The whaling treaty's rules, when combined with rules allowing quotas to be set by a three-quarter majority, allowed adoption of a 1982 ban on commercial whaling over the opposition of most whaling states (Epstein 2008). Negotiators have chosen to regulate many environmentally harmful behaviours directly but only to regulate the *trade* in endangered species, tropical timber and hazardous waste (Curlier and Andresen 2002; Khoo and Rau 2009; Nagtzaam 2009; Lucier and Gareau 2015). The types, specificity, ambitiousness and flexibility of institutional rules usually reflect both structural constraints that rule out certain options and strategic choices as well as framings designed to create winning coalitions, generate agreement and attract states to join (Newell et al. 2015). The 1911 fur seal treaty proved effective because negotiators rejected reciprocal exchanges of restraint in favour of Russian and US payments to Canada and Japan to completely halt their fur seal harvests (Barrett 2003). The appearance of a 1.5 degrees Celsius target and loss and damage clauses in the 2015 Paris Agreement (but not prior agreements) reflected enabling structural conditions that made their inclusion *possible* and the strategies of states, non-state actors and individuals to ensure that negotiators *actually included* those provisions in the final document. Those same influences help explain the shift from top-down obligations that bound only industrialized states under the Kyoto Protocol to the bottom-up pledge and review obligations of the Paris Agreement (Falkner 2016; Keohane and Oppenheimer 2016).

Finally, successful institutional formation requires that states join the institutions they negotiate. Both state and treaty characteristics help explain membership choices (Seelarbokus 2014b). Although most of the literature examines correlations between national economic and political traits and treaty membership, increasing attention is being paid to treaty design features such as legalization, formalization and obligational depth (Von Stein 2008; Perrin and Bernauer 2010;

Bernauer et al. 2013; Baccini and Urpelainen 2014; Seelarbokus 2014b; Spilker and Koubi 2016).

Explaining the Influence of Intergovernmental Institutions

States create, design and adapt intergovernmental institutions to reduce the environmental impacts of human behaviours. The factors and forces (noted in the previous section) that shape institutional emergence and design also shape whether, how and how much influence institutions have on state behaviour. The influence of intergovernmental environmental institutions, that is, their effectiveness, has been a major thread in earth system governance research for decades and has contributed significantly to broader international relations scholarship. Here we review recent contributions that build on a long tradition of scholarship by individuals and teams, including Thomas Bernauer, Edith Brown Weiss and Harold Jacobson, Abram Chayes and Antonia Chayes, George Downs, Peter Haas, Robert Keohane, Edward Miles, Ronald Mitchell, Arild Underdal, David Victor and Oran Young.

Such research initially responded to the neo-realist claim that intergovernmental institutions are, by their nature, ineffective: if states negotiate institutional provisions to reflect their interests, then only join those that fit their interests and are not subject to an overarching authority that enforces institutional commitments, then those institutions can have no independent influence on their behaviour (Strange 1983). However, over time, institutionalist scholars developed compelling theory and convincing evidence that, despite these considerations, intergovernmental institutions can alter state behaviours under identifiable circumstances.

A central debate has revolved around the distinction between compliance and effectiveness. Downs, Rocke and Barsoom (1996) claimed that states accept and fulfil 'shallow' obligations if they require little behavioural adjustment or, alternatively, accept but then renege on 'deep' obligations if they require too-costly behavioural adjustment. Scholars of intergovernmental environmental institutions take this claim seriously, accepting that states negotiate institutional obligations to reflect their narrow and short-term interests, reject membership in agreements that harm their interests and behave in ways that reflect their interests (Mitchell 2009). They have clarified that high compliance need not correspond with high effectiveness (Mitchell 2010; Breitmeier, Underdal and Young 2011; Spilker and Koubi 2016). But, scholars have also shown that these considerations do not preclude the *possibility* of institutions independently influencing state behaviour. States expend considerable effort to negotiate intergovernmental institutions, so that they *do* influence state behaviour. They design obligations and supporting provisions so that the benefits to member states exceed their costs and minimize risks. And scholars have documented many cases in which intergovernmental institutions have led states to engage in less environmentally harmful behaviour than they would have in the institutions' absence (Bernauer 2002, Mitchell 2009, Wettestad 2011).

Accurately assessing institutional influence requires overcoming problems of endogeneity. Institutional endogeneity makes it hard to separate the influence of an institution from the spurious correlation that arises because both the obligations a state negotiates and accepts and the behaviours in which it later engages are driven by their pre-institutional interests. To address endogeneity, scholars have sought to replace the legal terminology of compliance with a social science terminology of effectiveness. International relations scholars distinguish effectiveness as an institution's causal influence, which contrasts with the correlation between behaviours and legal standards that we expect to arise because of institutional endogeneity. Documenting effectiveness requires (a) identifying deviations of state behaviours from otherwise-similar 'no-regime counterfactuals' and (b) demonstrating (through process-tracing) that such deviations reflect the independent influence of the institution (Helm and Sprinz 2000; Ringquist and Kostadinova 2005; Dombrowsky 2008; Böhmelt and Pilster 2010). The Oslo-Potsdam solution (named after the locations of key scholars involved) argued, further, that meaningful and comparative measuring of effectiveness requires placing an institution on an (ordinal) scale ranging from a specified no-regime counterfactual to the 'collective optimum' of what that institution could have accomplished, given its membership, decision rules and internal allocation of power. Clearly, measuring effectiveness poses a demanding challenge that scholars must face head-on if they seek to assess the relative performance of institutions (e.g., Young 2001; Hovi, Sprinz and Underdal 2003; Young 2003; Mitchell 2006; Andresen 2013). Addressing endogeneity requires that scholars develop counterfactuals that 'net out' the effects of non-institutional variables and the effects of power, interests, capacities and the like on institutional design, institutional membership and state behaviour. Failing to do so will systematically overstate the influence of international institutions.

Earth system governance has attracted attention from a new generation of scholars of international law and economics as well as political science and international relations (Barrett 2007; Bodansky 2010; Victor 2011; Young 2017). Some have addressed less-studied yet timely questions. Some examine how non-governmental organizations and corporate actors increasingly finance, monitor and shame actors in ways that foster the effectiveness of intergovernmental institutions, rather than serving as alternatives to them (Gupta 2010; Hale 2016; van der Ven, Bernstein and Hoffmann 2017). Others highlight the power of norms and ideas (as distinct from self-interest) in promoting earth system governance (Pettenger 2007; Dryzek 2013). These and other innovative lines of inquiry, especially cross-disciplinary ones, provide more nuanced and sophisticated views of how institutions work and the actors and processes that promote their effectiveness.

Structures

Accurately assessing institutional effectiveness requires generating no-regime counterfactuals that identify how we expect states to behave in the absence of an institution by estimating the influence of pre- and non-institutional factors on state behaviour. For example, the incentives to generate externalities that produce tragedy-of-the-commons or upstream/downstream problems support a strong presumption that those states will continue their behaviours unless and until they join an institution designed to constrain them. Counterfactuals carefully specified to reflect those incentives and other influences on state behaviour provide a baseline, divergence from which we can interpret as plausible evidence of institutional influence. Such counterfactuals allow us to distinguish, for example, states that become greener over time because of an institution from those that do so for domestic political or economic reasons. Attention to endogeneity suggests that the leader states pushing institutional creation are likely to become greener over time anyway and that it will be the green behaviour of *important laggards* that have few other explanations that provide more compelling evidence of institutional influence. We note that institutions may have perverse effects, but that most scholars define *effectiveness* as behaviours that are in line with institutional goals.

A problem's structure helps identify both the type and difficulty of the task an institution faces (Miles et al. 2002). Some institutions must prompt states to halt a harmful behaviour; others must induce states to increase a beneficial behaviour already within their capacities; others must get states to pool financial or informational resources; and others must remedy the incapacities of some states to engage in desired behaviours. Some institutions must encourage adoption of alternatives that are numerous and cheap, while others must encourage adoption of alternatives that do not exist, are expensive or are not available to important actors. As the ozone and climate cases illustrate, the institutional task of reducing emissions is easier when cheap and profit-generating alternatives exist (as with chlorofluorocarbons) than when they do not (as with fossil fuels).

Material, ideational and normative considerations also should influence how we assess institutional effectiveness. Institutions face an easier task when powerful states support their objectives and are not engaged in the behaviours in question and a harder task when such states are committed to such behaviours and have strong counter-institutional interests. Powerful states sufficiently concerned about a problem to take unilateral pro-environmental action become examples and test sites for policies that weaker states may imitate even absent an intergovernmental institution. The size of such positive effects may depend on the framing of relations between weaker and stronger states: in climate change, both historical responsibility and economic asymmetries led developing countries to demand that industrialized states act first. Normative concerns also matter in other ways: institutions will find it easier to alter behaviours that are inconsistent with broadly accepted norms than those that align with such norms. Over time, the task of altering behaviour may ease as actors increasingly accept regulation as legitimate and appropriate.

Identifying one institution's influence requires that we also account for how other institutions influenced observed behaviours (see Chapter 7). One institution's influence may reflect synergies with other institutions (Young 2008; Keohane and Victor 2011; Oberthür and Stokke 2011; Stokke 2013). This can reflect either unintended interaction effects or self-conscious coordination among intergovernmental institutions (Andresen and Rosendal 2009). European air pollution may have declined in response to the Convention on Long-Range Transboundary Air Pollution and its protocols, European Union directives or some combination of both (Byrne 2015). Parsing the influence of multiple institutions requires distinguishing their additive, synergistic, conditional and alternative effects.

In general, accurate assessments of institutional influence require accounting for these and other systemic and state-specific factors as both alternatives to and conditioning factors on institutional influence. Features of the behaviours in question, the states involved in them and the material, normative and institutional landscapes mean that institutions differ in the challenges they face. Future researchers seeking to compare institutional effectiveness will need to take such differences into account.

Actors, Agency and Strategies

As with institutional creation and design, structural factors make institutional effectiveness more or less likely but operate through identifiable actors deploying particular strategies to induce states to adjust their behaviours. The efforts of such actors may serve as alternatives to institutional explanations or as the mechanisms through which institutions operate.

Over the past decade, scholars have improved theory, methods and evidence related to institutional influence (Underdal and Young 2004; Young 2011; Seelarbokus 2014a). Scholars have built new hypotheses and models on findings of early multi-treaty comparisons (Haas, Keohane and Levy 1993; Brown Weiss and Jacobson 1998; Victor, Raustiala and Skolnikoff 1998; Young 1999; Miles et al. 2002). The International Regimes Database (Breitmeier, Young and Zürn 2006) and Oslo-Seattle Database (Miles et al. 2002) have informed development of new datasets that cover more institutions, operationalize variables better and offer aggregate indices to capture empirical variation and facilitate hypothesis testing. Some have created datasets targeted on specific issues like climate change and rivers (Bernauer and Böhmelt 2013, 2014). The International Environmental Agreements (IEA) Database (IEA Database 2019) has been developed, maintained, linked and extended to cover all environmental treaties, all member states and numerous related variables. Scholars are increasingly conducting systematic analyses of a wide range of institutions by combining variables from these datasets with other readily available datasets.

Scholars have used both simple frequency counts and sophisticated econometrics to detect institutional effects in statistically significant differences between the pre- and post-treaty behaviours of member states or between member and non-member behaviour (Kim, Tanaka and Matsuoka 2017). Of numerous quantitative analyses of the protocols under the Convention on Long-Range Transboundary Air Pollution, most have found them to have little effect on emissions (Ringquist and Kostadinova 2005; Aakvik and Tjøtta 2011; Vollenweider 2013; Houghton and Naughton 2014; Byrne 2015). Scholars have also assessed intergovernmental institutions regulating carbon dioxide, ozone depleting substances, hazardous wastes, river pollution and other environmental problems (Dombrowsky 2008; Cullis-Suzuki and Pauly 2010; Berardo and Gerlak 2012; Myint 2012; Bodin and Österblom 2013; Kellenberg and Levinson 2014; Saleh and Abene 2016). Many of these studies have devised compelling strategies to address the challenges of endogeneity noted above (Downs, Rocke and Barsoom 1996; Bernauer et al. 2013). Statistical approaches have advantages in identifying an institution's average effect across all members and in allowing the analyst to isolate and compare an institution's influence on behaviour relative to other independent variables, after controlling for those variables.

Other scholars approach institutional effectiveness in set-theoretic terms (qualitative comparative analysis), looking for their influence in associations between *combinations* of necessary and sufficient conditions and positive environmental behaviours (Ragin 1987; Stokke 2012). One research team combined multiple techniques to compare findings from the International Regimes Database and the Oslo-Seattle Database (Breitmeier, Underdal and Young 2011). Despite the value of statistical techniques, qualitative comparative analysis proved more helpful in documenting that consensual knowledge was more commonly associated with effective regimes than any other variable. Scholars are increasingly deploying novel research strategies, including agent-based modelling, to assess the effectiveness of possible institutions that do not yet exist (Lempert, Scheffran and Sprinz 2009; Gerst et al. 2013; Nordhaus 2015; Hovi et al. 2017).

Alongside quantitative studies, qualitative studies, which dominated institutional effectiveness research in the 1990s, have continued to provide sophisticated, nuanced and compelling insight into the processes by which intergovernmental institutions influence behaviour (Hønneland and Stokke 2007; Brochmann and Hensel 2011; Victor 2011; Wettestad 2011; Young 2011; Mitchell and Zawahri 2015). Refuting the generally pessimistic conclusions of quantitative studies, qualitative assessments often find intergovernmental institutions to be at least moderately effective. In part, this divergence may reflect the fact that quantitative methods define effectiveness as an institution's *average* influence across all member states while qualitative methods seek evidence of their influence on some *specific states*. This suggests the need for future research to integrate divergent quantitative and qualitative findings, determining whether they are different but compatible assessments, reflect methodological artefacts or constitute irreconcilable claims about the same empirical record.

Despite sometimes-divergent findings, both quantitative and qualitative research support some claims about why some intergovernmental institutions are more effective than others. We know that problems prove less susceptible to resolution (more malign) when individual and collective costs and benefits diverge, when states have upstream-downstream rather than tragedy-of-the-commons incentives and when resolution requires a minimum number of contributors. Institutions are less likely to be effective in addressing malign problems and may become wholly ineffective if knowledge about the problem is weak or uncertain (Miles et al. 2002). The prospects for institutional effectiveness improve, however, when small homogeneous groups of states share interdependencies, have a source of leadership and can use pre-existing institutional capacities.

For institutions within a propitious context, scholars have made major progress in identifying links between institutional design and institutional influence. Treaty design involves trade-offs: more binding, precise, or ambitious treaties may deter some states from participating but lead those that join to larger behavioural adjustments (Böhmelt and Pilster 2010; Bernauer et al. 2013; Spilker and Koubi 2016). Certain designs can mitigate these trade-offs: institutions can convince recalcitrant states to join and meet institutional obligations by offering opportunities to receive side-payments or establish green reputations, or by reassuring them of reciprocation from other states if they make sacrifices for a collective goal (Baccini and Urpelainen 2012; Bernauer et al. 2013). Effectiveness is enhanced when institutions attract support from powerful states that take leadership roles, exercise behavioural restraint unilaterally or credibly threaten sanctions (Cirone and Urpelainen 2013; Sand 2013). We leave to Kalfagianni and colleagues (Chapter 4) a discussion of non-governmental organizations as independent sources of earth system governance, but we do note here the increasing frequency with which institutions involve non-governmental organizations in their implementation (Gulbrandsen and Andresen 2004; Betsill and Corell 2008).

A well-developed literature has demonstrated the contributions transparency makes to institutional effectiveness, by enhancing institutional legitimacy and by providing a foundation for sanctioning and shaming (Gupta 2008; Mason 2008; Haufler 2010). Support has accumulated for constructivist claims that effectiveness is a function of norms, legitimacy and equity as much as instrumental strategies for shaping behaviour (Andresen and Hey 2005; Breitmeier 2008; Epstein 2008; Brunnée and Toope 2010; Milkoreit 2015). Institutional promotion of scientific research expands and strengthens epistemic communities, which brings attention and support from national governments and universities, attracts media and public attention and may move domestic policy towards institutional goals. Corporations also shape institutional effectiveness. Their opposition can delay negotiations and reduce institutional influence while their support can do the opposite. Public commitments by chemical companies to eliminate the production of chlorofluorocarbons if such substances were shown to cause ozone depletion play an important role in most narratives of the Montreal Protocol's effectiveness (Parson 2003). Fossil fuel industry opposition has generally delayed carbon emission reductions, but some corporations' investments in renewable energy are helping reduce carbon emissions faster than they would otherwise. Corporate research and investment strategies, normative guidelines, production standards and certification and labelling programmes have contributed to the implementation and effectiveness of intergovernmental institutions addressing climate change, forestry and fisheries (Cashore et al. 2007; Auld 2014; Gulbrandsen 2014).

Conclusions and Future Directions

Our review identifies the many influences on the creation and design of intergovernmental institutions in earth system governance and on their effectiveness. With respect to institutional creation and design, structural factors (both material and socially constructed) condition whether, when, how easily and how states address a shared environmental problem. Scholars are enhancing our ability to diagnose environmental problems and to identify promising cures. Institutional emergence depends on structural conditions such as (a) the degree of shared, science-based knowledge about cause–effect relationships; (b) the nature of the behaviours that cause the environmental problem; and (c) the availability of attractive substitute behaviours. Within structural constraints and opportunities, states, existing institutions, non-governmental organizations and the private sector deploy strategies that either foster institutional creation in malign conditions or hinder it in benign conditions. The interaction of structural conditions and strategies determines whether and what type of institutions emerge. All these factors help explain variation in institutional emergence, as evident in the more rapid and effective efforts to address ozone depletion relative to climate change.

With respect to effectiveness, institutions succeed best when decision-makers and stakeholders identify designs that fit the problem, reflecting the structural constraints and opportunities, the power and incentives of important actors and a range of other important political, economic and social dynamics (Young 2002). Institutions become effective when proponents design them to alter behaviours within existing material, ideational and normative constraints. Solutions may require support from powerful actors and the availability of low-cost alternatives, but individuals and institutions often are crucial to identifying and directing efforts towards common interests (Miles et al. 2002: 450). The power and concern of the United States related to ozone depletion led to efforts that generated support among OECD countries and offered assistance to developing countries, leading to rapid and substantial emission reductions. By contrast, intergovernmental efforts on climate change have been far less effective because of a more complex and malign problem structure, greater resistance from powerful states and corporations and weaker capabilities among the many institutions in the climate change regime complex. Deepening our knowledge of why some intergovernmental institutions succeed while others do not will require scholars to continue researching complex interactions at and across the international, transnational, domestic and sub-national levels of earth system governance.

Future research opportunities abound. Scholars can assess what factors explain the creation and success (or failure) of hundreds of existing intergovernmental institutions in earth system governance. They can do so by building on already-robust analytic frameworks and methods. And the interdisciplinary skills of scholars of earth system governance position them well to make unique contributions to global efforts to address climate change, biodiversity loss, the hazardous waste trade, ocean and freshwater pollution and the myriad other environmental problems facing our planet.

References

Aakvik, A., & Sigve, T. (2011). Do collective actions clear common air? The effect of international environmental protocols on sulphur emissions. *European Journal of Political Economy*, 27 (2), 343–51.

Abbott, K. W., & Bernstein, S. (2015). The High-level Political Forum on Sustainable Development: Orchestration by default and design. *Global Policy*, 6 (3), 222–33.

Abbott, K. W., & Snidal, D. (2013). Taking responsive regulation transnational: Strategies for international organizations. *Regulation and Governance*, 7 (1), 95–113.

Allan, J. I., & Hadden, J. (2017). Exploring the framing power of NGOs in global climate politics. *Environmental Politics*, 26 (4), 600–20.
Andonova, L. B., & Mitchell, R. B. (2010). The rescaling of global environmental politics. *Annual Review of Environment and Resources*, 35 (1), 255–82.
Andresen, S. (2013). International regime effectiveness. In R. Falkner (ed.), *The handbook of global climate and environment policy* (pp. 304–19). Chichester: Wiley-Blackwell.
Andresen, S., & Hey, E. (2005). The effectiveness and legitimacy of international environmental institutions. *International Environmental Agreements*, 5 (3), 211–26.
Andresen, S., & Rosendal, K. (2009). The role of the United Nations Environment Programme in the coordination of multilateral environmental agreements. In F. Biermann, B. Siebenhüner, & A. Schreyögg (eds.), *International organizations in global environmental governance* (pp. 133–50). London: Routledge.
Auld, G. (2014). *Constructing private governance: The rise and evolution of forest, coffee and fisheries certification*. New Haven: Yale University Press.
Baccini, L., & Urpelainen, J. (2012). Strategic side payments: Preferential trading agreements, economic reform, and foreign aid. *Journal of Politics*, 74 (4), 932–49.
Baccini, L., & Urpelainen, J. (2014). Before ratification: Understanding the timing of international treaty effects on domestic policies. *International Studies Quarterly*, 58 (1), 29–43.
Barrett, S. (2003). *Environment and statecraft: The strategy of environmental treaty-making*. Oxford: Oxford University Press.
Barrett, S. (2007). *Why cooperate? The incentives to supply global public goods*. Oxford: Oxford University Press.
Berardo, R., & Gerlak, A. (2012). Conflict and cooperation along international rivers. *Global Environmental Politics*, 12 (1), 101–20.
Bernauer, T. (2002). Explaining success and failure in international river management. *Aquatic Sciences*, 64 (1), 1–19.
Bernauer, T., and Böhmelt, T. (2013). National climate policies in international comparison: The climate change cooperation index. *Environmental Science and Policy*, 25, 196–206.
Bernauer, T., and Böhmelt, T. (2014). Basins at risk: Predicting international river basin conflict and cooperation. *Global Environmental Politics*, 14 (4), 116–38.
Bernauer, T., Kalbhenn A., Koubi, V., & Spilker, G. (2013). Is there a 'depth versus participation' dilemma in international cooperation? *Review of International Organizations*, 8 (4), 477–97.
Betsill, M. M., Benney, T. M., & Gerlak, A. K. (eds.) (2019). *Agency in earth system governance*. Cambridge, UK: Cambridge University Press.
Betsill, M. M., & Corell, E. (2008). *NGO diplomacy: The influence of nongovernmental organizations in international environmental negotiations*. Cambridge, MA: The MIT Press.
Betzold, C., Bernauer, T., & Koubi, V. (2016). Press briefings in international climate change negotiations. *Environmental Communication*, 10 (5), 575–92.
Beunen, R., & Patterson, J. J. (2016). Analysing institutional change in environmental governance: Exploring the concept of 'institutional work'. *Journal of Environmental Planning and Management*, 62 (1), 12–29.
Biermann, F., & Pattberg, P. (eds.) (2012). *Global environmental governance reconsidered*. Cambridge, MA: The MIT Press.
Bodansky, D. (2010). *The art and craft of international environmental law*. Cambridge, MA: Harvard University Press.

Bodansky, D. (2015). Legally-binding vs. non-legally-binding instruments. In S. Barrett, C. Carraro, & J. de Melo (eds.), *Towards a workable and effective climate regime* (pp. 155–65). London: VoxEU.

Bodin, Ö., & Österblom, H. (2013). International fisheries regime effectiveness: Activities and resources of key actors in the southern ocean. *Global Environmental Change*, 23 (5), 948–56.

Böhmelt, T., & Betzold, C. (2013). The impact of environmental interest groups in international negotiations: Do ENGOs induce stronger environmental commitments? *International Environmental Agreements*, 13 (2), 127–51.

Böhmelt, T., & Pilster, U. H. (2010). International environmental regimes: Legalisation, flexibility and effectiveness. *Australian Journal of Political Science*, 45 (2), 245–60.

Breitmeier, H. (2008). *The legitimacy of international regimes*. Aldershot: Ashgate.

Breitmeier, H., Underdal, A., & Young, O. R. (2011). The effectiveness of international environmental regimes: Comparing and contrasting findings from quantitative research. *International Studies Review*, 13 (4), 579–605.

Breitmeier, H., Young, O. R., & Zürn, M. (2006). *Analyzing international environmental regimes: From case study to database*. Cambridge, MA: The MIT Press.

Brochmann, M., & Hensel, P. R. (2011). The effectiveness of negotiations over international river claims. *International Studies Quarterly*, 55 (3), 859–82.

Brown Weiss, E., & Jacobson, H. K. (eds.) (1998). *Engaging countries: Strengthening compliance with international environmental accords*. Cambridge, MA: The MIT Press.

Brun, A. (2016). Conference diplomacy: The making of the Paris Agreement. *Politics and Governance*, 4 (3), 115–23.

Brunnée, J., Bodansky, D., & Hey, E. (eds.) (2007). *Oxford handbook of international environmental law*. Oxford: Oxford University Press.

Brunnée, J., & Toope, S. J. (2010). *Legitimacy and legality in international law: An interactional account*. Cambridge, UK: Cambridge University Press.

Byrne, A. (2015). The 1979 Convention on Long-Range Transboundary Air Pollution: Assessing its effectiveness as a multilateral environmental regime after 35 years. *Transnational Environmental Law*, 4 (1), 37–67.

Carpenter, R. C. (2007). Setting the advocacy agenda: Theorizing issue emergence and nonemergence in transnational advocacy networks. *International Studies Quarterly*, 51 (1), 99–120.

Cashore, B., Auld, G., Bernstein, S., & McDermott, C. (2007). Can non-state governance 'ratchet up' global environmental standards? Lessons from the forest sector. *Review of European Community and International Environmental Law*, 16 (2), 158–72.

Cirone, A. E., & Urpelainen, J. (2013). Trade sanctions in international environmental policy: Deterring or encouraging free riding? *Conflict Management and Peace Science*, 30 (4), 309–34.

Corneloup, I. de Agueda, & Mol, A. P. J. (2014). Small island developing states and international climate change negotiations: The power of moral 'leadership'. *International Environmental Agreements*, 14 (3), 281–97.

Cullis-Suzuki, S., & Pauly, D. (2010). Failing the high seas: A global evaluation of regional fisheries management organizations. *Marine Policy*, 34 (5), 1036–42.

Curlier, M., & Andresen, S. (2002). International trade in endangered species: The cites regime. In E. L. Miles, A. Underdal, S. Andresen, J. Wettestad, J. B. Skjærseth, & E. M. Charlin (eds.), *Environmental regime effectiveness: Confronting theory with evidence* (pp. 357–78). Cambridge, MA: The MIT Press.

Dimitrov, R. S. (2016). The Paris Agreement on climate change: Behind closed doors. *Global Environmental Politics*, 16 (3), 1–11.

Dombrowsky, I. (2007). *Conflict, cooperation and institutions in international water management: An economic analysis*. Northampton, MA: Edward Elgar.

Dombrowsky, I. (2008). Institutional design and regime effectiveness in transboundary river management: The elbe water quality regime. *Hydrology and Earth System Sciences*, 12, 223–38.

Downs, G. W., Rocke, D. M., & Barsoom, P. N. (1996). Is the good news about compliance good news about cooperation? *International Organization*, 50 (3), 379–406.

Dryzek, J. S. (2013). *The politics of the earth: Environmental discourses*. Oxford: Oxford University Press.

Epstein, C. (2008). *The power of words in international relations: Birth of an anti-whaling discourse*. Cambridge, MA: The MIT Press.

Falkner, R. (2016). The Paris Agreement and the new logic of international climate politics. *International Affairs*, 92 (5), 1107–25.

Gerst, M. D., Wang, P., Roventini, A., Fagiolo, G., Dosi, G., Howarth, R. B., & Borsuk, M. E. (2013). Agent-based modeling of climate policy: An introduction to the engage multi-level framework. *Environmental Modelling and Software*, 44, 62–75.

Gulbrandsen, L. H. (2014). Dynamic governance interactions: Evolutionary effects of state responses to non-state certificate programs. *Regulation and Governance*, 8 (1), 74–92.

Gulbrandsen, L. H., & Andresen, S. (2004). NGO influence in the implementation of the kyoto protocol: Compliance, flexibility mechanisms, and sinks. *Global Environmental Politics*, 4 (4), 54–75.

Gupta, A. (2008). Transparency under scrutiny: Information disclosure in global environmental governance. *Global Environmental Politics*, 8 (2), 1–7.

Gupta, J. (2010). A history of international climate change policy. *WIRE Climate Change*, 1 (5), 636–53.

Haas, P. M., Keohane, R. O., & Levy, M. A (eds.) (1993). *Institutions for the earth: Sources of effective international environmental protection*. Cambridge, MA: The MIT Press.

Hale, T. (2016). 'All hands on deck': The Paris Agreement and nonstate climate action. *Global Environmental Politics*, 16 (3), 12–22.

Haufler, V. (2010). Disclosure as governance: The extractive industries transparency initiative and resource management in the developing world. *Global Environmental Politics*, 10 (3), 53–73.

Helm, C., & Sprinz, D. F. (2000). Measuring the effectiveness of international environmental regimes. *Journal of Conflict Resolution*, 44 (5), 630–52.

Hønneland, G., & Stokke, O. S. (2007). *International cooperation and arctic governance: Regime effectiveness and northern region building*. New York: Routledge.

Houghton, K. A., & Naughton, H. T. (2014). International environmental agreement effectiveness: A review of empirical studies. In T. Eisenberg, & G. B. Ramello (eds.), *Comparative law and economics* (pp. 442–55). Oxford: Oxford University Press.

Hovi, J., Sprinz, D. F., Sælen, H., & Underdal, A. (2017). The club approach: A gateway to effective climate co-operation? *British Journal of Political Science*, 1–26.

Hovi, J., Sprinz, D. F., & Underdal, A. (2003). The Oslo-Potsdam solution to measuring regime effectiveness: Critique, response, and the road ahead. *Global Environmental Politics*, 3 (3), 74–96.

IEA Database (2019). *International Environmental Agreements Database Project (Version 2018.1)*. Available at: http://iea.uoregon.edu. Accessed: 4 March 2019.

Ivanova, M. H. (2010). UNEP in global environmental governance: Design, leadership, location. *Global Environmental Politics*, 10 (1), 30–59.

Jinnah, S. (2011). Marketing linkages: Secretariat governance of the climate-biodiversity interface. *Global Environmental Politics*, 11 (3), 23–43.

Jinnah, S. (2014). *Post-treaty politics: Secretariat influence in global environmental governance.* Cambridge, MA: The MIT Press.

Kellenberg, D., & Levinson, A. (2014). Waste of effort? International environmental agreements. *Journal of the Association of Environmental and Resource Economists*, 1 (1–2), 135–69.

Keohane, R. O., & Oppenheimer, M. (2016). Paris: Beyond the climate dead end through pledge and review? *Politics and Governance*, 49 (3), 142–51.

Keohane, R. O., & Victor, D. G. (2011). The regime complex for climate change. *Perspectives on Politics*, 9 (1), 7–23.

Khoo, S., & Rau, H. (2009). Movements, mobilities and the politics of hazardous waste. *Environmental Politics*, 18 (6), 960–80.

Kim, Y., Tanaka, K., & Matsuoka, S. (2017). Institutional mechanisms and the consequences of international environmental agreements. *Global Environmental Politics*, 17 (1), 77–98.

Koremenos, B., Lipson, C., & Snidal, D. (2001). The rational design of international institutions. *International Organization*, 55 (4), 761–99.

Krasner, S. D. (ed.) (1983). *International regimes.* Ithaca, NY: Cornell University Press.

Lempert, R., Scheffran, J., & Sprinz., D. F. (2009). Methods for long-term environmental policy challenges. *Global Environmental Politics*, 9 (3), 106–33.

Lucier, C. A., & Gareau, B. J. (2015). From waste to resources?: Interrogating 'race to the bottom' in the global environmental governance of the hazardous waste trade. *Journal of World-Systems Research*, 21 (2), 495–520.

Mason, M. (2008). Transparency for whom? Information disclosure and power in global environmental governance. *Global Environmental Politics*, 8 (2), 8–13.

Miles, E. L., Underdal, A., Andresen, S., Wettestad, J., Skjærseth, J. B., & Carlin, E. M. (eds.) (2002). *Environmental regime effectiveness: Confronting theory with evidence.* Cambridge, MA: The MIT Press.

Milkoreit, M. (2015). Hot deontology and cold consequentialism: An empirical exploration of ethical reasoning among climate change negotiators. *Climatic Change*, 130 (3), 397–409.

Mitchell, R. B. (2006). Problem structure, institutional design, and the relative effectiveness of international environmental agreements. *Global Environmental Politics*, 6 (3), 72–89.

Mitchell, R. B. (2009). The influence of international institutions: Institutional design, compliance, effectiveness, and endogeneity. In H. V. Milner (ed.), *Power, interdependence and non-state actors in world politics: Research frontiers* (pp. 66–83). Princeton, NJ: Princeton University Press.

Mitchell, R. B. (2010). *International politics and the environment.* London: SAGE.

Mitchell, R. B., Clark, W. C., Cash, D. W., & Dickson, N. (eds.) (2006). *Global environmental assessments: information and influence.* Cambridge, MA: The MIT Press.

Mitchell, S. M. L., & Zawahri, N. A. (2015). The effectiveness of treaty design in addressing water disputes. *Journal of Peace Research*, 52 (2), 187–200.

Myint, T. (2012). *Governing international rivers: Polycentric politics in the Mekong and the Rhine.* Northampton, MA: Edward Elgar.

Nagtzaam, G. J. (2009). *The international tropical timber organization and conservationist forestry norms: A bridge too far.* Melbourne: Monash University.

Newell, P., Bulkeley, H., Turner, K., Shaw, C., Caney, S., Shove, E., & Pidgeon, N. (2015). Governance traps in climate change politics: Re-framing the debate in terms of responsibilities and rights. *WIRE Climate Change*, 6 (6), 535–40.

Nordhaus, W. D. (2015). Climate clubs: Overcoming free-riding in international climate policy. *American Economic Review*, 105 (4), 1339–70.

Oberthür, S., & Stokke, O. S. (2011). *Managing institutional complexity: Regime interplay and global environmental change*. Cambridge, MA: The MIT Press.

Oreskes, N., & Conway, E. M. (2010). *Merchants of doubt: How a handful of scientists obscured the truth on issues from tobacco smoke to global warming*. New York: Bloomsbury Press.

Parson, E. A. (2003). *Protecting the ozone layer: Science and strategy*. Oxford: Oxford University Press.

Perrin, S., & Bernauer, T. (2010). International regime formation revisited: Explaining ratification behavior with respect to long range transboundary air pollution agreements in Europe. *European Union Politics*, 11 (3), 405–26.

Pettenger, M. E. (ed.) (2007). *The social construction of climate change: Power, knowledge, norms, discourses*. Aldershot: Ashgate.

Ragin, C. C. (1987). *The comparative method: Moving beyond qualitative and quantitative strategies*. Berkeley: University of California Press.

Ringquist, E., & Kostadinova, T. (2005). Assessing the effectiveness of international environmental agreements: The case of the 1985 Helsinki protocol. *American Journal of Political Science*, 49 (1), 86–102.

Saleh, P., & Abene, N. M. (2016). Africa and the problem of transboundary movement of hazardous waste: An assessment of the Bamako convention of 1991. *Journal of Law, Policy and Globalization*, 48, 47–53.

Sand, P. H. (2013). Enforcing cites: The rise and fall of trade sanctions. *Review of European Community and International Environmental Law*, 22 (3), 251–63.

Schroeder, H., & Lovell, H. (2012). The role of non-nation-state actors and side events in the international climate negotiations. *Climate Policy*, 12 (1), 23–37.

Seelarbokus, C. B. (2014a). Assessing the effectiveness of international environmental agreements (IEAs): Demystifying the issue of data unavailability. *SAGE Open*, 4 (1), 1–18.

Seelarbokus, C. B. (2014b). The influence of treaty design on the participation of developing and developed nations in international environmental agreements (IEAs). *African Journal of Political Science and International Relations*, 8 (8), 288–301.

Seto, K. C., Davis, S. J., Mitchell, R. B., Stokes, E. C., Unruh, G., & Ürge-Vorsatz, D. (2016). Carbon lock-in: Types, causes, and policy implications. *Annual Review of Environment and Resources*, 41, 425–52.

Shi, L. (1999). Successful use of the tacit acceptance procedure to effectuate progress in international maritime law. *University of San Francisco Maritime Law Journal*, 11 (2), 299–332.

Siebenhüner, B. (2008). Learning in international organizations in global environmental governance. *Global Environmental Politics*, 8 (4), 92–116.

Spilker, G., & Koubi, V. (2016). The effects of treaty legality and domestic institutional hurdles on environmental treaty ratification. *International Environmental Agreements*, 16 (2), 223–38.

Sprinz, D. F., & Vaahtoranta, T. (1994). The interest-based explanation of international environmental policy. *International Organization*, 48 (1), 77–105.

Stokke, O. S. (2012). *Disaggregating international regimes: A new approach to evaluation and comparison*. Cambridge, MA: The MIT Press.

Stokke, O. S. (2013). Regime interplay in Arctic shipping governance: Explaining regional niche selection. *International Environmental Agreements*, 13 (1), 65–85.

Strange, S. (1983). Cave! Hic dragones: A critique of regime analysis. In S. D. Krasner (ed.), *International regimes* (pp. 337–54). Ithaca, NY: Cornell University Press.

Underdal, A., & Hanf, K. (eds.) (2000). *International environmental agreements and domestic politics: The case of acid rain*. Aldershot: Ashgate.

Underdal, A., & Young, O. R. (eds.) (2004). *Regime consequences: Methodological challenges and research strategies*. Dordrecht: Kluwer Academic Publishers.

Urpelainen, J. (2013). A model of dynamic climate governance: Dream big, win small. *International Environmental Agreements*, 13 (2), 107–25.

Vabulas, F., & Snidal, D. (2013). Organization without delegation: Informal intergovernmental organizations and the spectrum of intergovernmental arrangements. *Review of International Organizations*, 8 (2), 193–220.

Van der Ven, H., Bernstein, S., & Hoffmann, M. (2017). Valuing the contributions of nonstate and subnational actors to climate governance. *Global Environmental Politics*, 17 (1), 1–20.

Vanhala, L., & Hestbaek, C. (2016). Framing climate change loss and damage in UNFCCC negotiations. *Global Environmental Politics*, 16 (4), 111–29.

Victor, D. G. (2011). *Global warming gridlock: Creating more effective strategies for protecting the planet*. Cambridge, UK: Cambridge University Press.

Victor, D. G., Raustiala, K., & Skolnikoff, E. B. (eds.) (1998). *The implementation and effectiveness of international environmental commitments: Theory and practice*. Cambridge, MA: The MIT Press.

Vollenweider, J. (2013). The effectiveness of international environmental agreements. *International Environmental Agreements*, 13 (3), 343–367.

Von Stein, J. (2008). The international law and politics of climate change: Ratification of the United Nations Framework Convention and the Kyoto protocol. *Journal of Conflict Resolution*, 52 (2), 243–68.

Wettestad, J. (2011). The improving effectiveness of CLRTAP: Due to a clever design? In R. Lidskog, & G. Sundqvist (eds.), *Governing the air: The dynamics of science, policy, and citizen interaction* (pp. 39–60). Cambridge, MA: The MIT Press.

Young, O. R. (1999). *Governance in world affairs*. Ithaca, NY: Cornell University Press.

Young, O. R. (2001). Inferences and indices: Evaluating the effectiveness of international environmental regimes. *Global Environmental Politics*, 1 (1), 99–121.

Young, O. R. (2002). *The institutional dimensions of environmental change: Fit, interplay, and scale*. Cambridge, MA: The MIT Press.

Young, O. R. (2003). Determining regime effectiveness: A commentary on the Oslo-Potsdam solution. *Global Environmental Politics*, 3 (3), 97–104.

Young, O. R. (2008). The architecture of global environmental governance: Bringing science to bear on policy. *Global Environmental Politics*, 8, 14–32.

Young, O. R. (2010). *Institutional dynamics: Emergent patterns in international environmental governance*. Cambridge, MA: The MIT Press.

Young, O. R. (2011). Effectiveness of international environmental regimes: Existing knowledge, cutting-edge themes, and research strategies. *Proceedings of the National Academy of Sciences of the United States of America*, 108 (50), 19853–60.

Young, O. R. (2017). *Governing complex systems: Social capital for the Anthropocene*. Cambridge, MA: The MIT Press.

Young, O. R. (ed.) (1999). *The effectiveness of international environmental regimes: Causal connections and behavioural mechanisms*. Cambridge, MA: The MIT Press.

3

International Bureaucracies

DOMINIQUE DE WIT, ABBY LINDSAY OSTOVAR, STEFFEN BAUER
AND SIKINA JINNAH

Earth system governance is marked by a diversity of actors engaging in affairs that were previously assumed to be solely the purview of governments. Recent scholarship has demonstrated, for example, the important roles played by non-governmental organizations, corporations and subnational actors, such as cities, in shaping the outcomes of global governance.[1] This still-growing literature has demonstrated how diverse actors perform a wide variety of governance functions. Some researchers have shown, for instance, how cities have come to differentiate themselves from states by setting their own greenhouse gas reduction targets and by steering local actors towards bottom-up climate governance (Gordon and Johnson 2017). Others explain how non-governmental organizations create networks that shape the trajectory of global climate politics (Hadden 2015). Similarly, corporate actors use private authority that is either delegated to them by governments or established through their entrepreneurial activities (Green 2013). For example, corporations have developed their own rules and persuaded others to follow, such as with greenhouse gas accounting standards. Yet, while all these non-governmental actors can influence international politics, they do not operate in a vacuum: their actions still respond to the institutional environment set by governments (Green 2013; see also Cutler, Haufler and Porter 1999; Pattberg 2007).

A growing body of literature on international bureaucracies suggests a similar agency and autonomy to be true for this group of actors as well as for the 'bureaucratic authority' that they muster (Barnett and Finnemore 2004; Bauer 2006; Biermann and Siebenhüner 2009; Jinnah 2014). This chapter, hence, reviews the rapidly expanding literature on international bureaucracies as key actors in earth system governance. We first conceptualize the term, and then present key research findings of the last decade, followed by suggestions for future research.

---

[1] See, for instance, Pattberg 2007; Betsill and Corell 2008; Andonova, Betsill and Bulkeley 2009; Biermann and Siebenhüner 2009; Green 2013; Bulkeley et al. 2014; Hickmann 2015; Chan, Brandi and Bauer 2016; Allan and Hadden 2017.

Conceptualization

The treaty secretariats and other international bureaucracies that we discuss in this chapter are all administrative entities of international treaties and organizations, sometimes also referred to as international public administrations. Although they all differ in terms of size, staff, budget and the institutions they serve, they are all agencies created by governments but fall outside the direct control of any single national government (Bauer, Andresen and Biermann 2012). Consequently, they are at the same time public actors and non-state actors (Bauer and Weinlich 2011). Among international bureaucracies, treaty secretariats are a specific type with a more focused mandate compared to the bureaucracies of larger international organizations. However, commonalities in mandate, practice and institutional design warrant considering all these actors together for the purposes of this chapter.

Long presumed to be epiphenomenal functionaries of their member states, international bureaucracies are increasingly recognized as significant actors (Bauer and Weinlich 2011; Bauer, Andresen and Biermann 2012; Jinnah 2014). Indeed, recent governance scholarship has found that under certain conditions, bureaucracies can exhibit varying degrees of autonomy, authority and influence (Biermann and Siebenhüner 2009). Recent studies also demonstrate how bureaucracies can have a significant impact on global governance even without formal decision-making powers. Such impacts can include agenda-setting, institution building and institutional interlinkages, implementation of international norms and rules and even shaping state preferences through development and provision of expertise.

In this chapter, we review recent scholarship on the activities of international bureaucracies in earth system governance, and show that their activities are not merely instrumental in nature. Some scholars refer to this as bureaucratic influence; others focus on the authority or autonomy that bureaucracies muster against governments and other actors in global governance. We focus on the distinct ways in which bureaucracies are emerging as actors in their own right and beyond their narrow functional mandates. We review findings that help us to understand how bureaucracies influence politics and from what sources they draw their capacity to influence.

Specifically, we identify three themes that help us to understand the sources of the influence of bureaucracies in global governance. In the first section below, we review scholarship that demonstrates how bureaucracies derive authority from structural characteristics of the international system. We focus on how the relative position of bureaucracies to other actors enables them to partake in political processes in the context of the dynamic institutional fragmentation and complexity of earth system governance (see Chapter 8). In the next section, we consider the influence of bureaucracies as derived from their ability to deliver specific functions. We look to the administrative and governance functions that allow bureaucracies to act as agenda setters, policy entrepreneurs and brokers of expertise in an increasingly diverse setting of global governance. Third, we look at the influence of bureaucracies as rooted in organizational autonomy, reflecting recent arguments that a degree of organizational autonomy is a minimum prerequisite for bureaucratic influence.

In line with the broader earth system governance approach to analysis, we take an interdisciplinary perspective and review scholarship from several fields, including global environmental politics, international relations, organizational studies, public administration, comparative politics and policy analysis. While we refer to some essential early contributions from the broader literature on international organization and bureaucracy, our primary focus is on recent contributions to the study of international bureaucracies. To that end, we build on Biermann and Siebenhüner's (2009) take on international bureaucracies as 'managers of global change' as a vantage point.

Reviewing this literature, we cover both treaty secretariats that support international treaties or agreements and the bureaucracies of larger international organizations that often have broader and more generic mandates. Both are international bureaucracies created by governments to pursue organizational objectives or to assist in policy implementation, and both are beyond direct formal control of any single national government. While mandate and scope can differ considerably, the bureaucratic nature and practice of both types warrants important insights from comparative analysis.

Sources of Influence of International Bureaucracies

The state-centric orthodoxy of most international relations scholarship resulted in neglecting bureaucracies as political actors (Hickmann and Elsässer 2018). Even liberal institutionalists, who have considered international organizations as meaningful actors, still gave short shrift to the bureaucracies of international organizations. Indeed, they often conflated them with the international organizations of which they are part (Barnett and Finnemore 1999; Reinalda and Verbeek 2003; Hawkins et al. 2006). When acknowledging the agency of international organizations, rationalist scholars have typically adopted a principal-agent perspective to explain how and why governments might delegate autonomy to international bureaucracies, and why bureaucracies enjoy autonomy in circumstances when their preferences deviate from those of their state principals (Elsig 2010; Johnson 2013; Johnson and Urpelainen 2014).

While applying a principal-agent framework to understanding the behaviour of international bureaucracies is straightforward in many ways, there are also several shortcomings. Crucially, principal-agent scholarship is focused on the delegation of authority as the primary, if not exclusive, source of autonomous behaviour of international bureaucracies (Hawkins et al. 2006). It also assumes a priori specific preferences of principals as well as agents (Bauer and Weinlich 2011). Moreover, in research practice the application of principal-agent theory typically stops short of scrutinizing the actual behaviour of bureaucracies, even though it is acknowledged that principal-agent analysis will need 'to understand agents in greater detail' (Hawkins and Jacoby 2006: 200).

Sociological institutionalist approaches to international organizations, such as Barnett and Finnemore's (2004) ground-breaking *Rules for the World*, consider delegation as only one of several potential sources of authority and are ontologically inclined to shed light into the 'black boxes' of international organizations and their bureaucracies. This allows scholars to develop a deeper understanding of the internal dynamics and sources of authority of international bureaucracies in relation to their external environment (Bauer and Weinlich 2011: 256). Consequently, preferences do not need to be assumed as a prerequisite to investigate the behaviour of bureaucracies and their sources of influence.

The scholarship on international bureaucracies recognizes that bureaucracies are not always and necessarily influential in policy formulation or governance implementation. Nevertheless, a growing number of in-depth case studies demonstrates how bureaucracies can be influential based on autonomous behaviour and when this is likely to be the case. For a critical appraisal of this literature, we now turn to the influence of bureaucracies as derived from the three sources that we identified above: international structure, the functions of bureaucracies, and organizational autonomy.

International Structure

The international system is increasingly characterized by polycentric, fragmented and complex systems of governance, wherein a multiplicity of state and non-state actors engages in polycentric, fragmented and generally complex institutions across levels and scales (e.g., Raustiala and Victor 2004; Zelli and van Asselt 2013; Jordan et al. 2015; Young 2017; Nicholson, Jinnah and Gillespie 2018). In earth system governance, increasing complexity and fragmentation are driven considerably by a proliferation of international institutions and treaties and the concomitant increase of bureaucracies that help manage these bodies (Bauer, Andresen and Biermann 2012). This refers to both the specialized agencies of the United Nations and other larger intergovernmental organizations with their bureaucracies, such as the World Trade Organization or the OECD, and the large and growing number of multilateral agreements that establish their own treaty secretariats.

Part of the complexity follows from the underlying nature of earth system problems. It has become apparent how intertwined technically separate issue areas such as climate change, forestry, land degradation and biodiversity really are – hence the very quest for integrated earth system governance (Biermann 2014: 15–45). To give one example: if international climate policy requires the consideration of land use change and threats to biodiversity, regime overlaps between the Convention on Biological Diversity and the United Nations Framework Convention on Climate Change are inevitable. This contributes to complex governance situations, wherein 'regime complexes' made up of various overlapping institutions work with varying degrees of coordination to govern aspects of the same or similar issues (see Chapter 7).

In turn, there is now much overlap between the mandates, membership, policies and objectives of the complex web of institutions that are engaged in earth system governance. Such overlap is evident across, for example, a host of biodiversity-related agreements (Andresen and Rosendal 2009), but also across institutions that operate in seemingly separate realms of politics, such as trade and environment (e.g., Gehring 2011; Jinnah and Lindsay 2015; Morin and Jinnah 2018; Jinnah and Morin 2020). Several scholars have pointed to these structural conditions as enabling the influence of bureaucracies. For example, overlaps between institutions have created a specific governance niche for 'overlap management' and 'orchestration', thereby creating a new space for bureaucracies and other actors to engage (Jinnah 2014; Abbott et al. 2016).

Indeed, the influence of international bureaucracies may become most visible where regimes overlap (Jinnah 2014). In these instances, governments may lack adequate expertise, capacity or the institutions for domestic cross-issue coordination and hence delegate these tasks to international bureaucracies. Whereas governments prove increasingly unable to manage the burgeoning labyrinth of treaties by themselves, treaty secretariats have been able to navigate complex settings by drawing from their unique networks, capacities and expertise to shape the decision-making of international organizations.

The participation of international bureaucracies in politics also comports with broader understandings of transnational governance (see Chapter 4). Transnational governance reflects the characteristics of contemporary policymaking, which is increasingly addressed by a system of interaction among supranational, national, regional and local levels of government, both within and beyond the state. Indeed, a rich literature has emerged to identify and analyze interactions among different groups of non-state and subnational actors in global governance (e.g., Stone and Ladi 2015; Chan, Brandi and Bauer 2016; Gordon and Johnson 2017; Roger, Hale and Andanova 2017; Hickmann and Elsässer 2018). For example, several scholars have highlighted an increasing hybridity of actors producing transnational governance initiatives (Hale and Roger 2014). The 2015 Paris Agreement on climate change provides a recent case in point as it might have become the most extensive framework for non-state actor engagement to date (Chan, Brandi and Bauer 2016; Dryzek 2017). Recognizing such institutional innovations is important for understanding the influence of international bureaucracies as it points to the extending networks of actors within which bureaucracies interact in addition to states and through which they can potentially exert influence. This is even more plausible when international bureaucracies are situated at the hub of such networks by virtue of providing their functions.

Several scholars have demonstrated, for example, how international bureaucracies cooperate with non-governmental organizations to build support for preferred policy outputs and to implement policies (Prideaux 2015; Jörgens et al. 2016). In some instances, this is motivated by the limited financial resources available to international bureaucracies. For instance, the secretariat of the Convention on Migratory Species increasingly commits non-governmental organizations to policy implementation efforts because of its limited financial resources (Prideaux 2015: 380). By fostering such collaborative efforts between the treaty secretariat, non-governmental organizations and governments, the secretariat can influence how implementation efforts eventually unfold and on which terms.

Other scholars have focused on how international bureaucracies indirectly influence policymaking prior to policy implementation. For example, international bureaucracies may try to build support for policy outputs by working with actors within broader transnational policy networks to exert pressure on negotiators from outside formal negotiations (Jörgens et al. 2016). They may also liaise with governments, for instance, maintaining close and potentially influential relations with the chairpersons of intergovernmental processes, including at conferences of the parties (Depledge 2007).

Some scholars suggest, for example, that the secretariat of the Convention on Biological Diversity holds its own views on the problems related to biodiversity conservation, policy options and framing needed to reach an agreement in multilateral negotiations. It may deliberately 'ruffle some feathers' with governments when providing them with policy-relevant information (Jörgens et al. 2016: 74–6; see also Siebenhüner 2009). It attempts to build support for its policy outputs by engaging with actors such as international organizations and other international bodies, including UNESCO, FAO, UNEP and the International Treaty on Plant Genetic Resources for Food and Agriculture. The secretariat to the biodiversity convention further gathers and distributes information, intending to engage other organizations in issue-specific debates to increase support for policy outcomes and to advance negotiations (Jinnah 2010, 2011; Jörgens et al. 2016: 78).

Recently, international bureaucracies have also been recognized as orchestrators in governance (see Chapter 11; see also Hickmann et al. 2019). Orchestration has become a quickly growing conceptual focus in global governance, referring to the strategy in which intergovernmental organizations and other actors work through intermediaries such as private actors, civil society and transnational and subnational actors to indirectly pursue governance goals (Hale and Roger 2014; Abbott et al. 2015). While scholars advancing the concept of orchestration have primarily focused on states and international organizations as orchestrators, international bureaucracies can play this role as well. Hickmann and Elsässer (2018), for instance, argue that the secretariats of the conventions on climate change, biodiversity and desertification have all acted as orchestrators to various bottom-up projects and programmes conducted by non-state actors. They find, for example, that the *Knowledge Hub* that the secretariat to the desertification convention has created, directs the attention of non-state actors to specific desertification issues; and its strategic communication platform, *Global Land Outlook*, brings together experts and partner organizations to identify common opportunities on issues of relevance to desertification policy (Hickmann and Elsässer 2018). Ultimately, the observed practice of orchestration that emerges in increasingly complex global governance effectively reaffirms the capacity of international bureaucracies to act as agenda setters, policy entrepreneurs or policy brokers (Jörgens et al. 2016).

Governance Functions

The specific governance functions of international bureaucracies are also important for understanding their ability to influence policy outcomes. There is a range of such functions. For example, international bureaucracies may undertake administrative functions in assisting with multilateral negotiations, policy formulation and policy implementation through gathering information, drafting proposals, overseeing state policy and treaty implementation and so on (Biermann 2014: 65–70). They thus provide sources of specialized expertise that is often beyond that of governments, as they act as a repository of information and conduct technical studies independent of governments (Jinnah 2014). Moreover, international bureaucracies are often tasked with awareness-building measures regarding their sphere of influence, and with coordination between states as well as with non-state actors (Hickmann and Elsässer 2018). Also, as noted above, scholars have argued that the coordination function of international bureaucracies can play out as 'orchestration' that guides governance interactions with non-state or subnational actors (Abbott and Snidal 2010; Hale and Roger 2014; Abbott et al. 2015).

Pertinent to this set of functions, empirical research suggests that international bureaucracies wield more influence when governance problems are highly technical and complex, including for example issues that concern more than one international regime or that address new or emerging issues (Jinnah 2014). In such contexts, international bureaucracies are particularly suited to capitalize on sources of authority such as organized expertise, institutional memory and their rational-legal foundations. Accordingly, the expertise of international bureaucracies has been identified as critical in understanding their capacity to exert influence across actors with divergent perspectives (Bauer 2006; Biermann and Siebenhüner 2009; Johnson 2013; Jinnah 2014; Johnson and Urpelainen 2014).

Importantly, the relations between international bureaucracies and governments vary according to the type of international organizations. Larger secretariats of international organizations typically have broad, far-reaching mandates and responsibilities, such as advising countries on policies. This contrasts with the mandates of specialized treaty secretariats, such as those of the biodiversity convention or the convention on trade in endangered species, which are 'typically discrete, aimed at carrying out a specific task, such as monitoring implementation or building a database for collecting and disseminating information' (Jinnah 2014: 15). Jinnah and others have delineated how the relationship between the secretariats and the member states of their respective international organizations differs between these two types: larger secretariats tend to be given more formal authority and 'delegated discretion' (Hawkins and Jacoby 2006), whereas treaty secretariats interact with member states mainly through regular meetings such as annual conferences of the parties (Jinnah 2014).

This suggests neither that governments always follow the expertise of international bureaucracies nor that such expertise is offered invariably. Indeed, qualitative and quantitative empirical analyses confirm that state preferences and their need for expertise circumscribe the ability of international bureaucracies to influence from the outset. Quantitative research by Johnson (2013), for instance, shows how international bureaucracies have influenced institutional design when governments need their expertise and when the issue at stake had low political salience for governments. Qualitative case studies in turn arrive at complementary conclusions. Jinnah (2014) finds international bureaucracies to be most likely to influence outcomes when the preferences of governments are weakly formed (that is, expertise is needed to form them, and they are at least perceived as lacking salience), and when the expertise of international bureaucracies is not substitutable with that of other available actors, such as non-governmental organizations or scientists.

During multilateral negotiations, the functions of international bureaucracies such as the drafting of proposals can be a source of influence as well (Biermann and Siebenhüner 2009). This pathway of influence can be particularly strong, depending on the relationship between the secretariat and the chairperson of a negotiation. Building on qualitative analysis of negotiations under the desertification convention, Depledge (2007: 46) argues that the two are 'locked into a mutually enforcing relationship and symbiotic division of labour, so that the role of one cannot be fully understood without reference to the other'. Through analyzing politically contentious climate negotiations, she traces how international bureaucracies can operate under a chairperson's 'veil of legitimacy': secretariat staff can lend their expertise to the chairpersons, who then use the information and proposals at their discretion. Depledge (2007: 65) concludes that '[i]n essence, Chairpersons wield their political authority to exert leadership, whereas the secretariat provides much of the intellectual resources needed to make that leadership effective'.

This demonstrates how specialized expertise and knowledge serve as sources of authority for international bureaucracies. Although such knowledge is often assumed to be technical and impartial, it can also be strategically deployed and politicized. When the issues at stake are not particularly contentious, international bureaucracies have also been found to use such strategies to expand their mandates (Reinalda and Verbeck 2003; Barnett and Finnemore 2004; Bauer 2009; Hall 2016). International bureaucracies have been shown to expand their mandate successfully into areas in which they originally did not have expertise themselves but harnessed the expertise of external actors (Littoz-Monnett 2017). In general, scholars have found the influence of international bureaucracies to be strongest when their expertise was perceived as being objective, impartial, fair or valid (Biermann and Siebenhüner 2009; Busch 2014).

The coordination functions of international bureaucracies can also be a source of influence. International bureaucracies are often tasked with maintaining contacts, both governmental and non-governmental, which enables them to identify and involve actors that they consider relevant. For instance, Hale (2016) discusses the climate regime's embrace of 'climate action' by subnational and non-state actors in the context of the Paris Agreement. He highlights the role of the secretariat of the climate convention in recognizing these actors and in facilitating their participation through its Momentum for Change initiative, rather than leaving them at the sidelines of the state-centred institutions of the climate convention (Hale 2016: 15).

Furthermore, the connections of international bureaucracies with key actors in policy networks enable them to infer influence in the policy process without explicitly stating their policy preferences. For example, the secretariat of the climate convention administered an online Twitter debate about the Lima Work Programme on Gender, which intended to better incorporate gender aspects in the climate regime. The centrality of the position of the secretariat amidst other key actors of that debate indicates an influential brokerage position within policy networks (Jörgens, Kolleck and Saerbeck 2016).

The governance functions that international bureaucracies carry out thus provide a source for two main types of their influence – cognitive and normative. They have in-depth expertise in their area, often exceeding the expertise of governments, and they have extensive procedural as well as substantive information. This enables them to act as 'knowledge brokers' and influence the knowledge and belief systems of governments and others (Biermann and Siebenhüner 2009). Additionally, through their active roles within multilateral agenda setting, policy development and policy implementation they wield normative influence, affecting international norms and treaties (Biermann and Siebenhüner 2009). This normative influence can be further strengthened through their coordinating functions as well.

Bureaucratic Autonomy

Although scholars of international relations have only recently come to study international bureaucracies as political actors, the degree to which bureaucracies can exert autonomous influence on politics has long been a key concern in other fields such as public administration scholarship and sociology (e.g., Beetham 1987; Gruber 1987). In trying to understand how international bureaucracies act autonomously from governments, recent literature has underscored the importance of core institutional characteristics. Centrally, scholars have argued that international bureaucracies need a degree of discretion over the interpretation of their mandate, their organizational structure and the resources at their disposal as a minimum prerequisite for acting autonomously (Ege 2017). Understanding their internal structures is important for understanding the relation between the mandates, resources and behaviour of international bureaucracies. Hence, it is relevant whether there are variations in influence between international bureaucracies following from discrepancies in formal arrangements and organizational behaviour (Knill, Eckhard and Grohs 2016). Here we argue that the autonomy that international bureaucracies enjoy is an important source of their influence, yet it does not flow linearly from their mandate and formal organizational provisions.

Recognizing that structural characteristics of international bureaucracies are preconditions for their actions, we follow Bauer and Ege (2017) in assessing how bureaucratic autonomy is a source of influence. We distinguish two main types of autonomy. First, 'autonomy of will' stems from characteristics in the bureaucratic administration and refers to the ability of international bureaucracies to form their own preferences independent of governments, which can have ideational influence particularly in the early stages of policymaking. Second, 'autonomy of action' is the ability to translate preferences into action as a result primarily of delegated authority and commensurate resources (Bauer and Ege 2017).

The concept of an autonomy of will builds on sociological literature that focuses on intra-organizational characteristics. It is in line with earlier research that emphasized the relevance of organizational cultures and internal procedures to explain variation in bureaucratic autonomy (Biermann and Siebenhüner 2009).

Biermann and Siebenhüner break this down into organizational expertise (process knowledge), organizational structure, organizational culture and leadership. For example, they found that the World Bank demonstrates strong leadership and high expertise, among other factors, which in turn enabled it to undertake extensive knowledge generation activities, to strongly influence agenda setting and to facilitate proactive initiatives (Biermann and Siebenhüner 2009). Bauer and Ege (2017) argue that the freedom of international bureaucracies to develop preferences independent of governments is contingent on their ability to develop their own preferences ('administrative differentiation') and their ability to act as a unified organizational entity ('administrative cohesion'). Yet, autonomy of will does not automatically translate into action or influence.

Autonomy of action implies a more relational concept of the bureaucracies of international organizations and their member states. It reflects a principal-agent perspective in that the autonomy of the agent (the international bureaucracy) depends on the powers and resources delegated by the principals (that is, member states) (Hooghe and Marks 2015; Ege 2017). Therefore, the ability of international bureaucracies to act depends on the length of their leash, and governments are assumed to be able to reduce this ability by curbing statutory powers or the human and financial resources that enable the international bureaucracy to act (Bauer and Ege 2017). The length of the leash can also depend on the political salience of the issue at stake. For instance, Busch (2009) illustrates how the member states of the climate convention effectively 'straitjacketed' its secretariat through constrictive formal and informal rules in the context of politically sensitive issues of climate mitigation. Governments at the time were wary of a strong climate secretariat that might advance an independent policy agenda that could have adverse consequences on wealth or prosperity (Busch 2009). Governments rather wanted the secretariat to assist in negotiations as their servant, rendering it hence unable to exercise leadership or to play any autonomous role (Busch 2009: 254).

That stood in stark contrast to the secretariat of the Convention on Biological Diversity, which was at the same time characterized as a 'lean shark' that was effectively advancing international negotiations and was perceived as neutral and trustworthy by governments (Siebenhüner 2009). Contrasting the limited role of the secretariat in climate negotiations with the more involved role of the secretariat of the biodiversity convention shows the limits of bureaucratic autonomy subject to political salience and perceptions of technical neutrality (see also Bauer, Busch and Siebenhüner 2009 for a comparative analysis of the secretariats of the conventions on climate, biodiversity and desertification).

Drawing on international public administration research, Knill and colleagues (2016) note that the organizational constraints placed upon bureaucracies by governments cannot alone explain administrative styles, and that the bureaucracies' perceptions of the roles of their international organizations affect their behaviour. Comparing administrative styles across three analytical dimensions, they show how two very different bureaucratic structures – one with a more autonomous structure and one with a weak bureaucracy – both have similar administrative styles. They find both bureaucracies to act as entrepreneurs (as opposed to servants) during policy initiation (the problem-definition and agenda-setting phase), to try to strategically optimize policy solutions during policy formulation (as opposed to merely executing member states' directives without questioning their task) and to apply mediating approaches (as opposed to interventionist approaches) during policy implementation (Knill, Eckhard and Grohs 2016: 1060).

Knill and colleagues conclude that international bureaucracies chose different approaches than would be expected solely on the basis of their mandates and structures. They demonstrate that even the bureaucracy with a weak mandate chose entrepreneurial and strategic approaches as it was responding to context and seeking to avoid irrelevance. Conversely, the institution with the stronger mandate was aware that its strength could come under scrutiny from powerful governments, if it did not show self-restraint. This indicates that political context and self-perception need to be taken into account in addition to mandates and organizational structures (Knill, Eckhard and Grohs 2016).

Overall the perceived impartiality, authority and legitimacy of international bureaucracies are the underpinnings of their autonomy. While the authority of international bureaucracies is in part delegated from principals, it also draws on other sources as elaborated above. Yet, while international bureaucracies may be impartial to specific governments' preferences, the bureaucracies are themselves not neutral (Depledge 2005: 48). The trust governments have in international bureaucracies and the legitimacy they grant to their actions builds largely on attributing to them the role of international civil servants who are credited as being politically neutral and unbiased (Dingwerth and Pattberg 2009). In essence, the autonomy of international bureaucracies is based on their aptitude at walking a tightrope between their own preferences, provided they have autonomy of will, and the leeway granted to them by governments.

Conclusions and Future Directions

International bureaucracies are here to stay as a distinct analytical category of actors in earth system governance. This is a timely adjustment to the state-centred orthodoxy in the study of international relations, which hitherto had largely neglected the bureaucratic underpinnings of international organization. Even as the field opened to recognizing a variety of actors that are not states in the context of globalization and emergent concepts of transnational and global governance, international bureaucracies have remained latecomers – academically speaking – relative to non-governmental organizations, multinational corporations and other non-state actors (for a comprehensive overview see Reinalda 2011).

As with the theorizing of international regimes and institutions (Zürn 1998), environmental governance research was at the forefront of investigating the relevance of international bureaucracies. This is hardly a coincidence given the prevalence of multilateral agreements, the degree of international organization in environmental governance and the imperative for transboundary cooperation in addressing the complex challenges of earth system transformation (Biermann 2014; Young 2017).

In consequence, scholarly scrutiny of the roles and relevance of international bureaucracies has covered considerable ground over the last decade. This research responded well to the analytical problems posed by the 2009 Earth System Governance Science and Implementation Plan, notably regarding architecture and agency, but also related issues such as accountability and legitimacy (Biermann et al. 2009). Today, it provides the social sciences with a solid foundation for further academic inquiry.

That being said, pertinent questions about our understanding of international bureaucracies as actors and robust explanations of their behaviour remain to be addressed. Causal pathways regarding the effectiveness or legitimacy of international bureaucracies and their actions still need to be explored and to be tested. Normative considerations about the contributions of international bureaucracies to equity and justice in global governance also loom large.

This chapter can merely provide a brief overview of the state of the art in governance research on international bureaucracies and secretariats. Against this background we conclude by proposing four avenues for further research on international bureaucracies and their relevance for earth system governance. We see four overarching challenges: remaining gaps in the existing research; the lack of an integrative theory; the need for a more systematic account of the organizational turn in the study of international relations; and a more thorough reflection of the changing context of world politics and its implications for multilateral governance.

(1) First, with regard to empirical as well as conceptual gaps in our understanding of international bureaucracies, three approaches seem particularly promising: (a) comparative analysis of the agency of international bureaucracies; (b) conceptual cross-fertilization of theories of international relations and organizational studies, notably regarding organizational learning and institutional change within international bureaucracies; and (c) honing research methods to empirically better assess the effectiveness and legitimacy of international bureaucracies as autonomous actors.

(2) Second, regarding theory-building and the organizational turn, there remains a striking disconnect between research on public administration and on

international relations. Addressing this should be conducive to developing an integrated theory on bureaucratic influence in international organization and policymaking. Along similar lines, several scholars have pointed to a gap between the study of international organizations and the sociology of organizations (e.g., Bauer and Weinlich 2011; Jönsson 2017), the latter including organization studies with a more managerial focus as it is prevalent in business administration research (Jönsson 2017: 61).

(3) Third, one promising further vantage point to improve our empirical yardsticks in assessing the agency of international bureaucracies and their influence, will be paying more attention to the timing of bureaucratic interventions and variation in influence across phases of the policy cycle. A growing body of literature suggests that the influence of international bureaucracies may be waning in phases of policy formulation and implementation, relative to issues of institutional design, as political salience increases (Eckhard and Ege 2016; see also Elsig 2010; Johnson 2013; Weinlich 2014).

(4) Finally, and linking back to the structures of the international system, research on international bureaucracies must not ignore the dynamics of change in world politics and the apparent crisis of multilateralism. Notwithstanding the persistence of international organizations, the emergence of populist governments and 'our-nation-first' movements around the globe will not leave the environment of international bureaucracies unaffected. Conventional wisdom in the study of international relations would suggest a decrease in the autonomy of international bureaucracies when national power politics gain prevalence. Yet, it is an open question whether the shifting context in multilateral governance may not open new pathways of influence for international bureaucracies in view of multilateral processes that keep going on even as powerful nations turn their backs. Ironically, the governance functions that we attributed to international bureaucracies – notably organizational expertise and knowledge as well as their rational-legal authority – may gain currency as self-referential adversaries of multilateral cooperation resort to 'post-truth' politics and 'alternative facts'. Conversely, the questioning of the scientific rigor and findings, for instance regarding anthropogenic climate change or new technologies, that is advanced on the Internet and in social media may constrain the reach of these more traditional institutional configurations of earth system governance as it undermines the foundations of their mandates (Earth System Governance Project 2018; see also Burch et al. 2019).

The concomitant trend towards goal setting in global governance, as opposed to conventional rule-making and regulation, might prove a case in point for bureaucratic authority (Chapter 12). Here again, the study of earth system governance seems well suited as an empirical testing ground. Already, 'goal setting in the Anthropocene' is being considered as the 'ultimate challenge for planetary stewardship' (Young et al. 2017: 53). To that end, the research that we reviewed suggests that international bureaucracies stand ready to play their part.

References

Abbott, K. W., Genschel, P., Snidal, D., & Zangl, B. (eds.) (2015). *International organizations as orchestrators*. Cambridge, UK: Cambridge University Press.

Abbott, K. W., Genschel, P., Snidal, D., & Zangl, B. (2016). Two logics of indirect governance: Delegation and orchestration. *British Journal of Political Science*, 46 (4), 719–29.

Abbott, K. W., & Snidal, D. (2010). International regulation without international government: Improving IO performance through orchestration. *Review of International Organizations*, 5 (3), 315–44.

Allan, J. I., & Hadden, J. (2017). Exploring the framing power of NGOs in global climate politics. *Global Environmental Politics*, 26 (4), 600–20.

Andonova, L. B., Betsill, M. M., & Bulkeley, H. (2009). Transnational climate governance. *Global Environmental Politics*, 9 (2), 52–73.

Andresen, S., & Rosendal, K. (2009). The role of the United Nations Environment Programme in the coordination of multilateral environmental agreements. In F. Biermann, B. Siebenhüner, & A. Schreyögg (eds.), *International organisations in global environmental governance* (pp. 133–50). London: Routledge.

Barnett, M. N., & Finnemore, M. (1999). The politics, power, and pathologies of international organizations. *International Studies Quarterly*, 53 (4), 699–732.

Barnett, M. N., & Finnemore, M. (2004). *Rules for the world: International organizations in global politics*. Ithaca, NY: Cornell University Press.

Bauer, M. W., & Ege, J. (2017). A matter of will and action: The bureaucratic autonomy of international public administrations. In M. W. Bauer, C. Knill, & S. Eckhard (eds.), *International bureaucracy: Challenges and lessons for public administration research* (pp. 13–41). London: Palgrave Macmillan.

Bauer, S., Andresen, S., & Biermann, F. (2012). International bureaucracies. In F. Biermann, & P. Pattberg (eds.), *Global environmental governance reconsidered* (pp. 27–44). Cambridge, MA: The MIT Press.

Bauer, S., Busch, P. O., & Siebenhüner, B. (2009). Treaty secretariats in global environmental governance. In F. Biermann, B. Siebenhüner, & A. Schreyögg (eds.), *International organisations in global environmental governance* (pp. 174–91). London: Routledge.

Bauer, S. (2009). The secretariat of the United Nations Environment Programme: Tangled up in blue. In F. Biermann, & B. Siebenhüner (eds.), *Managers of global change: The influence of international environmental bureaucracies* (pp. 169–202). Cambridge, MA: The MIT Press.

Bauer, S. (2006). Does bureaucracy really matter? The authority of intergovernmental treaty secretariats in global environmental politics. *Global Environmental Politics*, 6 (1), 23–49.

Bauer, S., & Weinlich, S. (2011). International bureaucracies: Organizing world politics. In B. Reinalda (ed.), *The Ashgate research companion to non-state actors* (pp. 251–62). Farnham: Ashgate.

Betsill, M. M., & Corell, E. (2008). *NGO diplomacy: The influence of nongovernmental organizations in international environmental negotiations*. Cambridge, MA: The MIT Press.

Beetham, D. (1987). *Bureaucracy*. Minneapolis: University of Minnesota Press.

Biermann, F., & Siebenhüner, B. (2009). *Managers of global change: The influence of international environmental bureaucracies*. Cambridge, MA: The MIT Press.

Biermann, F., Betsill, M. M., Gupta, J. et al. (2009). *Earth system governance: People, places and the planet. Science and implementation plan of the Earth System Governance Project*. Bonn: Earth System Governance Project.

Biermann, F. (2014). *Earth system governance: World politics in the Anthropocene*. Cambridge, MA: The MIT Press.

Bulkeley, H., Andonova, L. B., Betsill, M. M. et al. (2014). *Transnational climate change governance*. Cambridge, UK: Cambridge University Press.

Burch, S., Gupta, A., Inoue, C. Y. A. et al. (2019). New directions in earth system governance research. *Earth System Governance*, 1.

Busch, P. O. (2009). The climate secretariat: Making a living in a straitjacket. In F. Biermann, & B. Siebenhüner (eds.), *Managers of global change: The influence of international environmental bureaucracies* (pp. 75–99). Cambridge, MA: The MIT Press.

Busch, P. O. (2014). Independent influence of international public administrations: Contours and future directions of an emerging research strand. In S. Kim, S. Ashley, & W. H. Lambright (eds.), *Public administration in the context of global governance* (pp. 45–62). Cheltenham: Edward Elgar.

Chan, S., Brandi, C., & Bauer, S. (2016). Aligning transnational climate action with international climate governance: The road from Paris. *Review of European Community and International Environmental Law*, 25 (2), 238–47.

Cutler, C., Haufler, V., & Porter, T. (1999). *Private authority in international affairs*. Albany: State University of New York Press.

Depledge, J. (2005). *The organization of global negotiations: Constructing the climate change regime*. London: Earthscan.

Depledge, J. (2007). A special relationship: Chairpersons and the secretariat in the climate change negotiations. *Global Environmental Politics*, 7 (1), 45–68.

Dingwerth, K., & Pattberg, P. (2009). Actors, arenas, and issues in global governance. In J. Whitman (ed.), *Palgrave advances in global governance* (pp. 41–65). London: Palgrave Macmillan.

Dryzek, J. S. (2017). The meanings of life for non-state actors in climate politics. *Environmental Politics*, 26 (4), 789–99.

Earth System Governance Project (2018). *Earth system governance: Science and implementation plan of the Earth System Governance Project*. Utrecht: Earth System Governance Project.

Eckhard, S., & Ege, J. (2016). International bureaucracies and their influence on policy-making: A review of empirical evidence. *Journal of European Public Policy*, 23 (7), 960–78.

Ege, J. (2017). Comparing the autonomy of international public administrations: An ideal-type approach. *Public Administration*, 95 (3), 555–70.

Elsig, M. (2010). The World Trade Organization at work: Performance in a member-driven milieu. *Review of International Organizations*, 5 (3), 345–63.

Gehring, T. (2011). The institutional complex of trade and environment: Toward an interlocking governance structure and a division of labor. In S. Oberthür & O. Schram Stokke (eds.), *Managing institutional complexity: Regime interplay and global environmental change* (pp. 227–54). Cambridge, MA: The MIT Press.

Gordon, D. J., & Johnson, C. A. (2017). The orchestration of global urban climate governance: Conducting power in the post-Paris climate regime. *Environmental Politics*, 26 (4), 694–714.

Green, J. F. (2013). *Rethinking private authority: Agents and entrepreneurs in global environmental governance*. Princeton, NJ: Princeton University Press.

Gruber, J. E. (1987). *Controlling bureaucracies*. Berkeley: University of California Press.

Hadden, J. (2015). *Networks in contention*. Cambridge, UK: Cambridge University Press.

Hale, T. (2016). All hands on deck: The Paris Agreement and non-state climate action. *Global Environmental Politics*, 16 (3), 12–21.

Hale, T., & Roger, C. (2014). Orchestration and transnational governance. *Review of International Organization*, 9 (1), 59–82.

Hall, N. (2016). *Displacement, development, and climate change: International organizations moving beyond their mandates*. New York: Routledge.

Hawkins, D. G., Lake, D. A., Nielson, D. L., & Tierney, M. J. (eds.) (2006). Delegation under anarchy: States, international organizations and principal-agent theory. In D. G. Hawkins, D. A. Lake, D. L. Nielson, & M. J. Tierney (eds.), *Delegation and agency in international organizations* (pp. 3–38). Cambridge, UK: Cambridge University Press.

Hawkins, D. G., & Jacoby, W. (2006). How agents matter. In D. G. Hawkins, D. A. Lake, D. L. Nielson, & M. J. Tierney (eds.), *Delegation and agency in international organizations* (pp. 199–228). Cambridge, UK: Cambridge University Press.

Hickmann, T. (2015). *Rethinking authority in global climate governance: How transnational climate initiatives relate to the international climate regime*. London: Routledge.

Hickmann, T., & Elsässer, J. (2018). *New alliances in global environmental governance: Intergovernmental treaty secretariats and sub- and non-state actors*. Paper presented at the General Conference of the European Consortium for Political Research, Hamburg.

Hickmann, T., Widerberg, O., Lederer, M., & Pattberg, P. (2019). The United Nations Framework Convention on Climate Change Secretariat as an orchestrator in global climate policymaking. *International Review of Administrative Sciences*, in press.

Hooghe, L., & Marks, G. (2015). Delegation and pooling in international organizations. *Review of International Organizations*, 10 (3), 305–28.

Jinnah, S., & Morin, J. F. (Forthcoming 2020). *Greening through trade: How American trade policy has promoted environmental protection abroad*. Cambridge, MA: The MIT Press.

Jinnah, S., & Lindsay, A. (2015). Secretariat influence on trade-environment politics: NAFTA's Commission on Environmental Cooperation. *Review of Policy Research*, 32 (1), 124–45.

Jinnah, S. (2010). Overlap management in the World Trade Organization: Secretariat influence on trade-environment politics. *Global Environmental Politics*, 11 (2), 54–79.

Jinnah, S. (2011). Marketing linkages: Secretariat governance of the climate-biodiversity interface. *Global Environmental Politics*, 11 (3), 23–43.

Jinnah, S. (2014). *Post-treaty politics: Secretariat influence in global environmental governance*. Cambridge, MA: The MIT Press.

Johnson, T. (2013). Looking beyond states: Openings for international bureaucrats to enter the institutional design process. *Review of International Organizations*, 8, 499–519.

Johnson, T., & Urpelainen, J. (2014). International bureaucrats and the formation of intergovernmental organizations: Institutional design discretion sweetens the pot. *International Organization* 68, 177–209.

Jönsson C. (2017). IR paradigms and inter-organizational theory: Situating the research program within the discipline. In J. Koops, & R. Biermann (eds.), *Palgrave handbook of inter-organizational relations in world politics* (pp. 49–62). London: Palgrave Macmillan.

Jordan, A. J., Huitema, D., Hildén, M. et al. (2015). Emergence of polycentric climate governance and its future prospects. *Nature Climate Change*, 5, 977–82.

Jörgens, H., Kolleck, N., & Saerbeck, B. (2016).Exploring the hidden influence of international treaty secretariats: Using social network analysis to analyse the Twitter debate on the 'Lima Work Programme on Gender'. *Journal of European Public Policy*, 23 (7), 979–98.

Jörgens, H., Kolleck, N., Saerbeck, B., & Well, M. (2016). Orchestrating (bio-)diversity: The secretariat of the Convention on Biological Diversity as an attention-seeking bureaucracy. In M. Bauer, C. Knill, & S. Eckhard (eds.), *International Bureaucracy* (pp. 73–95). London: Palgrave Macmillan.

Knill, C., Eckhard, S., & Grohs, S. (2016). Administrative styles in the European Commission and the OSCE secretariat: Striking similarities despite different organizational settings. *Journal of European Public Policy*, 23 (7), 1–20.

Littoz-Monnet, A. (2017). Expert knowledge as a strategic resource: International bureaucrats and the shaping of bioethical standards. *International Studies Quarterly*, 0, 1–12.

Morin, J. F., & Jinnah, S. (2018). The untapped potential of preferential trade agreements for climate governance. *Environmental Politics*, 27 (3), 541–65.

Nicholson, S., Jinnah, S., & Gillespie, A. (2018). Solar radiation management: A proposal for immediate polycentric governance. *Climate Policy*, 18 (3), 322–34.

Pattberg, P. H. (2007). *Private Institutions and Global Governance: The New Politics of Environmental Sustainability*. Cheltenham: Edward Elgar.

Prideaux, M. (2015). Wildlife NGOs: From adversaries to collaborators. *Global Policy*, 6 (4), 379–88.

Raustiala, K., & Victor, D. G. (2004). The regime complex for plant genetic resources. *International Organization*, 58 (2), 277–309.

Reinalda, B., & Verbeek, B. (eds.) (2003). *Autonomous policy making by international organizations*. London: Routledge.

Reinalda, B. (2011). *The Ashgate research companion to non-state actors*. Aldershot: Ashgate.

Roger, C., Hale, T., & Andonova, L. B. (2017). The comparative politics of transnational climate governance. *International Interactions*, 43 (1), 1–25.

Siebenhüner, B. (2009). The biodiversity secretariat: Lean shark in troubled waters. In F. Biermann, & B. Siebenhüner (eds.), *Managers of global change: The influence of international environmental bureaucracies*. Cambridge, MA: The MIT Press.

Stone, D., & Ladi, S. (2015). Global policy and transnational administration. *Public Administration*, 93 (4), 839–55.

Weinlich, S. (2014). *The UN secretariat's influence on the evolution of peacekeeping*. London: Palgrave Macmillan.

Young, O. R. (2017). *Governing complex systems: Social capital for the Anthropocene*. Cambridge, MA: The MIT Press.

Young, O. R., Underdal, A., Kanie, N., & Kim, R. E. (2017). Goal setting in the Anthropocene: The ultimate challenge for planetary stewardship. In N. Kanie, & F. Biermann (eds.), *Governing through goals: Sustainable Development Goals as governance innovation* (pp. 53–74). Cambridge, MA: The MIT Press.

Zelli, F., & van Asselt, H. (2013). The institutional fragmentation of global environmental governance: Causes, consequences, and responses. *Global Environmental Politics*, 13 (3), 1–13.

Zürn, M. (1998). The rise of international environmental politics: A review of current research. *World Politics* 50 (4), 617–49.

4

Transnational Institutions and Networks

AGNI KALFAGIANNI, LENA PARTZSCH
AND OSCAR WIDERBERG

This chapter reviews the literature on transnational institutions and networks in earth system governance research over the past decade. We organize our review around the issues of emergence, effectiveness and legitimacy of transnational governance and outline key debates and controversies surrounding the shifting authority between public and private actors. We also identify major research lines and open questions and provide an outlook towards the most promising directions in future research. Our review is based on a set of articles selected by using keyword search from an original dataset containing all 2,521 articles (co)authored by fellows of the Earth System Governance Project up to November 2016 (found in Web of Science). This approach rendered around 200 articles in total. To include publications from the last two years (2017 and 2018) we extended our search by examining the research output of those fellows who identify transnational governance as a key area of their interest, yielding an additional 15 articles. In this review, we highlight key insights from this burgeoning literature.

Conceptualization

We define transnational governance as the transboundary institutional arrangements emerging from the interaction among private actors or between private and public actors that steer actors' behaviour in an issue-specific area (Falkner 2003). We consider private actors to include civil society organizations as well as business actors and their associations, and public actors to include states, international organizations and substate actors, such as cities and regions. Transnational governance is characterized by two main aspects. First, the state is not the main actor involved in rule making; instead, it is especially private actors that develop rules (Auld, Renckens and Cashore 2015). Second, transnational governance is voluntary and relies on market forces and public scrutiny to exert pressure on the target group and generate public benefits (Kalfagianni and Pattberg 2013). In the past decades, transnational governance has become particularly prominent in the sustainability domain, which connects environmental, social and economic dimensions of development (World Commission on Environment and Development 1987). Examples include individual codes of conduct and reporting standards to international certification programmes, public–private partnerships and city networks signifying that transnational governance varies widely in the forms it takes (e.g., Cashore 2016; Green and Auld 2017).

Transnational governance has been met with enthusiasm as well as critique. Early on, supporters underlined its contribution to public policy objectives, particularly in areas where governments may be unwilling or unable to regulate (Cashore, Auld and Newsom 2004). Critics contended, however, that transnational governance may circumvent or undermine public national and international law (O'Rourke 2003) and stressed its potentially pre-emptive character, especially, in the absence of effective monitoring mechanisms (Greven 2004). In this context, scholars continue to warn that transnational governance may be preventing stricter government rules from developing, thereby contributing to an overall 'race to the bottom' (Fuchs, Kalfagianni and Haringa 2011). Below, we explore the variation of transnational governance and debates around its effectiveness and legitimacy, particularly as they have developed over the past 10 years.

The Emergence of Transnational Governance

Transnational governance has become increasingly common for regulating global markets and other spheres of earth system governance. Private actors have begun to create, implement and enforce rules to address problems, such as global warming, deforestation and biodiversity loss, beyond the nation-state (Fuchs 2007; Green 2013). This section discusses how earth system governance researchers have analyzed the emergence of transnational governance by focusing on three key forms, namely: (a) voluntary certification programmes; (b) public–private partnerships; and (c) subnational networks among, for example, cities and regions.

Voluntary Certification Programmes

Voluntary certification programmes – organized and coordinated by private actors – are the most substantial effort of transnational governance. They are a market-based mechanism resting on the idea that if suppliers disclose information and certifiers guarantee the compliance to specific standards, buyers are willing to pay a higher price for their products (Gupta 2008). In turn, if buyers refrain from purchasing (non)certified products, they may endanger the financial viability of the supplier and of the certifier as well (Koenig-Archibugi and Macdonald 2013), thus providing an incentive towards sustainable market behaviour.

Certification programmes define *product* and *process* standards related to the conditions under which items are produced or traded (Gupta 2008; Koenig-Archibugi and Macdonald 2013). An example of product standards from the food sector is colour and size specifications of the final product. An example of process standards from the same sector is specifications regarding the process of producing a final product with organic rather than conventional methods. To the extent that they remain private and voluntary, certification programmes are compatible with the rules of the World Trade Organization. For example, governments can implement technical regulations based on certification, such as carbon-neutrality, as long as they do not discriminate against products with a different country of origin or form an unnecessary obstacle to free trade. However, 'green' unilateralism through transnational governance has been the subject of much heated debate: The Global South has accused firms and countries in the Global North of using environmental standards as 'trade weapons' (Quark 2013: 5) to retain their dominance in global markets.

The origins of voluntary certification at the transnational level can be traced to the forestry sector after the adoption of an international forests agreement failed at the 1992 United Nations Conference on Environment and Development (also known as the Rio Earth Summit, see Bartley 2007; Chan and Pattberg 2008). Since then, certification has diffused to multiple other sectors as diverse as fisheries, greenhouse gas emissions, mining, and tourism, to name a few. Scholars debate the reasons behind the emergence and diffusion of transnational governance in the form of voluntary certification. Early on, Haufler (2003), for example, focusing on business actors explains commitment to certification as a result of 'the twin threats': first, environmental activism that targets the public reputation of specific firms and industry sectors as a whole and, second, the 'shadow of hierarchy' or continuous negotiations among governments on command-and-control types of regulation (Haufler 2003: 248). As a response to these threats, business actors, particularly multinational corporations, care about and even export human rights and environmental norms by adopting relevant labelling and certification schemes. Similar observations have been made more recently by others (e.g., Bingen and Busch 2006; Bartley 2007; Hickmann 2017b; Bartley 2018).

Scholars have debated the motivation not only of businesses but also of non-governmental organizations to participate in transnational governance (Bartley 2007; Betsill and Corell 2008; Pattberg, Betsill and Dellas 2011). Since the 1990s, an increasing number of non-governmental organizations has changed from strategies of 'naming and shaming' to more collaborative approaches (Haufler 2003). Non-governmental organizations, such as the World Wide Fund for Nature, are participating in multi-stakeholder certification schemes to develop ethical market alternatives. Voluntary certification programmes led by non-governmental organizations tend to be more ambitious than business-led programmes. In this context while non-governmental organization-driven initiatives, such as the Forest Stewardship Council, are considered possibly to have been intrinsically motivated, business-driven initiatives are often considered to be much more responsive. In the forestry sector, after the Forest Stewardship Council had already been established, for instance, members of timber (processing) industries created their own certification, without participation of public actors or non-governmental organizations (Gulbrandsen 2014). Such industry-driven certification schemes were originally conceived to counter the influence of the more ambitious pioneer schemes (Dingwerth and Pattberg 2009). In particular, the Programme for the Endorsement of Forest Certification is seen as an industry response to the environmental non-governmental organization-based Forest Stewardship Council. The Programme for the Endorsement of Forest Certification is an umbrella organization that gathers 35 independent national forest certification schemes and is considerably less ambitious than the Forest Stewardship Council (Gulbrandsen 2014; McDermott, Irland and Pacheco 2015). However, critical voices also point out that non-governmental organizations engaging in certification may themselves result from and reproduce the (neoliberal) structures and discourses that generated sustainability concerns in the first place instead of initiating real transformational change (Tucker 2014).

The role of government is also widely debated in the emergence and diffusion of voluntary certification. Governments often actively support the implementation of certification through public procurement policies which request that state agencies or state-owned companies purchase only certified products (up to 20 per cent of total timber consumption, for example) (Gulbrandsen 2014). Likewise, governments also provide funds to assist market actors in adopting voluntary certification, such as New Zealand's Environmental Certification Fund, offering grants to fisheries to pay up to 75 per cent of the costs of certification with transnational private schemes (Washington and Ababouch 2011). Finally, governments may also co-opt voluntary certification under public management and ownership, thereby enabling its wider uptake and support. A prominent example here is the adoption of the Marine Stewardship Council guidelines by the Ecuadorian government as the basis for reforming its fisheries policy (Kalfagianni and Roche 2017).

Public–Private Partnerships

Besides governance through voluntary certification, private actors are also increasingly involved in international politics, for example, in the form of public–private or multi-stakeholder partnerships (Mert 2009; Pattberg and Widerberg 2016). Transnational partnerships can be defined as an institutionalized cooperation between state and non-state actors from different countries, with the aim to produce political output (Kaul 2006; Partzsch 2007). They are voluntary and largely based on the self-interest of each partner who independently joins. For the most part, they have a horizontal structure with public and private actors as equal partners. They set and enforce rules that the international community considers as binding (Börzel and Risse 2005).

Such partnerships came to prominence in the early 2000s (Whitfield 2001: 281). At the 2002 World Summit on Sustainable Development in Johannesburg, public–private partnerships were officially introduced as a new tool of global governance. Besides intergovernmental agreements, such as the Johannesburg Plan of Implementation, partnerships were presented as official summit outcomes. These partnerships broke with the purely inter-*governmental* system of sovereign states. In short time, more than 340 partnerships for sustainable development registered with the United Nations (Andonova and Levy 2003). An example of such transnational partnership is the Global Alliance for Improved Nutrition, set up in 2002, which works against vitamin and mineral deficiencies in developing countries. While the Global Alliance for Improved Nutrition received some funding from the US Agency for International Development, the Canadian International Development Agency and the German and Dutch governments, it was mainly set up and funded by the private Bill and Melinda Gates Foundation (Kaan and Liese 2011). Although the official status of such public–private partnerships in international law was never clarified (Partzsch 2007), the partnership approach is currently being emulated in many other issue areas of global governance, such health, water and climate change (Pattberg and Widerberg 2016). Partnerships are also key for the implementation of the recently adopted Sustainable Development Goals, a set of 17 global goals and 169 targets that aim to reduce poverty and inequality, foster environmental sustainability and develop resilient livelihoods for all by 2030 (Biermann, Kanie and Kim 2017; Kanie and Biermann 2017). Goal 17, more specifically, is about 'Partnerships for the Goals'.

Subnational Networks

In parallel to the emergence of 'private authority' (Green 2013), we also see a trend towards more decentralized public governance. While globalization first appeared to produce incentives for large-scale territorial governance, for example observed in the creation of the World Trade Organization, we increasingly witness also activities by subnational actors, such as regions and cities, in transnational governance (Bulkeley and Betsill 2013; Kronsell 2013).

As the number of transnational networks between cities has grown, an increasing number of earth system governance scholars has engaged with the topic. Already in 2004, Bulkeley and Betsill (2004: 472) called for more attention to be spent on 'networks of local government', claiming that they were a 'significant phenomenon in environmental politics'. The authors trace the emergence of subnational networks to the 1992 United Nations Conference on Environment and Development ('Rio Earth Summit'), including the development of a Local Agenda 21 suggesting that 'Cities ... are key sites in the production and management of energy use and waste production, through processes over which local authorities have a (varying) degree of influence' (Bulkeley and Betsill 2004: 477). Bansard and colleagues (2017: 229) suggest that a 'popular leitmotiv' has emerged wherein cities might have advantages over states in addressing sustainability issues, at least in terms of climate governance. Some of the city networks have become incredibly large. The C40 for instance, a climate change network, consists of 96 cities that produce 25 per cent of global GDP. The Global Covenant of Mayors engages over 9,000 cities, representing nearly 800 million people or 10 per cent of the global population.

In sum, certification programmes, partnerships and subnational networks are prominent types of transnational initiatives that occupy various institutional spaces in earth system governance. Few transnational institutions emerge in an institutional vacuum, however. They interact both with each other and with existing international institutions, creating 'transnational regime complexes' (Abbott 2014; Widerberg and Pattberg 2017). Accordingly, many earth system governance scholars now study the quality of the institutional structure framing the debate with terms such as fragmentation, polycentricism and orchestration (Zelli and van Asselt 2013; Jordan et al. 2015; Kleinschmit et al. 2018; see also Chapters 8 and 11). Emphasizing interrelational aspects of multisectoral governance arrangements, particularly those that involve public and private actors, Ponte and Daugbjerg (2015) speak of hybrid governance (also Kuyper, Linnér and Schroeder 2018). Accordingly, the literature on emergence is now moving away from studying individual transnational governance initiatives to the study of the entire transnational governance architecture.

The Effectiveness of Transnational Governance

According to the Earth System Governance Project's first science plan, institutional sprawl could influence the performance of both individual institutions and the system as a whole, recommending that 'research on architecture should ask how the performance of environmental institutions is affected by their embeddedness in larger architectures' (Biermann et al. 2010: 281). This section discusses how earth system governance researchers have evaluated effectiveness in transnational governance architectures. It reviews how the earth system governance literature has measured effectiveness, whether transnational governance has increased effectiveness and how to improve effectiveness in increasingly complex institutional architectures.

As Kramarz and Park observe, adding actors, rules and norms does not automatically have positive effects on the environment (2016: 1). The fifth Global Environment Outlook seems to support their argument, noting that '[e]fforts to slow the rate or extent of change ... have resulted in moderate successes but have not succeeded in reversing adverse environmental changes. Neither the scope of these nor their speed has abated in the past five years' (UNEP 2012: 6; see also UNEP 2019). Hence, despite pockets of success (e.g., reduction of ozone depleting substances, recovery of some fish stocks and increased forest cover), the overall state of the environment across land, water and the atmosphere is in worse shape than before the rise of transnational governance. Effectiveness, however, is a multifaceted concept, and few earth system governance scholars actually measure the environmental impacts of their research objects. Instead, other criteria for measuring effectiveness are more popular, such as quality and quantity of transnational rules and norms, changes in behaviour and consequences of interplay between transnational institutions. The findings suggest a more nuanced picture, different from the Global Environment Outlook.

Measuring and Explaining Effectiveness

Measuring effectiveness in earth system governance is complicated and subject to extensive scholarly debates (e.g., Miles et al. 2001; Breitmeier, Underdal and Young 2011; Young 2011). Measuring the impact of transnational institutions on environmental state indicators (e.g., biodiversity levels, greenhouse gas concentrations in the atmosphere or fish stocks) generally suffers from lack of data, absence of good state indicators or difficulty in establishing a causal link between an institution and the observed change in environmental state. Randomized controlled experiments, often considered the gold standard of measuring effectiveness in science, are not available to earth system governance researchers due to the difficulty of treating and controlling the real world in which transnational institutions operate. Instead governance scholars must revert to other conceptualizations of effectiveness and methodologies for establishing the influence of institutions.

How effectiveness is defined determines if something can be considered effective. In research on environmental institutions, effectiveness is commonly thought about as the extent to which institutions are able to alleviate or resolve the problems that motivated their creation (Young 2011). Beyond this broad conceptualization, effectiveness can be measured against some hypothetical ideal, such as a 'collective optimum' (Miles et al. 2001); against a (politically feasible) counterfactual scenario; or against an institution's own goals (Hale and Held 2011: 24; Bernstein and Cashore 2012: 586). Moreover, political scientists choose what to study by separating effectiveness into three different levels (Easton 1957). Output-level effectiveness pertains to the quality and quantity of commitments, rules, norms and decision-making procedures; outcome-level effectiveness concerns changes in behaviour by actors resulting from the outputs; and impact-level effectiveness measures the results of the behavioural changes by actors.

Relaxing the criteria for what to consider effective, from impact-level indicators to output-level indicators, changes the norm for what to consider effective. The earth system governance literature has primarily been occupied with output-level effectiveness, often studying the *potential* for an institution to have desired outcomes and impacts. For example, in their study on five transnational rule-setting organizations in the global governance of fisheries and aquaculture, Kalfagianni and Pattberg (2013: 125) operationalize 'conditions for effectiveness' as: problem structure; comprehensiveness and stringency of standards; quality of audits; access of relevant societal actors to decision-making venues and procedures; and uptake of standards by relevant actors (2013: 127). Chan and colleagues (2016) use a 'function-output fit' assessment, measuring whether transnational climate institutions produce outputs that are consistent with their self-stated functions (also Pattberg et al. 2012). This assessment framework thus operationalizes effectiveness as whether transnational initiatives carry out what they have set out to do, similar to international regime effectiveness research (see Chapter 2). The earth system governance scholarship converges around institutional design features, participation and institutional and problem context as determinants for transnational institutional effectiveness (Szulecki, Pattberg and Biermann 2011; Kalfagianni and Pattberg 2013; Pattberg and Widerberg 2016).

However, factors explaining the effectiveness of individual institutions may be insufficient to explain the effectiveness of the institutional architecture. There are, in fact, few architecture-level analyses of effectiveness in transnational governance, yet the initial results are sobering. Some projects have taken a large-*n* comparative approach, for instance on partnerships for sustainable development (Pattberg et al. 2012) and transnational climate partnerships (Pattberg 2010; Bulkeley et al. 2014). The findings suggest a limited impact of transnational institutions, even though more recent cases suggest some sustainability improvements (Chan et al. 2016).

Similarly, regarding voluntary certification, research from forestry reveals that transnational forest institutions in the Brazilian Amazon failed to prevent deforestation for three main reasons (McDermott, Irland and Pacheco 2015). First, higher prices for certified timber could not prevent the expansion of agriculture as a main driver of deforestation. Second, the costs of certification turned out to be too high for smallholders, excluding them from sustainable forest management, with additionally negative effects of market concentration. Third, the complexity of supply chains and products was higher for tropical timber export products than for European and North American timber products sold at domestic markets. As a consequence, forest certification schemes 'favor large producers and concentrated supply chains destined for external markets ..., while extensive legal requirements inhibit local benefit-capture' (McDermott, Irland and Pacheco 2015: 134). For global business, transnational labelling schemes for tropical timber harmonized environmental standards, and they have the same effect as a cartel: potentially harmful competition in voluntarily self-restricted markets is avoided (Holzinger and Sommerer 2011: 316). A lack of control, divergence of national rules and dependence on the certifiers are the main points of criticism, especially with regard to voluntary certification (Kleinschmit 2015: 85). As a consequence, we see growing criticism about the effectiveness of transnational institutions – especially in the forestry sector but also more generally.

Some authors have also criticized the neglect of economic and social consequences, especially in the Global South, in transnational institutions and their impact evaluations (McDermott, Irland and Pacheco 2015). For instance, studying Honduran community forestry groups producing furniture and kitchenware for Danish design markets, Nygren (2015) suggests that transnational governance, particularly in the form of certification, creates images of Southern community producers as authentic and exotic 'others', portrayed as people who cherish local traditions and toil for their living. The stories behind the products are crucial to the sale of certified products, and images of people are carefully selected. For example, people sawing timber barefoot or carrying planks on their shoulders on muddy slopes are not chosen. The terms of the participation of Southern producers in global markets are rarely considered, however, and the same holds for the distribution of benefits and constraints among actors involved. While the 'distance' between Northern consumers and Southern producers may be shortened, the transnational institutions in this case study do not challenge asymmetrical trade relations (Nygren 2015).

The studies reviewed above, however, put limited focus on the possibility that the sum of transnational governance institutions may be greater than its parts and that institutional embeddedness may influence the effectiveness of an institution on the architecture of transnational governance (but see Hoffmann 2011). Climate governance provides another illustrative example. For transnational climate institutions, reductions in greenhouse gas emissions have been a central indicator for effectiveness (Blok et al. 2012; UNEP 2015; Hsu et al. 2016). However, van der

Ven and colleagues (2016: 6) consider evaluation criteria such as greenhouse gas emissions, other goals or intended outputs too narrow, arguing that it could lead to situations where subpar performance 'lead[s] policy-makers to prematurely abandon good ideas because of short-term failures to "measure up"'. Instead, scaling (how initiatives create scalable solutions) and entrenchment (how initiatives create durable solutions) are important performance criteria, challenging evaluators to expand the conceptual toolbox beyond traditional effectiveness criteria (van der Ven, Bernstein and Hoffmann 2016). Scaling and entrenchment could be considered system-level indicators where the evaluation is less concerned about how an intervention affects a target population but rather how it affects the system. Such calls for taking a system-level perspective are echoed by Abbott and colleagues (2016) in their study on how emerging 'private transnational regulatory organizations' are interacting with public governance. According to the authors, private transnational regulatory organizations could reinforce and complement public governance; provide natural experiments for governments and international organizations to learn; and offer venues for private actors to participate in international governance (Abbott, Green and Keohane 2016: 272).

In sum, reviewing the literature, three observations can be made considering measuring and explaining effectiveness in governance architectures characterized by a high number of transnational institutions. First, while earth system governance research has generated an impressive amount of descriptive data on individual institutions, including under what conditions institutions perform best, there are few system-level studies, in particular, taking impact-level effectiveness into account. Second, the rise of transnational governance challenges researchers to consider new indicators for effectiveness that take system-level effects into account, such as scaling. Third, system-level evaluations require more focus on the interplay between transnational institutions, considering their effects on each other and on the established international institutions.

Improving Effectiveness

After ten years of research, could earth system governance research provide decision-makers with recommendations on how to improve transnational governance? Orsini and colleagues (2013) pose the question in terms of institutional structure: Should one prefer 'fragmented, centralized, or dense' network structures in global governance? When searching for an answer, it is sometimes hard to separate analytical from normative claims in the literature (Jordan and Huitema 2014). Scholars writing in the tradition of Ostrom (2010), describing governance architectures in terms of polycentricity, have been keen on emphasizing the benefits of polycentric systems, including improved policies through learning and innovation over time, and improved cooperation through more interaction and communication (Cole 2015). Others have been more hesitant about the benefits of polycentricity, claiming that there is limited proof that a polycentric structure, where transnational governance plays a significant role, would outperform a monocentric structure, where governments play a more important role (Jordan et al. 2015).

Some scholars are developing a middle path between the monocentric and polycentric systems perspectives by focusing on how multilateral institutions need to change to integrate transnational actors and institutions into their work. Abbott (2014) argues that the links between the various institutions and organizations should be strengthened and better coordinated. A stream of research focuses on how to create synergies between the expanding universe of transnational climate governance and the existing multilateral institutions, most notably the United Nations Framework Convention on Climate Change (Betsill et al. 2015; Hickmann 2017a). Abbott's ideas build on 'orchestration', defined as a mode of governance whereby international organizations use non-state and subnational actors as intermediaries to enhance global governance (Abbott and Snidal 2009). Orchestration seems to have played a role in the run-up to the Paris Agreement, at which international organizations mobilized non-state and subnational actors to create a positive momentum, instilling confidence among governments to negotiate an ambitious climate agreement (Chan et al. 2015; Hale 2016; Widerberg 2017; Hickmann et al. 2019). Andonova and Hoffmann (2012: 61) make a similar argument that '[i]nternational environmental regimes can establish platforms to promote broader accountability and integration of private and hybrid governance solutions that are up to the task, and facilitate their up-scaling and more equitable distribution'. According to the middle path, improving effectiveness is thus about linking the transnational and international levels of governance, finding ways to coordinate and cooperate between transnational institutions, and for the international system to harness positive transnational forces.

The Legitimacy of Transnational Governance

The increasing importance of private actors in earth system governance raises questions about its legitimacy. Scholars observe, for instance, an increasing power of transnational corporations in governance architectures (Fuchs 2007). Individuals, such as Bill and Melinda Gates as heads of their foundation (Partzsch 2017) as well as some celebrities and social entrepreneurs, also increasingly influence global governance (Partzsch 2014, 2018). As a result, some scholars raise concerns that private actors are eroding state sovereignty and bypassing regulation based on democratic decision-making, even in their countries of origin

(Quark 2013; Kramarz 2016). Simultaneously, others argue that private actors are democratizing global governance particularly through the involvement of civil society organizations and citizen initiatives providing a space for previously marginalized voices (Kuyper 2014). This section discusses how earth system governance research has evaluated the legitimacy of transnational governance. It reviews how earth system governance research has measured legitimacy, whether transnational governance is considered legitimate and how to improve legitimacy beyond the state.

Measuring and Explaining Legitimacy

Legitimacy is considered as the property of a situation or behaviour that is defined by a set of social norms as correct or appropriate (Scott 1995). It refers to the perception by the actors of the overall quality of the social order, which includes institutions, norms and rules (Risse 2002). In modern democracies, legitimacy tends to rely on two pillars (Scharpf 1997), one based on input-oriented arguments, that is, government by the people, and one based on output-oriented arguments, that is, government for the people. According to input-oriented arguments, legitimacy derives from democratic procedures and formal arrangements, whereas according to output-oriented arguments, legitimacy derives from the effectiveness of the specific governance institution in designing policies that promote the public good.

The literature identifies three broad criteria as important for examining the democratic legitimacy of transnational governance: participation, transparency and accountability (Porter and Ronit 2010). Participation relates to the types of actors included in decision-making. Participation can be examined from the point of view of the democratic principle of inclusion in decision-making (Kalfagianni and Pattberg 2013). According to this principle, those who have to abide by the law need to participate in its creation. In modern democracies, this principle is satisfied on the basis of representation rules through the electoral process. In transnational governance, a key question is the representativeness of a global public within governance architectures. In this context, scholars propose different criteria for participation, such as the scope and quality of participation (Dingwerth 2007). Scope requires that those affected by decisions need to participate in their creation (Dingwerth 2007; Scholte 2008), while quality requires that participants are actively engaging in decision-making rather than being simply consulted (Sénit, Biermann and Kalfagianni 2017). Further, a distinction is made between physical and discursive representation, the latter inquiring not into who participates in decision-making but whose discourses are represented and deliberated (Dryzek 2012; Dryzek and Pickering 2017).

Transparency, in turn, refers to the provision of timely, reliable and comprehensible information (Fuchs, Kalfagianni and Havinga 2011). Scholars distinguish transparency in terms of procedures, that is, the openness of governance processes such as decision-making or adjudication; outcome, that is, the openness about regulated or unregulated behaviour (Auld and Gulbrandsen 2010); and substance, that is, the access to and dissemination of information related to the substance of decision-making (Sénit, Biermann and Kalfagianni 2017). It is an important dimension of legitimacy that aims to enhance public scrutiny and visibility and eventually strengthen meaningful participation and accountability in transnational governance architectures (Fuchs, Kalfagianni and Havinga 2011). In relation to participation and accountability, transparency has attracted more attention in transnational governance. Thus, scholars observe that transparency has taken on the aspect of a new global norm signifying a procedural turn in global politics (Gupta 2008) and a trend towards governance by disclosure (Haufler 2010).

Accountability refers to the willingness to accept responsibility or to account for one's actions (Biermann and Gupta 2011). It is related to the notion of responsibility and answerability and, thus, sanctions and redress are considered core dimensions of accountability (Bäckstrand 2008). Accountability is a relational concept between two or more actors often thought of in the form of a principal holding an agent accountable (Widerberg and Pattberg 2017). Scholars distinguish between internal and external accountability. Internal accountability refers to the presence of mechanisms of responsibility when principals and agents are institutionally linked (Widerberg and Pattberg 2017), such as between the board members of an organization and their constituents (Fuchs, Kalfagianni and Havinga 2011). External accountability refers to the relationship between decision-makers and the public (Biermann and Gupta 2011). An important prerequisite of external accountability is the ability of the public to intervene and adjust the governance institution either directly or through representatives (Fuchs, Kalfagianni and Havinga 2011). Accountability in transnational governance differs from accountability in democratic state governance where the public has the potential to vote decision-makers out of office. Rather scholars identify different forms of accountability such as peer, reputational and market accountability and highlight the need to consider plural forms of accountability when it comes to transnational forms of governance (Bäckstrand 2008).

In short, measuring and explaining the legitimacy of transnational governance involves asking questions about who participates in the development of rules and norms for sustainability, how transparent the process is, the substance and outcomes of decision-making, and which mechanisms to use to hold decision-makers accountable. While a similar set of questions applies to evaluating the legitimacy of state-led governance, the exact criteria based on which legitimacy is operationalized for transnational forms of governance differs. Particularly problematic measurement aspects include identifying who the relevant public is and how to translate legal forms of accountability into mechanisms applicable to voluntary and less formalized institutional arrangements beyond the state.

When evaluating the legitimacy of transnational governance along the three criteria of participation, transparency and accountability, some authors are rather optimistic. Kuyper (2014), for instance, argues that as we move away from single regimes to 'regime complexes' (Raustiala and Victor 2004; see also Chapter 7), the opportunities for increased participation and accountability surge. Much of the literature, however, remains sceptical. In terms of participation, scholars note obstacles to the provision of equal opportunities to influence the norms and rules that govern sustainability, particularly for civil society, developing country actors and marginalized groups (Fuchs and Kalfagianni 2010). Transparency is also generally lacking, and its implementation does not always produce the desirable results. Scholars note, for instance, the often-selective nature of transparency in transnational institutions (Dingwerth and Eichinger 2010), the drawbacks of drowning in too much information and the potential of powerful actors controlling information disclosure to their benefit (Gupta 2010). Finally, accountability is also problematic in transnational governance. Scholars note, in particular, the lack of accountability and redress mechanisms (Hachez and Wouters 2011), while they are concerned especially with regard to accountability to the public (Fuchs, Kalfagianni and Havinga 2011).

Improving Legitimacy

Developing measures that provide legitimacy for transnational governance is admittedly difficult. It is a challenge to create a level playing field in participation, for instance. While several transnational institutions try to improve Southern participation and inclusion of marginalized groups in their structures, the involvement of the latter remains primarily consultative (Hachez and Wouters 2011). Indeed, such an endeavour would require support for those facing resource or collective action problems hindering their participation in the first place. To address this problem, some scholars underline the importance of developing online platforms and information and communication technologies that can enhance the participation of civil society in particular (Davies and Chandler 2012). The problem remains, however, how to ensure inclusion and how to guarantee that the voices of all will be recognized and heard equally. Research on the use of information and communication technologies during the consultation around the Sustainable Development Goals reveals that this cannot be a panacea and similar problems with onsite participation apply (Gellers 2016; Sénit, Kalfagianni and Biermann 2016).

Creating transparent transnational institutions is also challenging. Particularly in cases where corporate actors dominate, transparency as defined in this chapter is an incredibly difficult norm to establish (Haufler 2010). It is also difficult to create mechanisms for accountability and the global attribution of responsibility of transnational actors. Some directions are provided by Ruggie (2013), who emphasizes the importance of strengthening grievance mechanisms alongside educating and training the potentially affected actors in the use of such mechanisms for their own benefit. Other proposals include the explicit recognition of the affected public by public actors, thus creating the conditions for the inclusion of their interests when these are negatively impacted upon by transnational governance institutions (Kalfagianni 2015); the creation of innovative spaces for discursive representation (Stevenson 2014); and an effort towards incremental improvement within specific issue areas instead of adopting a one-size-fits-all approach to legitimation (Kuyper 2014).

Conclusions and Future Directions

Transnational governance – with all its promises and pitfalls – is not likely to wane soon. We expect that the issues underlined in this chapter will continue to inform earth system governance scholarship for years to come. However, the focus of attention may change.

While many scholars dealt with the emergence of new forms of transnational governance in the late 1990s and early 2000s, more recent research is primarily concerned with questions of effectiveness and legitimacy. However, many questions of emergence and development of transnational governance remain unaddressed. For example, scholars continue to be puzzled about why private actors are more committed in some areas, such as forestry, compared to others, such as mining (e.g., Hilson 2014). Also little differentiation is made between earlier initiatives, such as the Forest Stewardship Council, and later ones, such as the Programme for the Endorsement of Forest Certification (for an exception see Dingwerth and Pattberg 2009). Future research needs to explain better the variation among transnational initiatives both across sectors and over time. The role of public actors therein, particularly in terms of orchestration, also gains prominence. Recent scholarship pays increasing attention to the state in providing direct incentives for private actors to create and to commit to new forms of transnational governance, for example, through public procurement and other supportive policies (Gulbrandsen 2014; Sarfaty 2015).

For scholars studying the influence and effectiveness of transnational governance, it is becoming increasingly difficult to decompose the system into discrete units. Instead, effectiveness must be understood in a relational sense. For instance, Eberlein and colleagues (2014) suggest that system-level effects include the interaction between various elements in the architecture. Effectiveness is then not so much a matter of what the final impact is on the problem that an institution is aiming to solve, but instead how the elements influence each other's regulatory capacity and performance. It echoes Biermann and colleagues (2009) suggesting that all global governance architectures are fragmented but that there can be synergistic, cooperative or conflictive outcomes of the institutional interactions. Questions regarding performance for future generations of scholars of earth system governance should encompass how one initiative relates to other initiatives in the institutional architecture and how they influence each other. This in turn poses challenges for future research designs and methodologies, but there are promising avenues such as the use of network approaches for mapping and analyzing transnational governance (Widerberg 2016; Widerberg and Pattberg 2017).

For scholars studying legitimacy, focusing alone on the institutional design elements of transnational governance will prove inadequate. Instead, legitimacy needs to be examined more systemically, particularly in relation to power. Global power relations intersect with legitimacy concerns, including discrimination against vulnerable groups (Schlosberg and Collins 2014) and the privileged position of some interest groups (e.g., business) over others in the transnational arena. Where power is exercised to exclude minorities or marginalized groups, exclusion reduces diversity in representation in decision-making (Earth System Governance Project 2018; see also Burch et al. 2019). Addressing legitimacy concerns, then, means simultaneously addressing power inequalities. In this context, promising future research can include attention to the extent to which transnational governance helps to curb abuses of power; whether transparency-driven approaches empower smaller actors or perpetuate inequalities; and the extent to which power inequalities can be exposed through transnational governance initiatives.

References

Abbott, K. W. (2014). Strengthening the transnational regime complex for climate change. *Transnational Environmental Law*, 3 (1), 57–88.

Abbott, K. W., Green, J. F., & Keohane, R. O. (2016). Organizational ecology and institutional change in Global Governance. *International Organization*, 70 (2), 247–77.

Abbott, K. W., & Snidal, D. (2009). Strengthening international regulation through transnational new governance: Overcoming the orchestration deficit. *Vanderbilt Journal of Transnational Law*, 42 (2), 501–78.

Andonova, L. B., & Hoffmann, M. J. (2012). From Rio to Rio and beyond: Innovation in global environmental governance. *The Journal of Environment and Development*, 21 (1), 57–61.

Andonova, L. B., & Levy, M. A. (2003). Franchising global governance: Making sense of the Johannesburg Type Two partnerships. In O. S. Stokke, & Ø. B. Thommessen (eds.),

Yearbook of international cooperation on environment and development (pp. 19–31). London: Earthscan Publications.

Auld, G., & Gulbrandsen, L. (2010). Transparency in nonstate certification: Consequences for accountability and legitimacy. *Global Environmental Politics*, 10 (3), 97–119.

Auld, G., Renckens, S., & Cashore, B. (2015). Transnational private governance between the logics of empowerment and control. *Regulation and Governance*, 9 (2), 108–24.

Bansard, J., Pattberg, P., & Widerberg, O. (2017). Cities to the rescue? Assessing the performance of transnational municipal networks in global climate governance. *International Environmental Agreements*, 17 (2), 229–46.

Bäckstrand, K. (2008). Accountability of networked climate governance: The rise of transnational climate partnerships. *Global Environmental Politics*, 8 (3), 74–102.

Bartley, T. (2007). Institutional emergence in an era of globalization: The rise of transnational private regulation of labor and environmental conditions. *American Journal of Sociology*, 113 (2), 297.

Bartley, T. (2018). *Rules without rights: Land, labor, and private authority in the global economy*. Oxford: Oxford University Press.

Bernstein, S., & Cashore, B. (2012). Complex global governance and domestic policies: Four pathways of influence. *International Affairs*, 88 (3), 585–604.

Betsill, M. M., & Corell, E. (eds.) (2008). *NGO diplomacy*. Cambridge, MA: The MIT Press.

Betsill, M. M., Dubash, N., Paterson, M., van Asselt, H., Vihma, A., & Winkler, H. (2015). Building productive links between the UNFCCC and the broader global climate governance landscape. *Global Environmental Politics*, 15 (2), 1–10.

Biermann, F., Kanie, N., & Kim, R. E. (2017). Global governance by goal-setting: The novel approach of the UN Sustainable Development Goals. *Current Opinion in Environmental Sustainability*, 26–27, 26–31.

Biermann, F., & Gupta, A. (2011). Accountability and legitimacy in earth system governance: A research framework. *Ecological Economics*, 70, 1856–64.

Biermann, F., Betsill, M. M., Gupta, J. et al. (2010). Earth system governance: A research framework. *International Environmental Agreements*, 10 (4), 277–98.

Biermann, F., Pattberg, P., van Asselt, H., & Zelli, F. (2009). The fragmentation of global governance architectures: A framework for analysis. *Global Environmental Politics*, 9 (4), 14–40.

Bingen, J., & Busch, L. (eds.) (2006). *Agricultural standards: The shape of the global food and fiber system*. New York: Springer.

Blok, K., Höhne, N., van der Leun, K., & Harrison, N. (2012). Bridging the greenhouse-gas emissions gap. *Nature Climate Change*, 2 (7), 471–4.

Börzel, T., & Risse, T. (2005). Public–private partnerships: Effective and legitimate tools of transnational governance? In E. Grande, & L. W. Pauly (eds.), *Complex sovereignty: Reconstituting political authority in the twenty-first century* (pp. 195–206). Toronto: University of Toronto Press.

Breitmeier, H., Underdal, A., & Young, O. R. (2011). The effectiveness of international environmental regimes: Comparing and contrasting findings from quantitative research. *International Studies Review*, 13 (4), 579–605.

Bulkeley, H. A., & Betsill, M. M. (2013). Revisiting the urban politics of climate change. *Environmental Politics*, 22 (1), 136–54.

Bulkeley, H. A., & Betsill, M. M. (2004). Transnational networks and global environmental governance: The cities for climate protection program. *International Studies Quarterly*, 48 (2), 471–93.

Bulkeley, H., Andonova, L. B., Betsill, M. M. et al. (2014). *Transnational climate change governance*. Cambridge, UK: Cambridge University Press.

Burch, S., Gupta, A., Inoue, C. Y. A. et al. (2019). New directions in earth system governance research. *Earth System Governance*, 1.

Cashore, B. W. (2016). Cross-sector partnerships for NSMD global governance: Change pathways and strategic implications. *Annual Review of Social Partnerships*, 11, 88–92.

Cashore, B. W., Auld, G., & Newsom., D. (2004). *Governing through markets: Forest certification and the emergence of non-state authority*. New Haven, CT: Yale University Press.

Chan, S., Falkner, R., Goldberg, M., & van Asselt, H. (2016). Effective and geographically balanced? An output-based assessment of non-state climate actions. *Climate Policy*, 18 (1), 24–35.

Chan, S., & Pattberg, P. (2008). Private rule-making and the politics of accountability: Analyzing global forest governance. *Global Environmental Politics*, 8 (3), 103–21.

Chan, S., van Asselt, H., Hale, T. N. et al. (2015). Reinvigorating international climate policy: A comprehensive framework for effective nonstate action. *Global Policy*, 6 (4), 466–73.

Cole, N. H. (2015). Advantages of a polycentric approach to climate change policy. *Nature Climate Change*, 5, 114–18.

Davies, T., & Chandler, R. (2012). Online deliberation design: Choices, criteria, and evidence. In T. Nabatchi, M. Weiksner, J. Gastil, & M. Leighninger (eds.), *Democracy in motion: Evaluating the practice and impact of deliberative civic engagement* (pp. 103–34). Oxford: Oxford University Press.

Dingwerth, K. (2007). *The new transnationalism: Transnational governance and democratic legitimacy*. New York: Palgrave Macmillan.

Dingwerth, K., & Eichinger, M. (2010). Tamed transparency: How information disclosure under the global reporting initiative fails to empower. *Global Environmental Politics*, 10 (3), 74–96.

Dingwerth, K., & Pattberg, P. (2009). World politics and organizational fields: The Case of transnational sustainability governance. *European Journal of International Relations*, 15 (4), 707–43.

Dryzek, J. S. (2012). Global civil society: The progress of post-Westphalian politics. *Annual Review Political Science*, 15, 101–19.

Dryzek, J. S., & Pickering, J. (2017). Deliberation as a catalyst for reflexive environmental governance. *Ecological Economics*, 131, 353–60.

Earth System Governance (2018). *Earth system governance: Science and implementation plan of the Earth System Governance Project*. Utrecht: Earth System Governance Project.

Easton, D. (1957). An approach to the analysis of political systems. *World Politics*, 9 (3), 383–400.

Eberlein, B., Abbott, K. W., Black, J., Meidinger, E., & Wood, S. (2014). Transnational business governance interactions: Conceptualization and framework for analysis. *Regulation and Governance*, 8 (1), 1–21.

Falkner, R. (2003). Private environmental governance and international relations: Exploring the links. *Global Environmental Politics*, 3 (2), 72–87.

Fuchs, D. (2007). *Business power in global governance*. London: Lynne Rienner Publishers.

Fuchs, D., & Kalfagianni, A. (2010). The causes and consequences of private food governance. *Business and Politics*, 12 (3), 145–81.

Fuchs, D., Kalfagianni, A., & Havinga, T. (2011). Actors in private food governance: The legitimacy of retail standards and multistakeholder initiatives with civil society participation. *Agriculture and Human Values*, 28 (3), 353–67.

Gellers, C. J. (2016). Crowdsourcing global governance: Sustainable Development Goals, civil society and the pursuit of democratic legitimacy. *International Environmental Agreements*, 16, 415–32.

Green, J. F. (2013). *Rethinking private authority: Agents and entrepreneurs in global environmental governance*. Princeton, NJ: Princeton University Press.

Green, J. F., & Auld, G. (2017). Unbudling the regime complex: The effects of private authority. *Transnational Environmental Law*, 6 (2), 259–84.

Greven, T. (2004). Private, Staatliche und Uberstaatliche Interventionen zur Verankerung von Arbeitnehmerrechten. In H. Bass, & S. Melchers (eds.), *Neue Instrumente zur Sozialen und Ökologischen Gestaltung der Globalisierung: Codes of Conduct, Sozialklauseln, nachhaltige Investmentfonds* (pp. 139–371). Münster: LIT Verlag.

Gulbrandsen, L. H. (2014). Dynamic governance interactions: Evolutionary effects of state responses to non-state certification programs. *Regulation and Governance*, 8 (1), 74–92.

Gupta, A. (2008). Transparency under scrutiny: Information disclosure in global environmental governance. *Global Environmental Politics*, 8 (2), 1–7.

Gupta, A. (2010). Transparency as contested political terrain: Who knows what about the global GMO trade and why does it matter? *Global Environmental Politics*, 10 (3), 32–52.

Hachez, N., & Wouters, J. (2011). A glimpse at the democratic legitimacy of private standards: Assessing the public accountability of global G. A. P. *Journal of International Economic Law*, 14 (3), 677–710.

Haufler, V. (2003). Globalization and industry self regulation. In M. Kahler, & D. A. Lake (eds.), *Governance in a global economy: Political authority in transition* (pp. 226–52). Princeton, NJ: Princeton University Press.

Haufler, V. (2010). Disclosure as governance: The extractive industries transparency initiative and resource management in the developing world. *Global Environmental Politics*, 10 (3), 53–73.

Hale, T. N. (2016). 'All hands on deck': The Paris Agreement and nonstate climate action. *Global Environmental Politics*, 16 (3), 12–22.

Hale, T. N., & Held, D. (2011). *Handbook of transnational governance*. Cambridge, UK: Polity Press.

Hickmann, T. (2017a). The reconfiguration of authority in global climate governance. *International Studies Review*, 19 (3), 430–51.

Hickmann, T. (2017b). Voluntary global business initiatives and the international climate negotiations: A case study of the Greenhouse Gas Protocol. *Journal of Cleaner Production*, 169, 94–104.

Hickmann, T., Widerberg, O., Lederer, M., & Pattberg, P. (2019). The United Nations Framework Convention on Climate Change Secretariat as an orchestrator in global climate policymaking. *International Review of Administrative Sciences*, in press.

Hilson, G. (2014). 'Constructing' ethical mineral supply chains in Sub-Saharan Africa: The case of Malawian fair trade rubies. *Development and Change*, 45 (1), 53–78.

Hoffmann, M. J. (2011). *Climate governance at the crossroads: Experimenting with a global response after Kyoto*. Oxford: Oxford University Press.

Holzinger, K., & Sommerer, T. (2011). 'Race to the bottom' or 'race to Brussels'? Environmental competition in Europe. *Journal of Common Market Studies*, 49 (2), 315–39.

Hsu, A., Cheng, Y., Weinfurter, A., Xu, K., & Yick, C. (2016). Track climate pledges of cities and companies. *Nature*, 532 (7599), 303.

Jordan, A. J., & Huitema, D. (2014). Innovations in climate policy: The politics of invention, diffusion, and evaluation. *Environmental Politics*, 23 (5), 715–34.

Jordan, A. J., Huitema, D., Hildén, M. et al. (2015). Emergence of polycentric climate governance and its future prospects. *Nature Climate Change*, (5), 977–82.

Kalfagianni, A. (2015). 'Just Food': The normative obligations of private agrifood governance. *Global Environmental Change*, 31, 174–86.

Kalfagianni, A., & Pattberg, P. (2013). Fishing in muddy waters: Exploring the conditions for effective governance of fisheries and aquaculture. *Marine Policy*, 38, 124–32.

Kalfagianni, A., & Roche, T. (2017). Domestic policy responses to transnational private governance: The Marine Stewardship Council in Alaska, Australia and Ecuador. In T. Havinga, & P. Verbruggen (eds.), *The hybridisation of food governance* (pp. 240–71). Cheltenham: Edward Elgar.

Kanie, N., & Biermann, F. (eds.) (2017). *Governing through goals: Sustainable Development Goals as governance innovation*. Cambridge, MA: The MIT Press.

Kaan, C., & Liese, A. (2011). Public private partnerships in global food governance: Business engagement and legitimacy in the global fight against hunger and malnutrition. *Agriculture and Human Values*, 28 (3), 385–99.

Kaul, I. (2006). *Exploring the policy space between markets and states: Global public private partnerships*. Boulder: Lynne Rienner.

Kleinschmit, D. (2015). Internationale Waldpolitik – Prinzip Freiwilligkeit. In H. Leitschuh, G. Michelsen, U. E. Simonis, J. Sommer, & E. U. Von Weizsäcker (eds.), *Jahrbuch Ökologie 2016* (pp. 82–7). Stuttgart: Hirzel.

Kleinschmit, D., Secco, L., Sergent, A., Wallin, I., & Pülzl, H. (2018). Orchestrating forest policy making: Involvement of scientists and stakeholders in political processes. *Forest Policy and Economics*, 89, 1–106.

Koenig-Archibugi, M., & Macdonald, K. (2013). Accountability-by-proxy in transnational non-state governance. *Governance*, 23 (3), 499–522.

Kramarz, T. (2016). World Bank partnerships and the promise of democratic governance. *Environmental Policy and Governance*, 26, 3–15.

Kramarz, T., & Park, S. (2016). Accountability in global environmental governance: A meaningful tool for action? *Global Environmental Politics*, 16 (2), 1–21.

Kronsell, A. (2013). Legitimacy for climate policy: Politics and participation in the green city of Freiburg. *Local Environment*, 18 (8), 965–82.

Kuyper, J. W. (2014). Global democratization and international regime complexity. *European Journal of International Relations*, 20 (3), 620–46.

Kuyper, J. W., Linnér, B. O., & Schroeder, H. (2018). Non-state actors in hybrid global climate governance: Justice, Legitimacy, and Effectiveness in a Post-Paris Era. *Climate Change*, 9 (1), e497.

Mert, A. (2009). Partnerships for sustainable development as discursive practice: Shifts in discourses of environment and democracy. *Forest Policy and Economics*, 11 (5–6), 326–39.

McDermott, C. L., Irland, L. C., & Pacheco, P. (2015). Forest certification and legality initiatives in the Brazilian Amazon: Lessons for effective and equitable forest governance. *Forest Policy and Economics*, 50, 134–42.

Miles, E. L., Underdal, A., Andresen, S., Wettestad, J., Skjærseth, J. B., & Carlin, E. M. (2001). *Environmental regime effectiveness: Confronting theory with evidence*. Cambridge, MA: The MIT Press.

Nygren, A. (2015). Governance and images: Representations of certified southern producers in high-quality design markets. *Environmental Values*, 24 (3), 391–412.

Orsini, A., Morin, J. F., & Young, O. R. (2013). Regime complexes: A buzz, a boom, or a boost for global governance? *Global Governance*, 19 (1), 27–39.

O'Rourke, D. (2003). Outsourcing regulation: Analyzing nongovernmental systems of labor standards and monitoring. *The Policy Studies Journal*, 31 (1), 1–19.

Ostrom, E. (2010). Polycentric systems for coping with collective action and global environmental change. *Global Environmental Change*, 20 (4), 550–7.

Partzsch, L. (2007). *Global Governance in Partnerschaft: Die EU-Initiative 'Water for Life'*. Baden-Baden: Nomos.

Partzsch, L. (2014). *Die neue Macht von Individuen in der globalen Politik: Wandel durch Prominente, Philanthropen und social entrepreneurs*. Baden-Baden: Nomos.

Partzsch, L. (2017). Powerful individuals in a globalized world. *Global Policy*, 8 (1), 5–13.

Partzsch, L. (2018). Take action now: The legitimacy of celebrity power in international relations. *Global Governance*, 24 (2), 229–48.

Pattberg, P. (2010). Public–private partnerships in global climate governance. *Wiley Interdisciplinary Reviews, Climate Change*, 1 (2), 279–87.

Pattberg, P., Betsill, M. M., & Dellas, E. (eds.) (2011). Agency in earth system governance. *International Environmental Agreements*, 11 (1), 85–98.

Pattberg, P., Biermann, F., Chan, S., & Mert, A. (2012). *Public–private partnerships for sustainable development: Emergence, influence and legitimacy*. Cheltenham: Edward Elgar.

Pattberg, P., & Widerberg, O. (2016). Transnational multi-stakeholder partnerships for sustainable development: Conditions for success. *AMBIO*, 45 (1), 42–51.

Ponte, S., & Daugbjerg, C. (2015). Biofuel sustainability and the formation of transnational hybrid governance. *Environmental Politics*, 24 (1), 96–114.

Porter, T., & Ronit, K. (eds.) (2010). *The challenges of global business authority: Democratic renewal, stalemate or decay?* New York: SUNY Press.

Quark, A. (2013). *Global rivalries: Standards wars and the transnational cotton trade*. Chicago, IL: University of Chicago Press.

Raustiala, K., & Victor D. G. (2004). The regime complex for plant genetic resources. *International Organization*, 58 (2), 277–309.

Risse, T. (2002). Transnational actors and world politics. In W. Carlsnaes, T. Risse, & B. Simmons (eds.), *Handbook of international relations* (pp. 255–74). London: SAGE Publications.

Ruggie, J. (2013). *Just business: Multinational corporations and human rights*. New York: W. W. Norton and Company.

Sarfaty, G. A. (2015). Shining light on global supply chains. *Harvard International Law Journal*, 56 (2), 419–63.

Scharpf, F. W. (1997). *Games real actors play: Actor-centered institutionalism in policy research*. London: Routledge.

Schlosberg, D., & Collins, L. B. (2014). From environmental to climate justice: climate change and the discourse of environmental justice. *WIREs Climate Change*, 5 (3), 359–74.

Scholte, J. A. (2008). Defining globalisation. *The World Economy*, 31 (11), 1471–1502.

Scott, W. R. (1995). *Institutions and organizations*. Thousand Oaks: Sage Publications.

Sénit, C. A., Biermann, F., & Kalfagianni, A. (2017). The representativeness of global deliberation: A critical assessment of civil society consultations in the shaping of Sustainable Development Goals. *Global Policy*, 8 (1), 62–72.

Sénit, C. A., Kalfagianni, A., & Biermann, F. (2016). Cyber-democracy? Information and communication technologies in civil society consultations for sustainable development. *Global Governance*, 22 (4), 533–54.

Stevenson, H. (2014). Representing green radicalism: The limits of state-based representation in global climate governance. *Review of International Studies*, 40, 177–201.

Szulecki, K., Pattberg, P., & Biermann, F. (2011). Explaining variation in the effectiveness of transnational energy partnerships. *Governance*, 24 (4), 713–36.

Tucker, K. (2014). Participation and subjectification in global governance: NGOs, acceptable subjectivities and the WTO. *Millennium: Journal of International Studies*, 42 (2), 376–96.

UNEP (2012). *Global environmental outlook 5: Summary for policy makers*. Nairobi: United Nations Environment Programme.

UNEP (2015). *The emissions gap report 2015: Executive Summary*. Nairobi: United Nations Environment Programme.

UNEP (2019). *Global environment outlook 6: Healthy planet, healthy people*. Nairobi: United Nations Environment Programme.

Van der Ven, H., Bernstein, S., & Hoffmann, M. (2016). Valuing the contributions of nonstate and subnational actors to climate governance. *Global Environmental Politics*, 17 (1), 1–20.

Washington, S., & Ababouch, L. (2011). *Private standards and certification in fisheries and aquaculture: Current practice and emerging issues*. Rome: Food and Agriculture Organization of the United Nations.

Whitfield, D. (2001). *Public services or corporate welfare: Rethinking the nation state in the global economy*. London: Pluto Press.

Widerberg, O. (2016). Mapping institutional complexity in the Anthropocene: A network approach. In P. Pattberg, & F. Zelli (eds.), *Environmental politics and governance in the Anthropocene: Institutions and legitimacy in a complex world* (pp. 81–102). London: Routledge.

Widerberg, O. (2017). The 'black box' problem of orchestration: How to evaluate the performance of the Lima-Paris Action Agenda. *Environmental Politics*, 26 (4): 1–23.

Widerberg, O. and Pattberg, P. (2017). Accountability challenges in the transnational regime complex for climate change. *Review of Policy Research*, 34 (1), 68–87.

World Commission on Environment and Development (1987). *Our common future*. Oxford: Oxford University Press.

Young, O. R. (2011). Effectiveness of international environmental regimes: Existing knowledge, cutting-edge themes, and research strategies. *Proceedings of the National Academy of Sciences of the United States of America*, 108 (50), 19853–60.

Zelli, F., & van Asselt, H. (eds.) (2013). The institutional fragmentation of global environmental governance: Causes, consequences and responses. *Global Environmental Politics*, 13 (3), 1–13.

5

Institutional Architectures for Areas beyond National Jurisdiction

ORAN R. YOUNG

Writing in 1977, a team of analysts associated with the Brookings Institution in Washington, DC, observed that as the 'earth becomes more crowded, there is increasing international concern over the management of international "commons" – those realms that have remained outside the jurisdiction of any country' (Brown et al. 1977: 1). They went on to say that accelerating economic and political developments in these realms raise 'basic policy questions' regarding how international commons are used by whom and based on what rules of the game (Brown et al. 1977). While current terminology was not in use at that time, these analysts were directing attention to what we now think of as earth system governance with particular reference to human activities in areas beyond national jurisdiction (Biermann et al. 2009). So, we can ask: have we made progress in the intervening 40 years in addressing these basic policy questions, and have our efforts added significantly to the theory and practice of earth system governance?

It is clear at the outset that what earth system scientists have documented as the Great Acceleration, starting roughly in the middle of the twentieth century, has heightened the importance of these concerns (Steffen et al. 2004). We now live in a world in which human actions have become driving forces not only regarding matters of longstanding concern (for example, the depletion of ocean fish stocks) but also regarding new realms (for example, the use of cyberspace) and the fate of the planet itself (for example, the earth's climate system). How have our international and transnational governance systems evolved to address these challenges? Can we reach any solid conclusions regarding factors that determine the success or effectiveness of governance systems dealing with matters arising in areas beyond national jurisdiction, and more specifically regarding matters of architecture and institutional design, in this context? Are efforts to come to grips with this class of needs for governance raising fundamental questions about the character of international society as a social system whose members are sovereign states (Young

2016)? In this chapter, I assess the state of play regarding efforts to address these questions and to chart a course for the next phase of research in this realm.

Conceptualization

It is useful to begin with some distinctions among subsets of the general class of areas beyond national jurisdiction. There are, to begin with, what we have long treated as international spaces (Young 2011). These are material or physical systems that are relatively easy to locate in spatial or areal terms. Familiar examples include the high seas, the deep seabed and (for all practical purposes) Antarctica (Berkman et al. 2011).[1] Together, these international spaces cover almost half of the earth's surface. A second distinct subset includes what we now think of as earth systems or planetary support systems, which play critical roles in determining the condition of the planet as a whole and its suitability for human habitation (Rockström et al. 2009). These, too, are physical or material systems. But it is difficult or impossible to delimit them in any ordinary spatial sense (Steffen et al. 2015). The earth's climate system is a prominent example of this subset of the broader category of areas beyond national jurisdiction. But others involve biogeochemical flows like the nitrogen and phosphorus cycles and the earth's overall complement of living species or planetary biological diversity (Kolbert 2014; Wilson 2016). In contrast to these material systems, a third subset comprises intangible or virtual systems that are increasingly important functionally but that are not associated with spaces easily identifiable in physical or material terms. Perhaps the quintessential example is cyberspace, characterized as 'the notional environment in which communication over computer networks occurs' (Oxford English Dictionary). In terms of this tripartite classification of areas beyond national jurisdiction, outer space constitutes an outlier or a somewhat unusual case. We often characterize outer space as an international space, much like the high seas. But outer space has no meaningful boundaries; it extends beyond the confines of planet earth with no obvious outer limit. As well, outer space is used often as a medium for the transmission of information that has no material presence.

The usefulness of drawing these distinctions, as I shall endeavour to show in the substantive sections of this chapter, arises from the fact that each of the subsets of areas beyond national jurisdiction generates distinct needs for governance, giving rise to a variety of institutional architectures that yield insights for those seeking to create effective governance systems. Given the flood of literature on issues of

---

[1] While Antarctica is subject to unextinguished jurisdictional claims on the part of seven states, the regime that has grown up under the terms of the 1959 Antarctic Treaty treats the continent as an international space in practice.

governance in areas beyond national jurisdiction, it is impossible in a short chapter to provide a comprehensive review of where we stand regarding the governance of these realms. As a way forward, I will focus on insights relating to the architecture of governance systems from one key example exemplifying each of the subsets identified in the preceding paragraph: the governance of ocean spaces, climate governance and the governance of the Internet.

The governance of ocean spaces presents a case of a comprehensive and well-established constitutive foundation – the 1982 United Nations Convention on the Law of the Sea (UNCLOS) – together with a growing collection of operational arrangements under the terms of Implementing Agreements and related arrangements dealing with sectoral but overlapping issues (Oude Elferink 2005). The fundamental challenge is to find ways to weave the elements of the resultant regime complex into an integrated and coherent whole (Chapter 7), especially as we move towards the development of an international legally binding instrument on what we have come to know as biodiversity beyond national jurisdiction (Oberthür and Stokke 2011). For its part, the case of governing the earth's climate system offers an example of a growing need for governance that has not yielded to traditional top-down procedures featuring the negotiation of legally binding multilateral agreements highlighting regulatory measures and directing attention to issues of compliance. What is emerging instead is a bottom-up effort featuring a system known increasingly as pledge-and-review, in which the pledges that individual countries make are politically significant even though they do not impose performance obligations that are legally binding (Cherry, Hovi and McEvoy 2014). The governance of the Internet, by contrast, provides an example featuring an order that is decentralized and in large measure self-generating in its origins, though it is now plagued with challenges likely to require deliberate and coordinated intervention to come to terms with increasingly serious problems (for example, problems involving politically motivated hacking, bullying and identity theft) (Hayek 1973). A key question here concerns the division of labour between governments, corporations and actors located in civil society in the development and implementation of measures needed to address these problems (Brousseau, Marzouki and Meadel 2015).

Research Findings

International Spaces: Governing Marine Areas beyond National Jurisdiction

Oceans cover about 70 per cent of the earth's surface. Of this, roughly two-thirds lie beyond the outer boundaries of the Exclusive Economic Zones of coastal states or, in other words, in areas beyond national jurisdiction. This means that about 45 per cent of the planet's surface belongs to the category of international spaces. The status of the seabed is much the same but with one major exception. Under the provisions of Article 76 of UNCLOS, coastal states are permitted to exercise authority over the seabed beyond the limits of their Exclusive Economic Zones when they can demonstrate convincingly that the areas in question are natural prolongations of land lying within their jurisdiction. UNCLOS establishes a well-defined process for handling such matters that are ongoing at this time but likely to eventuate in a situation in which areas beyond national jurisdiction on the seabed are substantially (but not drastically) more limited than parallel areas of the water column and superjacent airspace.[2] The bottom line is clear: the sphere of marine international spaces is large and will remain so for the foreseeable future.

Regarding ocean governance, the point of departure is that UNCLOS provides a widely accepted constitutive foundation for the public order of the oceans. Whether it is appropriate to treat UNCLOS as an actual constitution for the oceans is a matter that has provoked lively debate (see various contributions to Oude Elferink 2005). But it is a constitutive arrangement in at least two important senses. UNCLOS sets forth a structure of concepts, principles, rights and rules applicable to all matters of ocean governance, and it establishes a mechanism (the International Tribunal on the Law of the Sea) authorized to interpret the meaning of these provisions regarding specific situations and to settle disputes relating to such matters. In addition, UNCLOS provides a platform on which to erect more operational regimes in the form of Implementing Agreements or similar legally binding arrangements dealing with specific issues (for example, the 1995 fish stocks agreement). Operational regimes relating to marine issues developed under other auspices (for example, the International Maritime Organization's convention on the prevention of pollution from ships) also are expected to be compatible with the constitutive provisions of UNCLOS.[3]

In general terms, this governance system treats the high seas and the so-called Area (the portion of the deep seabed lying beyond the boundaries of national jurisdiction) differently. A basic norm of the system is that the high seas should be open to entry for various purposes (for example, navigation, fishing and laying submarine cables) on the part of all members of the international community. Of course, the high seas cover a smaller area now than they did when the doctrine of freedom of the seas took root in the seventeenth century as a foundational norm underpinning maritime commerce. But the spirit of this doctrine is still alive and

---

[2] Among other things, Art. 76 creates the Commission on the Limits of the Continental Shelf with a mandate to provide scientific assessments of submissions that individual states make regarding seabed jurisdiction.

[3] The 1973/1978 International Convention for the Prevention of Pollution from Ships (MARPOL) predates UNCLOS. But the contents of this agreement are compatible with the constitutive framework of UNCLOS. See Zacher and Sutton 1996.

embedded in many provisions of UNCLOS. The Area, on the other hand, is treated in UNCLOS as part of the common heritage of mankind (now humanity). This means that those desiring to exploit the resources of the deep seabed are expected to operate under internationally agreed rules and to comply with provisions relating to benefit sharing. UNCLOS establishes the International Seabed Authority as the body charged with administering this regime on a day-to-day basis.

In reality, the differences between the operational regimes for the high seas and the deep seabed are not as sharp. There are numerous restrictions on the freedom of the high seas under a range of operational regimes. These include many of the regional fisheries management organizations, some regional seas agreements (for example, the Convention for the Protection of the Marine Environment of the North-East Atlantic), numerous International Maritime Organization regulations dealing with commercial navigation, the arrangement pertaining to whales and whaling and others. When it comes to matters of governance, the high seas today certainly do not bear much resemblance to our conventional image of the Wild West. Nor is the situation clear-cut with regard to the governance of the Area. For one thing, opposition to the regime for the Area has proven to be the single biggest stumbling block regarding ratification of UNCLOS on the part of the United States. One motivating factor behind the 1994 Implementing Agreement relating to the regime for the Area (Part XI of UNCLOS) was a desire to make the regime more palatable to opponents seeking to minimize restrictions on the freedom of private corporations to operate in this realm. Still, this adjustment has not satisfied those opposed to any significant restrictions on free enterprise in this realm. The issue has faded somewhat in recent years due to a general lack of commercial interest in proceeding with deep seabed mining. If (as some now anticipate) changing economic conditions or new technologies produce a revival of interest in the resources of the Area, however, undoubtedly disagreements regarding Part XI of UNCLOS will resurface.

The most prominent issue pertaining to the architecture of ocean governance today involves the effort to negotiate an international legally binding instrument within the framework of UNCLOS on biodiversity beyond national jurisdiction (Rochette et al. 2015). Developed over several years by an informal working group, this proposal took a major step forward in 2015 when the United Nations General Assembly passed a resolution (Resolution 69/292 of 19 June 2015) establishing a Preparatory Committee with a mandate to consider the content of such an agreement and to report back to the assembly on its progress by the end of 2017. On the strength of the resultant report, the General Assembly passed Resolution 72/249 on 24 December 2017 establishing an Intergovernmental Conference to negotiate a legally binding instrument on biodiversity beyond national jurisdiction and authorizing four negotiating sessions during the period 2018–2020. This process has developed considerable momentum; a first-order draft of the proposed agreement is now in circulation. It is premature at this stage to make confident predictions regarding the prospects for success in these negotiations, much less regarding the prospects for entry into force of such an agreement if the Intergovernmental Conference does reach agreement on the terms of a legally binding instrument on biodiversity beyond national jurisdiction. Nevertheless, this is an important development that could eventuate in the imposition of new restrictions on the scope of the traditional doctrine of freedom of the high seas.

A distinctive feature of the proposed agreement is that it encompasses four distinct elements that may not seem intuitively obvious as elements of a coherent a package: (a) arrangements dealing with bioprospecting and marine genetic resources more generally; (b) provisions relating to marine protected areas or what are characterized as area-based management tools; (c) procedural provisions relating to conduct of environmental impact assessments; and (d) arrangements relating to capacity building and technology transfer. Several major concerns arise in this context and will play important roles in determining the fate of the effort to agree on the terms of a legally binding instrument on biodiversity beyond national jurisdiction. So far, the proponents are largely members of the legal and NGO communities, with groups interested in fishing, shipping and mining adopting a more sceptical view. Will the proponents be able to build an effective coalition encompassing active participation from interested industries, including fishing, maritime commerce and bioprospecting? An agreement on biodiversity beyond national jurisdiction inevitably would interact in a variety of ways with existing regimes created to deal with other human activities. Is it feasible to work out mutually satisfactory relationships between an agreement on biodiversity beyond national jurisdiction and various regional fisheries management organizations, arrangements covering marine mammals and regulations under the International Maritime Organization pertaining to maritime commerce? Also, because the boundaries between Exclusive Economic Zones and areas beyond national jurisdiction are determined by geopolitical considerations rather than biophysical factors, a regime on biodiversity beyond national jurisdiction would need to be able to coordinate with national authorities in many parts of the world. What mechanisms would be effective in addressing this concern? Given the limited administrative capacity of the United Nations, it would be necessary to develop new operating arrangements capable of administering the procedural elements of a regime on biodiversity beyond national jurisdiction relating to matters like the conduct of environmental impact assessments and the handling of technology transfers. What would be the appropriate way to address such matters that would loom large in the process of moving an agreement on biodiversity beyond national jurisdiction from paper to practice?

What general observations regarding architecture can we draw from this discussion of maritime regimes as a prominent example of meeting needs for governance relating to international spaces? Three observations stand out. First, the distinction between constitutive arrangements and operational regimes is important. The role of UNCLOS in providing a solid platform on which to build regimes for specific issues in the realm of ocean governance is clearly helpful. Second, when it comes to the development of operational regimes dealing with specific oceans issues, institutional interplay is the order of the day. The initiative on biodiversity beyond national jurisdiction is not unusual in this regard. Finding ways to avoid tensions or even encourage synergies in interactions among sectoral regimes will be critical to their success (Oberthür and Stokke 2011). Third, the focus on specific maritime issues (for example, fish stocks, maritime commerce, deep seabed mining) tends to divert attention from profound challenges to the future of the ocean stemming from exogenous sources. The obvious example is climate change, which is responsible for the acidification of marine systems and for changes in ocean temperatures. But there are others including matters like the dispersion of plastic debris throughout the oceans, the expansion of hypoxic or dead zones and the destruction of coastal ecosystems. Efforts to address needs for ocean governance that ignore these larger anthropogenic drivers are unlikely to succeed.

Earth Systems: Governing the Planet's Climate System

We live today in a world of human-dominated systems and growing concern about the fate of what the science community calls a 'planet under pressure' (Steffen et al. 2004). The result is the development of holistic perspectives on the earth, a rising concern for planetary boundaries and an emerging awareness of the need to maintain 'a safe operating space for humanity' (Rockström et al. 2009). Without doubt, climate change is the paradigmatic example of this class of concerns involving systems that are material in character but extend beyond the bounds of what we are used to thinking of as international spaces. Climate change also presents a need for governance that has not yielded to mainstream approaches to international environmental governance and that has become a focus of experiments with new strategies. What lessons can we derive concerning the architecture of governance systems for areas beyond national jurisdiction from an examination of experience with efforts to create an effective climate regime (Dryzek, Norgaard and Schlosberg 2013)?

Human actions that generate emissions of greenhouse gases, including the burning of fossil fuels and the destruction of biomass, have become major drivers of change in the earth's climate system (Intergovernmental Panel on Climate Change 2014). The concentration of carbon dioxide, the principal greenhouse gas, in the atmosphere has risen from about 270 parts per million in preindustrial times to ~415 parts per million today; it continues to rise at a rate of 2–3 parts per million per year.[4] If we add the other greenhouse gases, the concentration in the atmosphere today is over 450 ppm carbon dioxide equivalent. This is a shift of massive proportions largely, though not exclusively, the result of anthropogenic forces. To be sure, not much is known at this stage about the likely impacts of this transformative change on the behaviour of the climate system. Nevertheless, the global scientific community agrees, with a remarkably high level of consensus, that the resultant changes will be dramatic, that they will manifest themselves on a planetary scale and that they are already detectable through a suite of indicators particularly in the high latitudes of both hemispheres (Oreskes 2004). Because the climate system has profound consequences for all human activities, the result is a growing need for governance that dwarfs most other needs for governance, at least regarding problems that are fundamentally biophysical in character.

Although they are obviously coupled, it is helpful to differentiate three separate components of this need for governance: (a) the problem of reducing emissions of greenhouse gases or what is often described as mitigation; (b) the problem of finding ways to prevent rising concentrations of greenhouse gases from changing the climate system or what may be called avoidance; and (c) the problem of adjusting to the impacts of changes in the climate system or what is often referred to as adaptation. Each component raises major challenges with regard to governance. But these challenges are by no means the same. Efforts to address each component have met with limited success so far, but for different reasons. Accordingly, it makes sense to look into the experience with mitigation, avoidance and adaptation in the search for insights about institutional architecture for areas beyond national jurisdiction.

Until recently, efforts to reduce emissions of greenhouse gases took the form of regulatory measures of the sort that have become familiar in a range of other issue areas (Dryzek, Norgaard and Schlosberg 2013). The 1992 United Nations Framework Convention on Climate Change established a legally binding constitutive base followed by the introduction of explicit targets and timetables in the 1997 Kyoto Protocol; a growing emphasis on issues of compliance and the role of what have become known as monitoring, reporting and verification practices; and an ongoing effort to strengthen regulatory requirements and extend them to additional member states. This effort sought to apply a governance strategy that had proven generally effective in dealing with stratospheric ozone depletion under the Montreal Protocol to the case of reducing greenhouse gas emissions (Parson 2003).

---

[4] The annual reports of the Global Carbon Project provide the most authoritative data on concentrations of greenhouse gases in the atmosphere – www.globalcarbonproject.org.

But the strategy turned out to be ineffective in the case of climate change, grinding slowly to a halt in the tortuous negotiations over the provisions of what is known as the Copenhagen Accord adopted at the eleventh hour of the conference of the parties in 2009. In the aftermath of this debacle, consensus emerged on the need to adopt a fundamentally different approach to the problem of reducing greenhouse gas emissions (Cherry, Hovi and McEvoy 2014).

The new strategy, reflected most clearly in the terms of the Paris Climate Agreement adopted at the conference of the parties in 2015, is often described as bottom-up in contrast to the top-down character of the previous strategy. The upshot is a system of pledge-and-review under which the parties agree to a collective goal and individual member countries pledge to fulfil Nationally Determined Contributions to the achievement of the goal. The pledges are not legally binding, but are subject to periodic reviews expected to produce a step-by-step process of strengthening or ratcheting up commitments regarding reductions in emissions of greenhouse gases over time. Even if all the initial Nationally Determined Contributions are fulfilled faithfully, temperatures at the earth's surface are likely to rise by 3–3.5 degrees Celsius, far in excess of the Paris Agreement's stated goal of holding increases to well under 2.0 degrees Celsius and ideally to 1.5 degrees Celsius. The success of this strategy, therefore, will depend critically on the willingness or ability of the parties to ratchet up their pledges through a process described in the 2015 agreement as a periodic 'global stocktake'. Whether this process is likely to succeed is a matter of considerable speculation at this stage (Young 2016). But our experience in other cases suggests that critical determinants of the outcome will include the (non)occurrence of climate shocks, the rise of a global social movement dedicated to addressing the problem of climate change and the emergence of effective leadership in the form of champions who are willing and able to keep this issue on the front burner in a variety of policy arenas.

One interesting observation regarding mitigation involves the interplay between the climate regime and other related regimes. So far, at least, the Montreal Protocol on ozone-depleting substances has proven considerably more effective than the Kyoto Protocol as a mechanism for reducing emissions of greenhouse gases (Velders et al. 2007). In a remarkable initiative, the meeting of the parties of the Montreal Protocol initiated in 2016 a phase-out of the production and consumption of hydrofluorocarbons, which are powerful greenhouse gases but not ozone-depleting substances (in fact, hydrofluorocarbons became popular precisely because they are not ozone-depleting). This leads to the broader observation that institutional interplay – including instances that produce negative as well as positive results – needs to be taken seriously in any assessment of institutional architecture relating to the problem of climate change (Keohane and Victor 2011).

The elusiveness of efforts to mitigate emissions and the onset of undeniable evidence of the impacts of climate change have directed the attention of growing numbers of observers to the question of whether there are ways to avoid the impacts of climate change that do not require far-reaching reductions in emissions of greenhouse gases. The result, predictably, is a search for technological solutions and, in this instance, a growing interest in what has become known as geoengineering (Royal Academy of Sciences 2009; National Research Council 2015). In its most popular form, generally described as solar radiation management, geoengineering envisions efforts to reduce radiative forcing at the earth's surface by injecting various substances into the earth's atmosphere to limit temperature increases by blocking or limiting the penetration of solar radiation (Keith 2013). While understanding of geoengineering remains limited at present, several features of this strategy are already becoming clear. Solar radiation management may be relatively inexpensive, well within the means of many states and even some non-state actors. Because this strategy would not reduce concentrations of greenhouse gases in the atmosphere, it would be essential to make a long-term commitment to the use of this strategy once we start down the road of solar radiation management. What is more, geoengineering would not alleviate the acidification of marine and terrestrial ecosystems, even if it were to succeed in limiting increases in temperature at the earth's surface.

From the point of view of governance, this is an explosive combination of characteristics. We are faced now with the need to make decisions about rules and procedures governing research on and potential deployment of various forms of solar radiation management. Should scientists be allowed to make their own decisions about research on various forms of geoengineering? Should individual states or non-state actors be allowed to make unilateral decisions about the deployment of solar radiation management systems? What are the requirements of good governance in this realm? And do we need to put in place an international geoengineering regime, or at least an agreed code of conduct, now rather than waiting until initiatives in this realm are more advanced to focus on the development of a suitable regime in this realm? There are no clear-cut answers to these questions. But the decisions we make may have profound consequences not only for human well-being but also for the future of the planet. This may be an area that will benefit from the development and dissemination of ethical or normative principles to guide the behaviour of all parties concerned rather than confining our attention to the more technical issues of devising rules and regulations (Young 2017).

If the impacts of climate change continue to grow, the need for adaptation strategies will become increasingly prominent. Adaptation does not lend itself to the development of general institutional responses. The actual impacts of climate change will vary greatly from place to place even within individual countries, and the ability of countries to deal effectively with needs for adaptation to the impacts of climate change varies enormously. From the perspective of governance, perhaps the most general concern about adaptation centres on the question of financing and, in particular, the need to provide financial assistance to countries that are likely to be affected first and foremost by climate change but lack the resources to launch effective adaptation strategies. This issue has been prominent in the climate negotiations since the beginning, giving rise to various flexibility mechanisms designed to increase the incentives of leaders of industrialized countries to assist victims of climate change and to numerous calls for new and additional financial assistance. The Green Climate Fund, established in 2010 under the climate convention, is now the principal vehicle for addressing these issues. Without going into detail, it seems accurate to conclude that progress to date in this realm has been severely limited.

Funding is not the only challenge of governance relating to adaptation to the impacts of climate change. Another issue that seems likely to become a lightning rod for intense controversy concerns the treatment of people displaced from their homes and in many cases from their countries of origin because of the impacts of climate change. While it is premature to make firm predictions, it is easy to imagine the numbers of 'climate refugees' swelling into the tens or hundreds of millions in the coming decades (Biermann and Boas 2008). In the case of low-lying small-island developing states, this may mean relocating entire populations. In other cases, the impacts of climate change are likely to interact with other problems to produce particularly toxic situations. Recent experiences with the treatment of people displaced from war-torn regions makes it clear that our capacity to deal with refugees is severely limited. The rise of populism and the 'new nationalism' in Europe and the United States will make this problem even more difficult to solve. From the point of view of governance, it may be that this is an area where national governments are severely constrained and the contributions of non-state actors will become increasingly important.

Cyberspace: Governing the Internet

Cyberspace is a medium for the transmission of data through either hardwired broadband access systems (for example, cable and digital subscriber lines) or wireless broadband access systems (for example, satellite broadband) that allow users to transmit and receive information electronically. A cascade of technological innovations has made cyberspace a prominent frontier in the expansion of human activities and led to the emergence of a wholly new genre of needs for governance extending beyond the reach of national authorities. Arguably, the most fundamental element of this expanding virtual world is the Internet, an electronic medium that not only allows for direct exchanges of email messages between and among users but also serves as an entry point for accessing search engines (for example, Google and Baidu), social networking sites (for example, Facebook, Twitter and WeChat) and content aggregation sites (for example, YouTube, Google Scholar and Wikipedia) (DeNardis 2014: 171). The extraordinary growth of these forms of communication over the last three decades highlights the success of the Internet as a means of connecting those desiring to communicate with one another or to access information available through services like sites on the World Wide Web.[5] In the last 25 years, the community of Internet users has grown from a base of just 10 million people in 1993 to upwards of 4 billion or over 50 per cent of the population of the planet today. Yet this explosive growth also has brought into focus a range of challenges that pose new and rapidly growing needs for governance and raise questions about the adequacy of the decentralized or distributed form of governance that has characterized the Internet since its inception.

Unlike the oceans, where UNCLOS provides a constitutive foundation or point of departure for the development of more focused regimes addressing a variety of human activities, there is no constitutive law of cyberspace or even a law of the atmosphere comprehensive enough to include provisions pertaining to the transmission of data electronically (for perspectives on the idea of a law of the atmosphere see International Law Commission 2016). Nor is there a more focused intergovernmental agreement spelling out the major provisions of a governance system or regime applicable to the Internet itself. Nevertheless, it would be a mistake to conclude that the Internet lacks a governance system. In fact, the essential features of a decentralized Internet governance system emerged at an early stage. They have remained fundamentally unchanged in form during the intervening years, though their content has expanded rapidly to keep pace with the extraordinary growth in the scope as well as the volume of electronic communication in recent years.

What makes Internet governance particularly interesting to those concerned with institutional architecture is that it is neither an international regime established under the terms of an intergovernmental agreement nor an ordinary market arrangement emerging spontaneously from interactions between producers and consumers of specific products. The system that has emerged is heterarchic rather than hierarchic (Brousseau, Marzouki and Meadel 2015: 16) taking the form, as DeNardis (2014: 23) observes, of 'distributed and networked multi-stakeholder

---

[5] Though the transmission of data electronically is virtual, the use of the Internet requires hardware (for example, computers) and infrastructure (for example, cables) that are material in nature and often located within the jurisdiction of sovereign states. This becomes a matter of considerable importance when governments seek to suppress information or otherwise control access to the Internet.

governance' in which a mix of corporate entities, non-profit organizations and public agencies interact in complex ways to devise and administer a set of guiding practices. As she goes on to say, the system is 'a contested space reflecting broader global power struggles' but one that has managed to adapt quickly to rapid technological, socio-economic and political changes (DeNardis 2014: 222).

Suggestive as they are, however, these descriptive statements are rather imprecise. On further examination, several specific features of the Internet governance system stand out. Together, they make it clear that the architecture of this regime differs from that of other international governance systems in significant ways. Internet governance involves not only many individual actors but also a diverse collection of actors located in distinct spheres. There is no established hierarchy in this realm; individual actors located in the public sector, the private sector and civil society play essential roles that are not superordinate or subordinate to one another. For the most part, the principal actors arrive at decisions through informal processes best described as a form of consensus building leading to the development of social practices that are effective without being legally binding. Few of the essential features of this governance system are embedded in formal intergovernmental agreements. While the system is highly dynamic, its essential features have remained relatively stable. Overall, Internet governance constitutes a form of 'governance without government' (Rosenau and Czempiel 1992) that relies on 'consensual, evolving and soft norms', even though public agencies do play roles of some significance in the resulting system (Rosenau and Czempiel 1992; Brousseau, Marzouki and Meadel 2015: 35).

The architecture of this governance system comprises two principal components known as (a) critical Internet resources and (b) transmission control protocols and Internet protocols. Critical Internet resources include Internet addresses, domain names and Autonomous System Numbers. Necessary for the Internet to remain operational, the critical Internet resources are administered by the Internet Corporation for Assigned Names and Numbers and its subsidiary the Internet Assigned Numbers Authority, which is linked contractually to the United States Department of Commerce, a matter of some controversy in the world of Internet governance. But the Internet Corporation for Assigned Names and Numbers and the Internet Assigned Numbers Authority have become international administrative arrangements in practice subject to review on the part of bodies like the United Nations Internet Governance Forum (Brousseau, Marzouki and Meadel 2015: 18). Transmission control protocols and Internet protocols constitute the rules of the game of the Internet. They are drafted by the Internet Engineering Task Force and formalized by the Internet Architecture Board operating 'under a not-for-profit, membership-oriented organization called the Internet Society (ISOC)' (DeNardis 2014: 69–70).

The resultant system has performed well in the sense that it has presided over the development of the Internet as a universal, resilient and adaptable arrangement for the electronic transmission of data. Close observers like DeNardis (2014: 230) have argued that the Internet's 'decentralized and distributed balance of power', which supports the organic growth of a system of established practices, accounts for its success. Whether or not this interpretation is correct, it seems fair to say that the Internet operates on the basis of a set of recognized practices that have normative force without being embedded in legally binding rules and that are able to adapt to changing circumstances in the absence of the difficulties associated with efforts to amend or revise legally binding instruments. The essential glue that makes the system work is self-interest. While the existing system may not be ideal for all participants, no one has a compelling reason to act in ways that would risk triggering a collapse of the system.

Nevertheless, this does not mean that Internet governance is an unmitigated success that can be counted on to find solutions to all the problems that have emerged in connection with the operation and growth of this use of cyberspace. In fact, there are no definitive solutions for many of these problems; the best we can hope for are management measures that prevent specific problems from wrecking the Internet and that are responsive to the emergence of new challenges. Among the most urgent of these problems are:

- Maintaining overall functionality in the face of disruptive actions featuring worms, viruses and the like;
- Preventing unauthorized parties from penetrating secure sites or deliberately spreading information intended to be misleading or deceptive;
- Minimizing uses of the Internet for purposes of espionage or terrorist activities;
- Securing intellectual property rights and preventing unauthorized publication or dissemination of documents for political purposes;
- Controlling Internet bullying, harassment, hate speech and reputational harm;
- Regulating control of access through blockages, denial of services and various forms of censorship using firewalls;
- Managing a range of new issues coming into focus in connection with the development of the Internet of everything; and
- Balancing the erosion of anonymity associated with the use of big data for advertising and commercial promotions.

It is unlikely that such problems will lead to a breakdown or collapse of the Internet as we know it, precipitating drastic changes in the architecture of Internet governance. Still, the growth of the Internet has precipitated a collection of challenges that will test the Internet governance system severely in the coming years, very likely leading to substantial changes in the character of this system in the foreseeable future. Whether these challenges can be addressed through the evolution of social norms and practices in contrast to the negotiation of explicit agreements is an open question.

What makes the case of the Internet governance system intriguing in the context of this analysis of institutional architecture is that it differs fundamentally from mainstream thinking about the creation and operation of international or transnational governance systems. As DeNardis (2014: 11) puts it, '[m]ost Internet governance functions have historically not been the domain of governments but have been executed through private ordering, technical design and new institutional forms'. The result is a heterarchic system that has not only produced a remarkable degree of order in the use of cyberspace but also demonstrated an impressive capacity to respond agilely to rapid technological and socio-economic changes. We cannot assume that the essential features of this system can be transferred directly to other realms featuring growing needs for governance in areas beyond national jurisdiction, and the system faces increasingly serious challenges on several fronts. But in a world of complex systems in which mainstream approaches to governance seem less and less effective, there are good reasons to ponder the experience of Internet governance in the search for fresh perspectives on governance that may prove helpful in coming to terms with emerging needs for governance in the Anthropocene.

Conclusions and Future Directions

When it comes to developing institutional architectures for the governance of areas beyond national jurisdiction, one size does not fit all. We need to go beyond panaceas to develop architectures that are well-matched to specific needs for governance (Ostrom and Turner 2007; Young, Webster et al. 2018). As a result, we will need to sharpen our skills at diagnosing needs for governance and matching these needs with appropriate institutional architectures (Young 2002; Ostrom and Turner 2007; Young 2008). Nevertheless, the examples I have explored do suggest some general observations about institutional architecture in settings in which there is no alternative to finding solutions in the form of governance without government.

The case of ocean governance is the most familiar in institutional terms. It features efforts to make progress by introducing multilateral agreements that are legally binding. Even in this case, however, we find ourselves seeking to manage a regime complex in ways that minimize problems of fragmentation and that may even produce synergies (Oberthür and Stokke 2011; see also Chapter 8). Moving on to climate governance, we encounter a case in which the familiar regulatory approach has failed, leading parties to shift towards a system of pledge-and-review in which individual participants in the governance system are asked to contribute to fulfilling a common goal in the absence of legally binding commitments (Cherry, Hovi and McEvoy 2014). Whether or not this strategy will prove successful remains to be seen. But it certainly reflects a way of addressing needs for governance going beyond the familiar regulatory approach. The case of Internet governance involves yet another strategy in which a variety of states and non-state actors engage in consensual efforts to develop social practices in the absence of formal, much less legally binding, agreements. It is undeniable that this strategy has proved successful in producing a form of governance without government. But it is also clear that there are growing challenges that may require more concerted action on the part of major players during the next phase of Internet governance.

To what extent are experiences with governance relating to the three subsets of areas beyond national jurisdiction transferable across issue areas? Would it be useful to work on the development of a constitutive law of the atmosphere analogous to UNCLOS (International Law Commission 2016)? Are bottom-up pledge-and-review systems likely to become increasingly prominent in dealing with the challenges of the Anthropocene? Does the Internet constitute a unique case in which self-generating consensual practices simply evolved along with the evolution of the technology underlying the Internet itself? Of course, there are no simple answers to these questions. We must develop a well-stocked toolkit and make a concerted effort to address needs for governance in a manner that is sensitive to the key features of individual cases. Even so, the experience with governance systems for areas beyond national jurisdiction provides some basis for optimism going forward. We have a collection of institutional tools that have been used to good advantage in addressing at least some major needs for governance in areas beyond national jurisdiction. If we proceed in a manner that is sensitive to the distinctive features of individual cases and avoid the temptation to adopt favoured approaches to governance regardless of the characteristics of individual cases, there is reason to believe that we can achieve some success in developing institutional architectures responsive to the needs associated with specific areas beyond national jurisdiction, despite the highly decentralized character of international society.

Three additional observations are noteworthy in this context. First, we need to pay attention to issues of process as well as to substance. It is well known that encouraging participation in regime formation can contribute substantially to the legitimacy of the institutional architectures that emerge and that ensuring transparency in the operations of regimes can engender a sense of trust among those who are subject to the requirements of governance systems (Young 2017). While these observations are generic in the sense that they apply to a wide range of issue areas, there is every reason to expect that they are applicable to institutions created to respond to needs for governance in areas beyond national jurisdiction. As the case of ocean governance makes clear, it is not easy to meet high standards regarding matters of participation and transparency. The experiment with a shift from a familiar regulatory approach to a more bottom-up participatory strategy in the case of climate governance will be interesting to follow in this regard.

Somewhat related comments are in order regarding what we may call discursive embeddedness. Institutions tend to work best when they are linked to underlying concepts, principles and broader narratives that make specific institutional architectures easy to comprehend and create a sense that they flow naturally from some common understanding of the nature of the problems at stake and the sources of solutions (Young 2017). A critical source of the effectiveness of UNCLOS lies in its success in articulating an emerging consensus acknowledging the rights of coastal states in their Exclusive Economic Zones, reaffirming the traditional rights of all in the high seas and introducing the idea of the common heritage of humanity regarding the resources of the Area. What is missing at this stage from the effort to devise an institutional architecture capable of addressing the problem of climate change is the development of a common narrative regarding this problem that could provide a discursive foundation for specific arrangements addressing mitigation, avoidance and adaptation.

Finally, thinking about institutional architectures for areas beyond national jurisdiction highlights the critical importance of balancing stability and agility in devising institutional architectures that prove effective in governing complex systems (Young 2017). Institutional architectures that lack stability or robustness cannot succeed in steering the behaviour of key players in specific issue areas. But as we move deeper into a world of complex systems characterized by non-linear change, bifurcations and emergent properties, there is a growing premium on creating governance systems that are agile or nimble in responding to changes in the issue areas they address. This is one reason for scepticism about whether the development of an international legally binding instrument on biodiversity beyond national jurisdiction constitutes the best approach to governing emerging activities affecting marine systems beyond national jurisdiction. It is also an explanation for the appeal of the somewhat fluid and consensual social practices that are a prominent feature of Internet governance, though the rising challenges in this realm may spawn strong pressures to introduce more formal and enforceable architectural arrangements relating to Internet governance during the near future. The bottom line in this regard is clear. Finding ways to enhance the agility needed to address the challenges of complex systems without undermining the effectiveness of current architectures is emerging as a key to the success of all forms of earth system governance.

References

Berkman, P. A., Lang, M. A., Walton, D. W. H., & Young, O. R. (2011). *Science diplomacy: Antarctica, science and the governance of international spaces*. Washington, DC: Smithsonian Institution Scholarly Press.

Biermann, F., Betsill, M. M., Gupta, J. et al. (2009). *Earth system governance: People, places and the planet. Science and implementation plan of the Earth System Governance Project*. Bonn: Earth System Governance Project.

Biermann, F., & Boas, I. (2008). Protecting climate refugees: The case for a global protocol. *Environment*, 50 (6): 8–16.

Brousseau, E., Marzouki, M., & Meadel, C. (eds.) (2015). *Governance, regulations and powers on the internet*. Cambridge, UK: Cambridge University Press.

Brown, S., Cornell, N. W., Fabian, L. L., & Brown Weiss, E. (1977). *Regimes for the ocean, outer space and weather*. Washington, DC: Brookings Institution.

Cherry, T. L., Hovi, J., & McEvoy, D. M. (eds.) (2014). *Toward a. new climate agreement: Conflict, resolution and governance*. London: Routledge.

DeNardis, L. (2014). *The global war for internet governance*. New Haven: Yale University Press.

Dryzek, J. S., Norgaard, R. B., & Schlosberg, D. (eds.) (2013). *The Oxford handbook of climate change and society*. Oxford: Oxford University Press.

Hayek, F. A. (1973). *Rules and order: Law, legislation and liberty*. Chicago, IL: University of Chicago Press.

Intergovernmental Panel on Climate Change (2014). *Climate change 2014: Synthesis report. Contribution of Working Groups I, II and III to the Fifth Assessment Report of the Intergovernmental Panel on Climate Change*. Geneva: Intergovernmental Panel on Climate Change.

International Law Commission (2016). *Third report on the protection of the atmosphere*. UN Doc. A/CN.4/692.

Keith, D. (2013). *A case for climate engineering*. Cambridge, MA: The MIT Press.

Keohane, R. O., & Victor, D. G. (2011). The regime complex for climate. *Perspectives on Politics*, 9, 7–23.

Kolbert, E. (2014). *The sixth extinction: An unnatural history*. New York: Henry Holt.

National Research Council (2015). *Climate intervention: Reflecting sunlight to cool earth*. Washington, DC: National Academies Press.

Oberthür, S., & Stokke, O. S. (eds.) (2011). *Managing institutional complexity: Regime interplay and global environmental change*. Cambridge, MA: The MIT Press.

Oreskes, N. (2004). The scientific consensus on climate change. *Science*, 306, 1686.

Ostrom, E., & Turner, II, B. L. (eds.) (2007). A diagnostic approach to going beyond panaceas. *Proceedings of the National Academy of Sciences of the United States of America*, 104, 15181–7.

Oude Elferink, A. G. (ed.) (2005). *Stability and change in the law of the sea: The role of the LOS convention*. Leiden: Martinus Nijhoff.

Parson, E. A. (2003). *Protecting the ozone layer: Science and strategy*. Oxford: Oxford University Press.

Rochette, J., Wright, G., Gjerde, K. M., Greiber, T., Unger, S., & Spadone, A. (2015). *A new chapter for the high seas?* Potsdam: Institute for Advanced Sustainability Studies.

Rockström, J., Steffen, W., Noone, K. et al. (2009). A safe operating space for humanity. *Nature*, 461, 472–5.

Rosenau, J. N., & Czempiel, E. O. (1992). *Governance without government: Order and change in world politics*. Cambridge, UK: Cambridge University Press.

Royal Society (2009). *Geoengineering the climate: Science, governance and uncertainty*. London: The Royal Society.

Steffen, W., Sanderson, A., Tyson, P. D. et al. (2004). *Global change and the earth system: A planet under pressure*. Heidelberg: Spring Verlag.

Steffen, W., Richardson, K., Rockström, J. et al. (2015). Planetary boundaries: Guiding human development on a changing planet. *Science*, 347, 1259855.

Velders, G. J. M., Andersen, S. O., Daniel, J. S., Fahey, D. W., & McFarland, M. (2007). The importance of the Montreal Protocol in protecting climate. *Proceedings of the National Academy of Sciences of the United States of America*, 104, 4814–19.

Wilson, E. O. (2016). *Half-Earth: Our planet's fight for life*. New York: Liveright.

Young, O. R. (2002). *The institutional dimensions of climate change: Fit, interplay, and scale*. Cambridge, MA: The MIT Press.

Young, O. R. (2008). Building regimes for socioecological systems: Institutional diagnostics. In O. R. Young, L. A. King, & H. Schroeder (eds.), *Institutions and environmental change: Principal findings, applications and research frontiers* (pp. 115–44). Cambridge, MA: The MIT Press.

Young, O. R. (2011). Governing international spaces: Antarctica and beyond. In P. A. Berkman, M. A. Lang, D. W. H. Walton, & O. R. Young (eds.), *Science diplomacy* (pp. 287–94). Washington, DC: Smithsonian Institution Scholarly Press.

Young, O. R. (2016). International relations in the Anthropocene. In K. Booth, & T. Erskine (eds.), *International relations theory today* (pp. 231–49). Cambridge, UK: Polity Press.

Young, O. R. (2017). *Governing complex systems: Social capital for the Anthropocene*. Cambridge, MA: The MIT Press.

Young, O. R., Webster, D. G. et al. (2018). Moving beyond panaceas in fisheries governance. *Proceedings of the Academy of Sciences of the United States of America*, 115, 9065–73.

Zacher, M. W., & Sutton, B. A. (1996). *Governing global networks: International regimes for transportation and communications*. Cambridge, UK: Cambridge University Press.

Part II

Core Structural Features

6

Institutional Interlinkages

THOMAS HICKMANN, HARRO VAN ASSELT, SEBASTIAN OBERTHÜR, LISA
SANDERINK, OSCAR WIDERBERG AND FARIBORZ ZELLI

Next to the myriad of existing intergovernmental institutions (Chapter 2), numerous new governance initiatives have emerged to tackle transboundary environmental challenges (Chapter 4). These initiatives include informal clubs of like-minded national governments such as the Climate and Clean Air Coalition, private certification schemes such as the Carbon Trust Standard and multi-stakeholder forums such as the Roundtable on Sustainable Palm Oil. This growing array of governance initiatives has considerably increased the institutional complexity of global environmental policymaking (Oberthür and Stokke 2011). Moreover, this proliferation of institutions causes more institutional interlinkages.

As institutions and their interlinkages have boomed over the past few years, so has the scholarship on institutional interlinkages. Of particular interest here is the expansion of interlinkages across different governance levels and scales. Earlier studies primarily (though not exclusively) focused on *horizontal* interlinkages between intergovernmental institutions operating at the same level of governance. More recently, however, scholars have devoted growing attention to *vertical* interlinkages across supranational, international, national and subnational layers of authority, as well as to *transnational* interlinkages between institutions set up by state and non-state actors. Against this backdrop, this chapter examines the extent to which present concepts and typologies of institutional interlinkages can capture the various interlinkages between different kinds of institutions in the evolving architecture of earth system governance.

The chapter proceeds as follows. First, we elaborate what we mean by institutional interlinkages and distinguish this term from related concepts covered in this book. We then review and synthesize the literature on institutional interlinkages and highlight key findings with relevance for earth system governance research. Finally, we identify gaps in our knowledge on institutional interlinkages and point to promising future research directions.

Conceptualization

Institutional interlinkages can be broadly understood as formal or informal connections between two institutions and their associated policy processes. Whereas the recent debate on the Anthropocene has drawn attention to the interconnectedness of different ecosystems and the inherent complexity of the earth system, governance has likewise become increasingly multifaceted and entangled (Biermann 2014; Pattberg and Zelli 2016; Hickmann et al. 2019). The responses to earth system changes are governed through explicit and implicit rule systems that operate at various levels and involve a broad range of actors with different motivations. Because of this governance complexity, an increasingly dense web of interlinked institutions addresses transboundary environmental problems. While earth system scientists have shed light on the biophysical interlinkages between environmental issues such as climate change, biodiversity loss or desertification, earth system governance scholars have focused on the interlinkages between the institutions that aim to tackle these challenges.

In their effort to study the connections and relations between institutions, scholars have proposed several terms, including institutional interaction, institutional interplay and institutional overlap (e.g., Zelli and van Asselt 2010; Brosig 2011; Oberthür and Gehring 2011; Oberthür and Stokke 2011; Van de Graaf and De Ville 2013; Betsill et al. 2015). In this chapter, we stick to the term institutional interlinkages, unless authors that we cite explicitly employ a different term. In line with the terminology used in this volume, we understand the dyadic interlinkages between two institutions as a key microscopic structural feature of the overall global governance landscape. In other words, institutional interlinkages are perceived here as the most basic building blocks or units of analysis in current scholarship on institutional architectures. The study of institutional interlinkages is thus a logical starting point for investigating the broader institutional setting of earth system governance.

Institutional interlinkages can be distinguished from related concepts addressed in this book, such as regime complexes (Chapter 7) and governance fragmentation (Chapter 8), both of which capture the relationships between institutions at a much higher analytical level. The concept of regime complexes stands at the meso level and emphasizes the interconnectedness and entanglement of three or more institutions within a larger governance architecture (Chapter 7; see also Orsini, Morin and Young 2013). The concept of governance fragmentation brings in a macro-level perspective, allowing for a comparison of different types and degrees of fragmentation across policy domains (Chapter 8; see also Zelli and van Asselt 2013). Depending on the nature of the dyadic relationships between individual components of the governance architecture in a given area, scholars can assess whether the respective field is characterized by conflictive, cooperative or synergistic fragmentation (Biermann et al. 2009).

Other concepts covered in this book such as interplay management, policy integration and orchestration are also related to the concept of institutional interlinkages. Although the three concepts have varying connotations, they all involve and propose certain forms of direct or indirect steering in response to the plethora of institutions and their interlinkages. They can best be seen as frameworks to cope with increasing institutional complexity and entail, hence, a normative dimension. The identified options for policymakers range from setting hierarchical guidelines and creating coordinating or centralized institutions, to collective decision-making in the individual institutions or using intermediaries for achieving policy goals.

Over the past decades, the global environmental politics literature has shifted its focus from individual international regimes (Krasner 1983) to institutional interlinkages and complexes (see Chapter 1). In the mid-1990s, scholars concerned with international environmental policymaking started to highlight the importance of understanding such intricate relationships of international institutions (e.g., Herr and Chia 1995; Young 1996; Von Moltke 1997). Subsequently, several research projects provided important insights in this respect, including the Inter-Linkages Initiative of the United Nations University (Chambers 2001, 2008), the Institutional Interaction Project (Oberthür and Gehring 2006b, 2011; Gehring and Oberthür 2009), the Institutional Dimensions of Global Environmental Change Project (Young 2002; Young, King and Schroeder 2008) and the Global Governance Project (e.g., van Asselt, Gupta and Biermann 2005; Biermann et al. 2009; Zelli 2011).

In these research projects and other earlier studies on institutional interlinkages, the conceptual and empirical emphasis was mainly placed on *horizontal* interlinkages, that is, linkages between institutions at the same level of governance. There has been, for instance, a particularly strong focus on interlinkages between different international environmental regimes (Oberthür 2001; J. Kim 2004), between international environmental regimes and international economic institutions like the World Trade Organization (Young, King and Schroeder 2008; Zelli and van Asselt 2010) and between the international climate regime and other international organizations such as the International Civil Aviation Organization and the International Maritime Organization (Oberthür 2003, 2006). Together with a few influential conceptual works (e.g., Young 1996; Stokke 2001; Gehring and Oberthür 2009), these empirical studies laid the foundation for our current understanding of institutional interlinkages in global governance.

However, these studies shed light on only one part of the connections between institutions, given the myriad interlinkages across different levels and scales. To fill this gap, several scholars have focused on new types of institutional interlinkages beyond the horizontal dimension. They investigate *vertical* interlinkages between institutions operating at different levels of governance. Several of these studies deal with and examine the multilevel governance system of the European Union or other regional regulatory schemes with distinct competencies in the realm of environmental politics (e.g., Selin and VanDeveer 2003; Oberthür and Gehring 2006b; Balsiger and VanDeveer 2012; Kluvánková-Oravská and Chobotová 2012; O'Neill 2013; Lindstad et al. 2015). Furthermore, scholars have started to analyze *transnational* interlinkages involving different kinds of public and private, as well as hybrid institutions (e.g., Bulkeley et al. 2014; Green 2014; Hale and Roger 2014; Betsill et al. 2015; Andonova 2017; Hickmann 2017b; Roger, Hale and Andonova 2017).

We now turn to synthesizing the literature on institutional interlinkages in more detail. Following an overview of existing typologies, we highlight key findings on the underlying reasons for interlinkages and their consequences, before discussing existing theoretical approaches and summarizing empirical studies of institutional interlinkages with a focus on scholarship from 2007 until today.

Research Findings

Typologies of Interlinkages

Scholars have proposed different typologies to categorize the various institutional interlinkages in global environmental policymaking, most of them prior to the publication of the 2009 Earth System Governance Science and Implementation Plan (Young 1996; Rosendal 2001; Stokke 2001; Young 2002; Oberthür and Gehring 2006a). Already in 2003, Henrik Selin and Stacy VanDeveer lamented that 'the literature on linkages remains littered with proposed taxonomies of linkages and little agreement regarding their utility for advancing understanding of the implications of such linkages' (2003: 14). In a seminal article, Young (1996: 2–7) distinguished four different types of interlinkages between the elements of international regimes: embedded, nested, clustered and overlapping institutions. The different types describe how the institutional units could be intentionally or unintentionally connected in terms of functional or political impacts.

Building upon Young's typology, scholars introduced other types of interlinkages, such as utilitarian, normative and ideational interplay (Stokke 2001: 10–11). Oberthür and Gehring (2006a, 2009) have proposed a typology consisting of four causal mechanisms operating at three levels of effectiveness of governance institutions. 'Cognitive interaction' and 'interaction through commitment' operate at the output level. Regarding these causal mechanisms, collective knowledge or specific commitments generated under one institution may shape decisions in another.

'Behavioural interaction' refers to inter-institutional influence at the outcome level. In this sense, behavioural changes of relevant actors induced in the domain of one institution at the implementation level affect the behaviour of relevant actors in the domain of another institution. Finally, 'impact-level interaction' occurs where effects on the ultimate target of governance of one institution (e.g., the climate system) influence the ultimate target of another governance institution (e.g., biodiversity or desertification).

The typologies for institutional interaction that we discussed so far were created for analyzing interlinkages between intergovernmental regimes or their elements. Over the past decade, however, several researchers have added typologies for interactions between transnational institutions that include non-state and subnational actors, and state-based institutions (e.g., Abbott and Snidal 2009; Abbott 2012; Green 2013). Eberlein and colleagues (2014), for instance, introduce a dynamic approach to examining transnational business-governance interactions. Their typology comprises six dimensions of interaction and six components of regulatory governance (Eberlein et al. 2014: 3). The dynamic aspect of this typology consists of mapping institutional interactions over time, that is, across the regulatory governance process similar to the policy cycle: starting from agenda setting and rule-formation towards implementation, enforcement, monitoring and evaluation. This typology has been used to study transnational environmental governance (e.g., Gulbrandsen 2014; Overdevest and Zeitlin 2014).

Underlying Reasons for Interlinkages

We can identify at least two drivers of the increase in the triggering of the causal mechanisms introduced by Gehring and Oberthür (2009) and hence the growth of institutional interlinkages.

First, the growth of the number of institutions in earth system governance increased institutional density and hence augmented the potential for overlaps and interlinkages. In this respect, it is noteworthy that many environmental institutions are dynamic in nature, that is, they feature decision-making systems – for example, in the form of conferences of the parties to multilateral environmental agreements – that continue to produce relevant norms and rules beyond their initial creation, which further increases the potential for institutional interlinkages. For instance, while the provisions of the United Nations Framework Convention on Climate Change that are related to biodiversity remain few and far between, subsequent rule development with regard to forest carbon sinks resulted in further interactions with the international biodiversity regime (van Asselt 2014).

Second, politics and more precisely actors' strategies and interests have been an important driver, also with respect to the institutional growth that we mentioned. As research on actor strategies has shown, there can be important reasons for actors to establish new international institutions and to use and develop interlinkages strategically (Alter and Meunier 2009; Van de Graaf and De Ville 2013).

In this context, the evolving research on interplay management and orchestration (see Chapters 10 and 11) is also highly relevant for understanding the dynamics of institutional interlinkages. Eventually, institutional interlinkages are shaped by the decisions that actors collectively take within each of the interacting institutions as well as potentially in a coordinated or overarching way (Chapter 13). When these decisions are taken consciously in order to purposefully or even strategically alter the interlinkages, they qualify as interplay management or orchestration. Consequently, the decisions of (collective) actors that trigger and form interlinkages and those intended to shape them overlap and often become inseparable. This underscores the fact that interplay management and orchestration themselves create institutional interlinkages.

As with the typologies of institutional interlinkages discussed above, a set of causal mechanisms has been identified based on interlinkages between intergovernmental institutions. The question thus arises whether these mechanisms can also be applied to interlinkages that involve other types of institutions, such as international bureaucracies (Biermann and Siebenhüner 2009; Biermann and Koops 2017a) or transnational arrangements that have been highlighted in the polycentric framing of global governance (Jordan et al. 2018). Research findings on intergovernmental institutions and the relevant drivers (knowledge, norms or commitments, behaviour and impact) appear potentially relevant also for this broader field of institutional interlinkages. Yet the importance of the causal mechanisms may vary in different subfields, and the mechanisms themselves may need to be complemented by new cause-and-effect relationships.

Consequences of Interlinkages

Interlinkages can have various consequences. As a starting point, previous studies have suggested that institutional interlinkages can result in a conflict between the two institutions involved, or in synergy between them, or have neutral or indeterminate effects (Oberthür and Gehring 2006a: 46). Although the focus of many studies has been on potential conflicts, several cases of institutional interlinkages may, in fact, result in synergies. Gehring and Oberthür (2006: 318), for instance, found that more than 60 per cent of their sample of 163 cases of interactions between different kinds of institutions resulted in synergy. Furthermore, the nature of the relationship between two institutions may change over time and move, for example, from conflictive towards synergistic.

The concepts of conflict and synergy, however, remain under-explored. With reference to the literature on norm conflicts in international law, van Asselt (2014) proposes to distinguish between more narrowly defined 'norm conflicts' and broader 'policy conflicts'. Norm conflicts are incompatibilities between the norms (obligations, permissions and prohibitions) of two treaties, meaning that a party cannot comply with one norm without violating the other (Vranes 2009). Such conflicts pose particular problems from the perspective of international law; addressing them may require recourse to conflict resolution mechanisms that specify the priority of one norm over another (Chapter 13).

However, many situations in which environmental institutions are in tension with each other or with other non-environmental institutions may not be captured by such a definition. For instance, while no rule in the climate regime explicitly obliges or permits a party to implement projects that have adverse impacts on biodiversity, the economic incentives provided through the climate regime's market mechanisms could result in such impacts (van Asselt 2014). Resolving such policy conflicts does not necessarily require establishing a hierarchy between two norms but can still lead to detrimental outcomes. This may be the case because the goals of two different institutions are at odds with each other since different principles and concepts are adhered to, or because opposing economic incentives are provided.

Other scholars have also offered broader conceptualizations of 'conflict'. Pulkowski (2014) distinguishes various types of regime conflicts, namely conflicts of legal rules, conflicts of (policy) goals and conflicts resulting from (inter-) institutional conflict and power struggle. Zelli (2010) follows sociologists like Simmel (1992) and Dahrendorf (1968) in his understanding of institutional interlinkages as conflicts. Conflicts in this sense do not genuinely or solely take place among institutions, but rather reflect positional differences among actors who constitute these institutions. Thus, the essence of a positional difference – and thereby of the institutional interlinkage – relates to the overlapping or contested issues among two or more institutions.

The notion of 'synergy' has surprisingly drawn less attention than that of 'conflict'. Broadly speaking, it can refer to a situation in which the aggregate effects of two institutions are larger than the sum of effects produced on their own (Rosendal 2001), or in which the individual effects are at least complementary with each other. This, in turn, suggests that some measure of regime effectiveness is needed to determine whether we can speak of a synergy. This raises methodological challenges, especially when it comes to impact-level effectiveness.

The concepts of conflict and synergy arguably offer an overly narrow framing of the various types of consequences of institutional interlinkages, with the notions primarily linked to problem-solving effectiveness (Oberthür and Gehring 2006a).

Yet the consequences of institutional interlinkages may also be understood in other terms, for instance focusing on the efficiency of global policymaking, the distributional effects or the effects on legitimacy and accountability of institutions involved in the interaction. Moreover, conflict and synergy denote specific relationships at the expense of others, for instance competition, coordination and convergence (Eberlein et al. 2014). Another type of consequence concerns the division of labour between international institutions. Gehring and Faude (2013, 2014) suggest that inter-institutional competition leads to the specialization of institutions, with each institution fulfilling its own specific niche.

Theorizing Interlinkages

Following the pioneering sets of causal mechanisms developed by Stokke (2001), Oberthür and Gehring (2006b) and Rosendal (2001), several deeper explanatory approaches have been developed over the last decade. Often building upon classical regime theory and related institutionalist perspectives, these accounts have sought to address and fill the theoretical gap that several observers had identified in the research on institutional interlinkages (Chambers, Kim and ten Have 2008; Young 2008). Three trends seem noteworthy regarding the focus of this chapter.

First, there is a move away from classical power-based explanations towards interest-based, cognitivist, critical and discursive approaches. Early power-based explanations drew on tenets of hegemonic stability theory and instrumental multilateralism. They basically argued that hegemonic governments play a crucial role not only in the generation and design of a single international institution, but also in causing overlaps and rivalry among institutions. Such rivalry may be used to weaken or strengthen an incumbent institution or to open opportunities for 'forum shopping' (Raustiala and Victor 2004). Later studies expanded or modified this reasoning and showed that non-hegemonic actors also use their resources to shape or navigate institutional interlinkages (Alter and Meunier 2009; Helfer 2009; Orsini, Morin and Young 2013).

In the same vein, interest- and knowledge-based explanations of institutional interlinkages gained more prominence. These include, for instance, Van de Graaf's (2013) analysis of the establishment of the International Renewable Energy Agency. Drawing on neoliberal institutionalism, he posits that domestic preferences may lead to an institutional hedging strategy, whereby governments deliberately create overlapping institutions. Morse and Keohane (2014) termed this strategy 'competitive regime creation' in an era of 'contested multilateralism'. Furthermore, scholars began exploiting critical and discursive theories to understand institutional overlaps. Based on the notion of dominant liberal environmentalism (Bernstein 2002), Zelli and colleagues (2013) hold that a prevalence of global norms that promote economic efficiency and environmental improvements through market-based mechanisms can partly explain the development of institutional interlinkages in the fields of biological diversity, biosafety, forestry, climate change and trade. Other scholars draw on different strands of discursive institutionalism by Arts and Buizer (2009), Schmidt (2008, 2017) or Hajer (1995) to understand how underlying discourses shape institutional interlinkages.

Second, another theoretical trend is a growing consideration of new 'spheres of authority' (Rosenau 1997: 41). Although the first wave of inter-organizational studies dates back to the 1970s (Gordenker and Sanders 1978; Hanf and Scharpf 1978), it was not until the early 2010s that a larger number of scholars scrutinized the role of transnational and private actors within institutional interlinkages in a theory-driven and systematic manner (e.g., Green 2014; Dingwerth and Green 2015; Hickmann 2016; Biermann and Koops 2017b). Their approaches acknowledge that interlinkages consist of many sites of political authority and that 'liquid authority' – meaning transnational, non-state, non-electoral authority – is replacing and/or supplementing traditional 'solid' sovereign authority (Krisch 2017). According to Hickmann (2017a), this development does not necessarily generate a shift of authority away from intergovernmental institutions and following Bäckstrand, Zelli and Schleifer (2018: 340), it 'implies reconfigurations of the functions of central institutions in a changing authoritative landscape'.

Third, theoretical accounts of institutional interlinkages increasingly adopt insights from disciplines other than political science and international relations. Abbott and colleagues (2016) refer to organizational ecology theories and their concepts of density, resources and niches to hypothesize with regard to the trajectories of institutional constellations in general and specific institutions therein. Likewise, social network analysis has become popular among international relations scholars, offering transparent and replicable measures to identify privileged actors within institutional overlaps (e.g., Kim 2013; Widerberg 2016; Hollway et al. 2017). While these and other complexity approaches might particularly lend themselves to analyzing the meso and macro levels, they can also offer novel explanations for the emergence and developments of dyadic institutional interlinkages.

Next to employing theories from different disciplines, a major unexplored territory in the theoretical literature is the formulation of more fundamental research questions. It is striking that the vast majority of approaches have sought to explain or understand a relatively concise set of aspects – first and foremost the emergence and shape of interlinkages, their synergistic or conflictive nature and the roles that specific institutions play. More far-reaching consequences of institutional interlinkages such as impacts on legitimacy, accountability, effectiveness and justice have largely remained under the radar of most theory-driven propositions.

Empirical Study of Interlinkages

Regarding the empirical study of institutional interlinkages, there has been a strong focus on global climate politics (e.g., Green 2008; Zelli 2011; van Asselt 2014; Betsill et al. 2015; Hickmann 2017a; Pattberg et al. 2018). This policy domain is of particular interest for studying institutional interlinkages because of the cross-cutting nature of climate change and the steadily growing number of institutions that address the problem of climate change directly or indirectly. Accordingly, several scholars concentrate their analyses on the connections between institutions dealing with climate change and use this field as a testing field or laboratory for investigating institutional interlinkages in depth in order to draw more general conclusions on current trends in global policymaking (e.g., Abbott, Green and Keohane 2016).

The focus of many scholars on global climate governance, however, does not mean that other areas of earth system governance have been neglected altogether. Several scholars have, for instance, examined the interlinkages between international environmental and economic institutions (e.g., Oberthür and Gehring 2006b; Jinnah 2010; Zelli and van Asselt 2010; Zelli, Gupta and van Asselt 2013; Jinnah 2014). In this area, scholars have focused on the relations between multilateral environmental agreements and the World Trade Organization. Moreover, scholars have put considerable efforts into analyzing the interlinkages across different biodiversity-related institutions (e.g., Caddell 2013; Oberthür and Pożarowska 2013). In addition, there is a growing body of literature about the interactions between environmental institutions and those operating in the field of human rights (e.g., Schapper and Lederer 2014) and security politics (e.g., De Grenade et al. 2016).

Two other trends in the empirical study of institutional interlinkages stand out. First, scholars have paid increasing attention to interactions within and beyond the field of global energy governance (e.g., Colgan, Keohane and Van de Graaf 2012; Van de Graaf 2013; Lesage and Van de Graaf 2016). This field is marked by a similarly large growth of institutions as the global climate policy domain. Moreover, the overlap between energy and climate governance is particularly strong when compared to other domains (Sanderink et al. 2017). Second, the adoption of the 2030 Agenda for Sustainable Development has given rise to numerous studies of complex connections and overlaps between institutions at different levels and scales that aim to achieve the 17 Sustainable Development Goals (e.g., Biermann, Kanie and Kim 2017; Lima et al. 2017; Tosun and Leininger 2017; see also Chapter 12).

In sum, while intergovernmental institutions remain important in earth system governance, numerous other institutions have lately been established that work at different governmental levels (e.g., supranational organizations like the European Union or local environmental agencies) and across national boundaries (e.g., transnational city networks). These institutions may have some potential to fill the regulatory gap and help attain sustainable development in their jurisdictions and constituencies, but such effects have so far not been studied in enough detail. In this regard, the recent increase in vertical and transnational institutional interlinkages still needs to be fully digested and further empirical studies on these trends are warranted to better understand the broader earth system governance landscape.

Conclusions and Future Directions

This chapter has reviewed and synthesized the scholarship on institutional interlinkages and highlighted key findings with regard to earth system governance research. After providing a basic definition of the term institutional interlinkages, we looked back at the origins of this research strand. Then, we presented an overview of typologies, discussed the reasons for interlinkages as well as their consequences and recapitulated theoretical approaches and the current state of research on institutional interlinkages with a focus on literature that has been released since 2007. Based on the previous sections, we now stress remaining gaps in this research area, before we point to promising future research directions.

We see three gaps with regard to research on institutional interlinkages.

(1) First, concepts and typologies of institutional interlinkages sometimes stand next to each other without referring to and building on each other. While all these concepts and typologies place emphasis on different aspects of institutional interlinkages and are employed in different contexts, they have many commonalities and could benefit from mutual awareness and consideration. This would also enhance conceptual clarity, allowing for consistency across empirical analyses and possibly enable better communication to policymakers and practitioners. In this regard, the present edited volume lays some groundwork to increase precision of a sometimes-confusing research area with many similar concepts and competing typologies.

(2) Second, research communities concerned with institutional interlinkages remain disconnected. Although efforts have been made to bring together international law and international relations research on the issue (e.g., Young 2011; Pulkowski 2014; van Asselt 2014), the debates on the fragmentation of international law and norm conflicts are still largely overlooked in international relations scholarship. Conversely, international law scholars have by and large refrained from applying concepts, typologies and theories on institutional interlinkages when studying norm conflicts. Other examples of such disconnects concern research in public administration and public policy, as well as the emerging research strand on

inter-organizational relations (Biermann 2008; Biermann and Koops 2017a). A better collaboration between (sub-)disciplines would permit ideas to navigate across specialized themes and facilitate cross-fertilization and innovation (Morin and Orsini 2013: 562).

(3) Third, there are still blind spots regarding our empirical knowledge on institutional interlinkages. This is mainly due to the rise of the transnational and vertical dimension of institutional interlinkages. Some issue areas are well studied, whereas others are basically neglected. Examples of under-researched areas include the interactions between institutions that aim to regulate the use of chemicals or those that seek to reduce marine plastics. Regarding vertical interlinkages, only very few studies explore regional–global institutional interlinkages such as the interplay between regional water agreements and international sea conventions. Regarding transnational interlinkages, we lack studies that look beyond usual suspects and investigate interactions between less prominent institutions such as those aiming at promoting carbon pricing, avoiding land degradation or protecting endangered species.

Next to these gaps, we see three rewarding avenues for further research.

(1) First, there is much potential for studying the (inter-)connections between and beyond institutions operating in the fields of global energy and climate governance. While studies have extensively analyzed interlinkages between the international climate regime and trade as well as biodiversity institutions, the interplay between institutions operating in the climate-energy nexus merits further attention (Van de Graaf and Colgan 2016). Both the Paris Agreement from 2015 and the 2030 Agenda for Sustainable Development emphasize the close link between global energy systems and a changing climate and the need for an integrated approach. While the United Nations Framework Convention on Climate Change constitutes the center of global climate governance, global energy governance is more fragmented and lacks a core. It is hence important to compare across these policy domains and explore the evolving 'climate-energy nexus' (Sanderink et al. 2017).

(2) Second, such a 'nexus approach' can be fruitful to investigate interlinkages across different Sustainable Development Goals. In fact, the 2030 Agenda for Sustainable Development allows for the analysis of new types of institutional interlinkages in the context of specific goals and their 169 sub-targets (Weitz, Nilsson and Davis 2014; Boas, Biermann and Kanie 2016). With the adoption of such an unprecedented overarching policy framework for sustainable development, scholars can examine how new interlinkages between institutions aiming to foster sustainable development emerge and others get strengthened. In this regard, a particularly interesting topic is the interplay between international bureaucracies and transnational institutions that mobilize advocacy, create

demonstration effects, or otherwise pressure national governments for generating transformative shifts towards sustainable development (Hickmann and Elsässer 2018).

(3) Third, a crucial theoretical and practical question concerns the implications of a changing approach to earth system governance in times of a severe crisis of multilateralism. In general, there are two different scholarly perspectives. Some take a positive stance on the increasingly fragmented governance architecture and its effects pointing to an emerging polycentric governance system. Others remain more sceptical about the lack of coordination and coherence between and across institutions. While there might be some room for a middle ground between these two positions, we need more empirical studies that build on rigorous theory and methodology. These studies need to start from a sound and thorough understanding of the dyadic interlinkages between two institutions as a key 'microscopic' structural feature of the overall governance architecture.

References

Abbott, K. W. (2012). Engaging the public and the private in global sustainability governance. *International Affairs*, 88 (3), 543–64.

Abbott, K. W., Green, J. F., & Keohane, R. O. (2016). Organizational ecology and institutional change in global governance. *International Organization*, 70 (2), 247–77.

Abbott, K. W., & Snidal, D. (2009). Strengthening international regulation through transnational new governance: Overcoming the orchestration deficit. *Vanderbilt Journal of Transnational Law*, 42 (2), 501–78.

Alter, K. J., & Meunier, S. (2009). The politics of international regime complexity. *Perspectives on Politics*, 7 (1), 13–24.

Andonova, L. B. (2017). *Governance entrepreneurs: International organizations and the rise of global public–private partnerships*. Cambridge, MA: Cambridge University Press.

Arts, B., & Buizer, M. (2009). Forests, discourses, institutions: A discursive-institutional analysis of global forest governance. *Forest Policy and Economics*, 11 (5–6), 240–7.

Bäckstrand, K., Zelli, F., & Schleifer, P. (2018). Legitimacy and accountability in polycentric climate governance. In A. Jordan, D. Huitema, H. van Asselt, & J. Forster (eds.), *Governing climate change: Polycentricity in action?* (pp. 338–56). Cambridge, UK: Cambridge University Press.

Balsiger, J., & VanDeveer, S. D. (2012). Navigating regional environmental governance. Global Environmental Politics, 12 (3), 1–17.

Bernstein, S. (2002). Liberal environmentalism and global environmental governance. *Global Environmental Politics*, 2 (3), 1–16.

Betsill, M. M., Dubash, N., Paterson, M., van Asselt, H., Vihma, A., & Winkler, H. (2015). Building productive links between the UNFCCC and the broader global climate governance landscape. *Global Environmental Politics*, 15 (2), 1–10.

Biermann, F. (2014). *Earth system governance: World politics in the Anthropocene*. Cambridge, MA: The MIT Press.

Biermann, F., Kanie, N., & Kim, R. E. (2017). Global governance by goal-setting: The novel approach of the UN Sustainable Development Goals. *Current Opinion in Environmental Sustainability*, 26, 26–31.

Biermann, F. Pattberg, P., van Asselt, H., & Zelli, F. (2009). The fragmentation of global governance architectures: A framework for analysis. *Global Environmental Politics*, 9 (4), 14–40.

Biermann, F., & Siebenhüner, B. (eds.) (2009). *Managers of global change: The influence of international environmental bureaucracies*. Cambridge, MA: The MIT Press.

Biermann, R. (2008). Towards a theory of inter-organizational networking. *Review of International Organizations*, 3 (2), 151–77.

Biermann, R., & Koops, J. A. (2017a). Studying relations among international organizations in world politics: Core concepts and challenges. In R. Biermann & J. A. Koops (eds.), *Palgrave handbook of inter-organizational relations in world politics* (pp. 1–46). Basingstoke: Palgrave.

Biermann, R., & Koops, J. A. (eds.) (2017b). *Palgrave handbook of inter-organizational relations in world politics*. Basingstoke: Palgrave Macmillan.

Boas, I., Biermann, F., & Kanie, N. (2016). Cross-sectoral strategies in global sustainability governance: towards a nexus approach. *International Environmental Agreements*, 16 (3), 449–64.

Brosig, M. (2011). Overlap and interplay between international organisations: Theories and approaches. *South African Journal of International Affairs*, 18 (2), 147–67.

Bulkeley, H., Andonova, L., Betsill, M. M. et al. (2014). *Transnational climate change governance*. Cambridge, UK: Cambridge University Press.

Caddell, R. (2013). Inter-treaty cooperation, biodiversity conservation and the trade in endangered species. *Review of European, Comparative & International Environmental Law*, 22 (3), 264–80.

Chambers, W. B. (ed.) (2001). *Inter-Linkages: The Kyoto Protocol and the international trade and investment regimes*. Tokyo: United Nations University Press.

Chambers, W. B. (2008). *Interlinkages and the effectiveness of international environmental agreements*. Tokyo: United Nations University Press.

Chambers, W. B., Kim, J. A., & ten Have, C. (2008). Institutional interplay and the governance of biosafety. In O. R. Young, C. W. Bradnee, J. A. Kim, & C. ten Have (eds.), *Institutional interplay: Biosafety and trade* (pp. 3–19). Tokyo: United Nations University Press.

Colgan, J. D., Keohane, R. O., & Van de Graaf, T. (2012). Punctuated equilibrium in the energy regime complex. *Review of International Organizations*, 7 (2), 117–43.

Dahrendorf, R. (1968). Zu einer Theorie des sozialen Konflikts. In W. Zapf (ed.), *Theorien sozialen Wandels* (pp. 108–23). Berlin: Kiepenheuer & Witsch.

De Grenade, R., House-Peters, L., Scott, C., Thapa, B., Mills-Novoa, M., Gerlak, A., & Verbist, K. (2016). The nexus: Reconsidering environmental security and adaptive capacity. *Current Opinion in Environmental Sustainability*, 21, 15–21.

Dingwerth, K., & Green, J. F. (2015). Transnationalism. In K. Bäckstrand, & E. Lövbrand (eds.), *Research handbook on climate governance* (pp. 153–63). Cheltenham: Edward Elgar.

Eberlein, B., Abbott, K. W., Black, J., Meidinger, E., & Wood, S. (2014). Transnational business governance interactions: Conceptualization and framework for analysis. *Regulation & Governance*, 8 (1), 1–21.

Gehring, T., & Faude, B. (2013). The dynamics of regime complexes: Microfoundations and systemic effects. *Global Governance*, 19 (1), 119–30.

Gehring, T., & Faude, B. (2014). A theory of emerging order within institutional complexes: How competition among regulatory international institutions leads to institutional adaptation and division of labor. *Review of International Organizations*, 9 (4), 471–98.

Gehring, T., & Oberthür, S. (2006). Empirical analysis and ideal types of institutional interaction. In S. Oberthür, & T. Gehring (eds.), *Institutional interaction in global environmental governance: Synergy and conflict among international and EU policies* (pp. 307–71). Cambridge, MA: The MIT Press.

Gehring, T., & Oberthür, S. (2009). The causal mechanisms of interaction between international institutions. *European Journal of International Relations*, 15 (1), 125–56.

Gordenker, L., & Sanders, P., A. (1978). Organization theory and international organization. In P. Taylor, & A. J. R. Groom (eds.), *International organization: A conceptual approach* (pp. 84–107). London: Pinter.

Green, J. F. (2008). Delegation and accountability in the clean development mechanism: The new authority of non-state actors. *Journal of International Law and International Relations*, 4 (2), 21–51.

Green, J. F. (2013). Order out of chaos: Public and private rules for managing carbon. *Global Environmental Politics*, 13 (2), 1–25.

Green, J. F. (2014). *Rethinking private authority: Agents and entrepreneurs in global environmental governance*. Princeton, NJ: Princeton University Press.

Gulbrandsen, L. H. (2014). Dynamic governance interactions: Evolutionary effects of state responses to non-state certification programs. *Regulation & Governance*, 8 (1), 74–92.

Hajer, M. A. (1995). *The politics of environmental discourse: Ecological modernization and the policy process*. Oxford: Clarendon Press.

Hale, T., & Roger, C. (2014). Orchestration and transnational climate governance. *Review of International Organizations*, 9 (1), 59–82.

Hanf, K., & Scharpf, F. W. (1978). *Interorganizational policy making: Limits to coordination and central control*. Thousand Oaks: Sage.

Helfer, L. (2009). Regime shifting in the international intellectual property system. *Perspectives on Politics*, 7 (1), 39–44.

Herr, R. A., & Chia, E. (1995). The concept of regime overlap: Towards identification and assessment. In D. Bruce (ed.), *Overlapping maritime regimes: An initial reconnaissance* (pp. 11–26). Hobart: Antarctic Climate and Ecosystems Cooperative Research Centre.

Hickmann, T. (2016). *Rethinking authority in global climate governance: How transnational climate initiatives relate to the international climate regime*. London: Routledge.

Hickmann, T. (2017a). The reconfiguration of authority in global climate governance. *International Studies Review*, 19 (3), 430–51.

Hickmann, T. (2017b). Voluntary global business initiatives and the international climate negotiations: A case study of the greenhouse gas protocol. *Journal of Cleaner Production*, 169, 94–104.

Hickmann, T., & Elsässer, J. (2018). *New alliances in global environmental governance: Intergovernmental treaty secretariats and sub- and non-state actors*. Paper presented at the General Conference of the European Consortium for Political Research, Hamburg.

Hickmann, T., Partzsch, L., Pattberg, P., & Weiland, S. (eds.) (2019). *The Anthropocene debate and political science*. New York: Routledge.

Hollway, J., Lomi, A., Pallotti, F., & Stadtfeld, C. (2017). Multilevel social spaces: The network dynamics of organizational fields. *Network Science*, 5 (2), 187–212.

Jinnah, S. (2010). Overlap management in the World Trade Organization: Secretariat influence on trade-environment politics. *Global Environmental Politics*, 10 (2), 54–79.

Jinnah, S. (2014). *Post-treaty politics: Secretariat influence in global environmental governance*. Cambridge, MA: The MIT Press.

Jordan, A., Huitema, D., van Asselt, H., & Forster, J. (eds.) (2018). *Governing climate change: Polycentricity in action?* Cambridge, UK: Cambridge University Press.

Kim, J. A. (2004). Regime interplay: The case of biodiversity and climate change. *Global Environmental Change*, 14 (4), 315–24.

Kim, R. E. (2013). The emergent network structure of the multilateral environmental agreement system. *Global Environmental Change*, 23 (5), 980–91.

Kluvánková-Oravská, T., & Chobotová, V. (2012). Regional governance arrangements. In F. Biermann, & P. Pattberg (eds.), *Global environmental governance reconsidered* (pp. 219–35). Cambridge, MA: The MIT Press.

Krasner, S. D. (ed.) (1983). *International regimes*. Ithaca, NY: Cornell University Press.

Krisch, N. (2017). Liquid authority in global governance. *International Theory*, 9 (2), 237–60.

Lesage, D., & Van de Graaf, T. (2016). *Global energy governance in a multipolar world*. New York: Routledge.

Lima, M. G. B., Kissinger, G., Visseren-Hamakers, I. J., Braña-Varela, J., & Gupta, A. (2017). The Sustainable Development Goals and REDD+: Assessing institutional interactions and the pursuit of synergies. *International Environmental Agreements*, 17 (4), 589–606.

Lindstad, B., Pistorius, T., Ferranti, F., Dominguez, G., Gorriz-Mifsud, E., Kurttila, M., Leban, V., Navarro, P., Peters, D. M., Pezdevsek Malovrh, S., Prokofieva, I., Scuck, A., Solberg, B., Viiri, H., Zadnik Stirn, L., & Krc, J. (2015). Forest-based bioenergy policies in five European countries: An explorative study of interactions with national and EU policies. *Biomass and Bioenergy*, 80, 102–13.

Morin, J. F., & Orsini, A. (2013). Insights from global environmental governance. *International Studies Review*, 15 (4), 562–5.

Morse, J. C., & Keohane, R. O. (2014). Contested multilateralism. *Review of International Organizations*, 9 (4), 385–412.

O'Neill, K. (2013). Vertical linkages and scale. *International Studies Review*, 15 (4), 571–3.

Oberthür, S. (2001). Linkages between the Montreal and Kyoto Protocols: Enhancing synergies between protecting the ozone layer and the global climate. *International Environmental Agreements*, 1 (3), 357–77.

Oberthür, S. (2003). Institutional interaction to address greenhouse gas emissions from international transport: ICAO, IMO and the Kyoto Protocol. *Climate Policy*, 3 (3), 191–205.

Oberthür, S. (2006). The climate change regime: Interactions with ICAO, IMO, and the EU burden-sharing agreement. In S. Oberthür, & T. Gehring (eds.), *Institutional interaction in global environmental governance: Synergy and conflict among international and EU policies* (pp. 53–77). Cambridge, MA: The MIT Press.

Oberthür, S., & Gehring, T. (2006a). Conceptual foundations of institutional interaction. In S. Oberthür, & T. Gehring (eds.), *Institutional interaction in global environmental governance: Synergy and conflict among international and EU policies*. Cambridge, MA: The MIT Press.

Oberthür, S., & Gehring, T. (eds.) (2006b). *Institutional interaction in global environmental governance: Synergy and conflict among international and EU policies*. Cambridge, MA: The MIT Press.

Oberthür, S., & Gehring, T. (2011). Institutional interaction: Ten years of scholarly development. In S. Oberthür, & O. S. Stokke (eds.), *Managing institutional complexity: Regime interplay and global environmental change* (pp. 25–58). Cambridge, MA: The MIT Press.

Oberthür, S., & Pożarowska, J. (2013). Managing institutional complexity and fragmentation: The Nagoya protocol and the global governance of genetic resources. *Global Environmental Politics*, 13 (3), 100–18.

Oberthür, S., & Stokke, O. S. (eds.) (2011). *Managing institutional complexity: Regime interplay and global environmental change*. Cambridge, MA: The MIT Press.

Orsini, A., Morin, J. F., & Young, O. R. (2013). Regime complexes: A buzz, a boom or a boost for global governance? *Global Governance*, 19 (1), 27–39.

Overdevest, C., & Zeitlin, J. (2014). Assembling an experimentalist regime: Transnational governance interactions in the forest sector. *Regulation & Governance*, 8 (1), 22–48.

Pattberg, P., Chan, S., Sanderink, L., & Widerberg, O. (2018). Linkages: Understanding their role in polycentric governance. In A. Jordan, D. Huitema, H. van Asselt, & J. Forster (eds.), *Governing climate change: Polycentricity in action?* (pp. 169–87). Cambridge, UK: Cambridge University Press.

Pattberg, P., & Zelli, F. (eds.) (2016). *Environmental politics and governance in the Anthropocene: Institutions and legitimacy in a complex world*. London: Routledge.

Pulkowski, D. (2014). *The law and politics of international regime conflict*. Oxford: Oxford University Press.

Raustiala, K., & Victor, D. G. (2004). The regime complex for plant genetic resources. *International Organization*, 58 (2), 277–309.

Roger, C., Hale, T., & Andonova, L. (2017). The comparative politics of transnational climate governance. *International Interactions*, 43 (1), 1–25.

Rosenau, J. N. (1997). *Along the domestic-foreign frontier: Exploring governance in a turbulent world*. Cambridge, UK: Cambridge University Press.

Rosendal, G. K. (2001). Impacts of overlapping international regimes: The case of biodiversity. *Global Governance*, 7 (1), 95–117.

Sanderink, L., Widerberg, O., Kristensen, K., & Pattberg, P. (2017). *Mapping the institutional architecture of the climate-energy nexus*. Amsterdam: Institute for Environmental Studies.

Schapper, A., & Lederer, M. (2014). Introduction: Human rights and climate change: Mapping institutional inter-linkages. *Cambridge Review of International Affairs*, 27 (4), 666–79.

Schmidt, V. A. (2008). Discursive institutionalism: The explanatory power of ideas and discourse. *Annual Review of Political Science*, 11 (1), 303–26.

Schmidt, V. A. (2017). Theorizing ideas and discourse in political science: Intersubjectivity, neo-institutionalisms, and the power of ideas. *Critical Review*, 29 (2), 248–63.

Selin, H., & VanDeveer, S. D. (2003). Mapping institutional linkages in European air pollution politics. *Global Environmental Politics*, 3 (3), 14–46.

Simmel, G. (1992). *Soziologie. Untersuchungen über die Formen der Vergesellschaftung* (Gesamtausgabe Band 11). Frankfurt am Main: Suhrkamp.

Stokke, O. S. (2001). *The interplay of international regimes: Putting effectiveness theory to work*. Lysaker: The Fridtjof Nansen Institute.

Tosun, J., & Leininger, J. (2017). Governing the interlinkages between the Sustainable Development Goals: Approaches to attain policy integration. *Global Challenges*, 1 (9), 1700036.

van Asselt, H. (2014). *The fragmentation of global climate governance: Consequences and management of regime interactions*. Cheltenham: Edward Elgar.

van Asselt, H., Gupta, J., & Biermann, F. (2005). Advancing the climate agenda: Exploiting material and institutional linkages to develop a menu of policy options. *Review of European Community and International Environmental Law*, 14 (3), 255–64.

Van de Graaf, T. (2013). Fragmentation in global energy governance: Explaining the creation of IRENA. *Global Environmental Politics*, 13 (3), 14–33.

Van de Graaf, T., & Colgan, J. (2016). Global energy governance: A review and research agenda. *Palgrave Communications*, 2, 15047.

Van de Graaf, T., & De Ville, F. (2013). Regime complexes and interplay management. *International Studies Review*, 15 (4), 568–71.
Von Moltke, K. (1997). Institutional interactions: The structure of regimes for trade and the environment. In O. R. Young (ed.), *Global governance: Drawing insights from the environmental experience* (pp. 247–72). Cambridge, MA: The MIT Press.
Vranes, E. (2009). Climate change and the WTO: EU emission trading and the WTO disciplines on trade in goods, services and investment protection. *Journal of World Trade*, 43 (4), 707.
Weitz, N., Nilsson, M., & Davis, M. (2014). A nexus approach to the post-2015 agenda: Formulating integrated water, energy, and food SDGs. *Review of International Affairs*, 34 (2), 37–50.
Widerberg, O. (2016). Mapping institutional complexity in the Anthropocene: A network approach. In P. Pattberg, & F. Zelli (eds.), *Environmental politics and governance in the Anthropocene: Institutions and legitimacy in a complex world* (pp. 81–102). London: Routledge.
Young, M. A. (2011). *Trading fish, saving fish: The interaction between regimes in international law.* Cambridge, UK: Cambridge University Press.
Young, O. R. (1996). Institutional linkages in international society: Polar perspectives. *Global Governance*, 2 (1), 1–24.
Young, O. R. (2002). *The institutional dimensions of environmental change: Fit, interplay, and scale.* Cambridge, MA: The MIT Press.
Young, O. R. (2008). Deriving insights from the case of the WTO and the Cartagena Protocol. In O. R. Young, C. W. Bradnee, J. A. Kim, & C. ten Have (eds.), *Institutional interplay: Biosafety and trade* (pp. 131–58). Tokyo: United Nations University Press.
Young, O. R., King, L. A., & Schroeder, H. (eds.) (2008). *Institutions and environmental change: Principal findings, applications, and research frontiers.* Cambridge, MA: The MIT Press.
Zelli, F. (2010). *Conflicts among international regimes on environmental issues: A theory-driven Analysis.* Tübingen: Eberhard-Karls University.
Zelli, F. (2011). The fragmentation of the global climate governance architecture. *Wiley Interdisciplinary Reviews: Climate Change*, 2 (2), 255–70.
Zelli, F., Gupta, A., & van Asselt, H. (2013). Institutional interactions at the crossroads of trade and environment: The dominance of liberal environmentalism? *Global Governance*, 19 (1), 105–18.
Zelli, F., & van Asselt, H. (2010). The overlap between the UN climate regime and the World Trade Organization: Lessons for post-2012 climate governance. In F. Biermann, P. Pattberg, & F. Zelli (eds.), *Global climate governance beyond 2012: Architecture, agency and adaptation* (pp. 79–96). Cambridge, UK: Cambridge University Press.
Zelli, F., & van Asselt, H. (2013). Introduction: The institutional fragmentation of global environmental governance: Causes, consequences, and responses. *Global Environmental Politics*, 13 (3), 1–13.

7

Regime Complexes

LAURA GÓMEZ-MERA, JEAN-FRÉDÉRIC MORIN
AND THIJS VAN DE GRAAF

The introduction of the concept of 'regime complexes' was a key theoretical innovation. It emerged from the pioneering work of scholars like Young (1996) and Aggarwal (1998), who pointed out early on that some international institutions are embedded within broader institutional frameworks. These institutional frameworks have been called 'clusters of regimes' (Stokke 1997; Rosendal 2001; Oberthür 2002), 'conglomerate regimes' (Leebron 2002; Helfer 2004), 'correlated regimes' (Sprinz 2000) and 'networks of regimes' (Underdal and Young 2004). They are now widely referred to as regime complexes, a term coined by Raustiala and Victor in their seminal 2004 article (Raustiala and Victor 2004). Since then, Google Scholar reports more than 4,200 scientific publications on regime complexes; the concept has become a central element of the theoretical repertoire of global governance.

The broad appeal of the concept of regime complexes arises from the recognition that international institutions are not created in a vacuum and do not develop in isolation from each other (Biermann et al. 2009a: 31). For example, a wide set of institutions governs climate change, including intergovernmental agreements, development banks, international scientific panels, transnational private regulations, agencies specialized on energy, free trade agreements and international networks of cities (Keohane and Victor 2011; Abbott 2012; see also Chapter 5). The concept of regime complexes calls anyone who aspires to understand the creation, evolution, implementation or effectiveness of a particular institution to take into account its broader institutional environment. Indeed, the institutional density and overlaps characterizing regime complexes have been documented in diverse areas of global governance, including trade and investment, security and human rights, among others. However, the concept of regime complexes 'was first raised in the context of the global environment' (Raustiala 2012: 9) and researchers in global environmental politics 'have so far produced the largest volume of writings on the subject' (Van de Graaf and De Ville 2013: 7).

At least four factors explain the concentration of the regime complex literature in the field of environmental governance. Firstly, the global environmental governance architecture is particularly fragmented due to the absence of a centralized world environment organization (Chapters 8; Chapter 13). Secondly, environmental institutions have proliferated in recent years to the point of creating an exceptionally high level of institutional density (Chapter 2). Thirdly, several scholars of environmental politics are particularly concerned with institutions' effectiveness, drawing their attention to the spill-overs and externalities resulting from other international institutions (J. Kim 2004; Johnson and Urpelainen 2012). Lastly, some of these scholars are keen to find an adaptive governance system that could correct the institutional mismatch between stable political institutions and changing biophysical and socio-economic systems (Galaz et al. 2008; Young 2010; Kim and Mackey 2014).

This chapter reviews this literature on regime complexes in earth system governance. The first section clarifies the definition of regime complex and distinguishes it from similar concepts. The following three sections look respectively at the emergence, the development and the consequences of regime complexes. The fifth section surveys the different methods used in the regime complex literature. Finally, the last section discusses future directions for research on environmental regime complexes.

Conceptualization

Raustiala and Victor (2004: 279) define a regime complex as 'an array of partially overlapping and non-hierarchical institutions governing a particular issue-area'. As such, a regime complex is located at a meso level of organization. It goes beyond a discrete institution or even the mere linkage between two institutions (Chapter 6). A regime complex encompasses several distinct institutions (Orsini, Morin and Young 2013: 30). Yet, a regime complex is located a lower level than the governance architecture taken as a whole, which stands at the macro level of governance (Chapter 1; see also Biermann et al. 2009b). The loosely coupled elements of a regime complex are related to the same issue area and often share some normative principles (Zelli, Gupta and van Asselt 2013). Thus, a regime complex can usefully be conceptualized as an open system, sufficiently held together to be recognizable but not completely detached from the rest of global governance.

As a system, a regime complex is made of units and connections. Mapping a regime complex, therefore, requires identifying these units and characterizing these connections. This task, however, often proves to be challenging, as the definitions of both units and connections remain contentious.

First, regarding the units of a regime complex, Raustiala and Victor (2004) argue that 'elemental regimes' are explicit legal agreements. This legalistic understanding of elemental regimes facilitates the mapping of regime complexes, but it leaves several analysts unsatisfied. It excludes institutions such as implicit norms, guidelines, clubs, private regulations and transnational initiatives, which many analysts consider as important elements within a regime complex (Abbott 2012; Green 2013; Hickmann 2015; Green and Auld 2017; Widerberg and Pattberg 2017).

An alternative is to rely on Krasner's classic definition of an international regime, as a set 'of implicit and explicit principles, norms, rules and decision-making procedures around which actor expectations converge in a given area of international relations' (Krasner 1982: 186). This definition, however, is also problematic in the context of a regime complex (Orsini, Morin and Young 2013). It creates ambiguity in the level of analysis as the elemental regimes constituting a regime complex can themselves be sets of various instruments. If an institution is made of institutions, which are themselves made of institutions, how can one know which of these is the regime complex? For example, should we consider the set of institutions governing endangered species as a regime complex, made of elemental regimes such as the whaling regime and the Atlantic tuna regime, or should we consider the set of institutions on endangered species as an elemental regime itself within the broader biodiversity complex? The most reasonable answer to this question is that the two positions can be valid, depending on the research question at hand. The labels 'regime' and 'regime complex' are heuristic constructs that do not exist independently from the analyst. Their scale and scope are socially constructed. Thus, debates as to whether an institution on endangered species is a regime or a regime complex are futile if unrelated to a specific research question. The label 'regime complex' is appropriate at any level of analysis as long as institutions under study are analyzed as a set rather than as unconnected units or a cohesive block.

The second constitutive component of a regime complex is the connections linking the different constitutive elements. It is clear from the definition provided by Raustiala and Victor (2004) that these connections do not arise out of any form of legal hierarchy (Chapter 13). Instead, they emerge from partial overlaps over a given issue area. These overlaps can be at the normative or the impact level. In the regime complex for genetic resources, for example, it is one thing to argue whether the private property rights protected by the Agreement on Trade-Related Aspects of Intellectual Property Rights are consistent with indigenous communities' rights over genetic resources as recognized in the Nagoya Protocol to the Convention on Biological Diversity. Yet, it is another thing to ask whether patent examination impacts the effectiveness of the Nagoya Protocol. Overlaps can also be conflicting or synergic. In the case of genetic resources, while some actors see a conflict between the Nagoya Protocol and the Agreement on Trade-Related Aspects of Intellectual Property Rights, other stakeholders claim that they are in a synergic relationship. In many cases, in fact, actors argue over the nature of the connections linking the various elements of a regime complex, making them particularly unstable.

Equally contentious are the consequences of partial membership overlaps. A regime complex can be made of plurilateral and transnational institutions, with public, private or hybrid membership (Green and Auld 2017). These overlapping memberships add a vertical dimension to a regime complex's thematic horizontal dimension (Morin, Pauwelyn and Hollway 2017). Thus, a regime complex is composed neither of parallel regimes with a clear division of labour, nor of nested regimes embedded within each other like Russian dolls (Young 1996; Aggarwal 1998). A regime complex is messier than these neatly organized ideal types.

Moreover, the ambiguous and contested nature of overlaps between elemental regimes makes regime complexes particularly dynamic. As actors try to address inconsistencies and reduce negative spillovers within a regime complex, they can alter regimes or create new ones. Because of these actions, the institutional architecture of a regime complex at one point in time will most likely have a different shape later. Accordingly, time is an important third dimension that must be included in the mapping of a regime complex (Anderson 2002).

This time dimension has drawn some analysts to use complex system theory to shed light on the evolution and expansion of regime complexes (Alter and Meunier 2009; Green 2013; Kim 2013; Kim and Mackey 2014; Meunier and Morin 2015; Morin, Pauwelyn and Hollway 2017). Alter and Meunier (2009) coined the term 'regime complexity' to express this marriage of regime complexes and complex system theory. However, the bulk of the literature on regime complexes is rooted in mainstream institutionalist thinking and the input from complex systems theory remains marginal. For most analysts, a regime complex refers neither to a theory nor to an attribute. It is merely a system of loosely coupled institutions.

Regime complexes can nevertheless be compared to each other for theory building and theory testing purposes. Drezner (2009) suggests comparing complexes according to their degree of vulnerability to regulatory capture. Gehring and Oberthür (2009) point out that some regime complexes encompass competitive relationships while other regime complexes are characterized by cooperative relationships. These attributes, however, more appropriately describe elemental regimes (which vary in vulnerability to regulatory capture) or connections between two regimes (competitive or cooperative) rather than a regime complex as a whole. A useful variable truly attached to a regime complex rather than its constitutive elements and connections is its degree of integration. As Keohane and Victor (2011) argue, a regime complex is situated somewhere in between the two extremes of fully integrated regime and completely fragmented collection of institutions. Locating regime complexes in this integration–fragmentation continuum emerges as a promising focus for further study (Chapter 8; see also Morin and Orsini 2014). As the next sections discuss, the degree of integration can be approached either as a dependent variable, calling for explanations, or as an independent variable, pointing towards consequences for global governance.

Research Findings

Causes and Origins

The creation of an intricate regime complex might appear as a counterintuitive anomaly. Conventional wisdom would expect states to use or modify an existing institution, rather than to pay the high transaction costs associated with the creation and management of overlapping institutions (Van de Graaf 2013: 15). Yet, regime complexes are increasingly frequent and many of them are even expanding over time. This observation, however, only appears counterintuitive if one views regime complexes as the intentional consequence of states' concerted efforts. In reality, most regime complexes are the unintentional results of successive interactions.

First, divergence of interests is one main reason that explains the emergence and existence of overlapping institutions (Keohane and Victor 2011). When actors that are crucial in an area have strong but divergent preferences, they will be unlikely to converge around a single institution. They will be more likely to collaborate with like-minded countries to create institutions that are limited in scope and membership. This is especially the case for environmental problems that do not require a global concerted effort to extract gains from cooperation, such as the regulation of dangerous waste or genetically modified organisms. Powerful actors might then consider that the benefits of a comprehensive and universal regime do not offset the concessions it requires from them (Rabitz 2016).

Even when an integrated regime is created, a regime complex can still emerge from it. As interests are not fixed but vary over time, some states might find themselves dissatisfied with the established regime. A coalition of dissatisfied states could then engage in regime shifting, which is the 'attempt to alter the status quo ante by moving treaty negotiations, law-making initiatives, or standard setting activities from one international venue to another' (Helfer 2004: 14). By shifting the debates to another regime, the challengers create a feedback effect in the first regime. Alternatively, they can engage in 'competitive regime creation' by creating de novo an institution that more closely represents their interests (Morse and Keohane 2014) and ideas (Oh and Matsuoka 2017). This is what Germany and other countries that were dissatisfied with the International Energy Agency did by creating the International Renewable Energy Agency (Van de Graaf 2013; Urpelainen and Van de Graaf 2015).

Second, in addition to competitive regime creation, regime complexes can also result from the creation of new linkages between existing regimes. Linkages can be created strategically to increase the gains from cooperation and create additional incentives for compliance (Leebron 2002). For example, the United States is able to extract more precise and more enforceable commitments on forestry and endangered species when these issues are negotiated in the context of a trade agreement (Jinnah 2011b). Linkages can also result from the recognition that one regime impedes the effectiveness of another regime and efforts are deployed to reduce these unintended negative spill-overs. When countries banned certain ozone-depleting substances and adopted substitutes that are potent greenhouses gases, the ozone regime was amended to better take into account norms from the climate regime (Johnson and Urpelainen 2012).

Governments are not the only creators of regime complexes. Important actors in the creation and expansion of regime complexes are private actors who establish institutions of their own (Abbott 2012; Green 2013; Green and Auld 2017) and advocate for new connections between existing institutions (Orsini 2013; Orsini 2016; Gómez-Mera 2017). They also include international organizations, which may create subsidiary organizations (Johnson 2014) and actively promote linkages among them (Jinnah 2011a; Gómez-Mera 2016).

In fact, research has found that the creation of regime complexes often results from the coalescence of different factors and the interaction of various actors (Keohane and Victor 2011; Van de Graaf 2013). Far from creating pressures for a more centralized and integrated institutional architecture, the proliferation of actors and institutions in global governance seems to lay a fertile ground for even more regime complexes. In this sense, systems of institutions have the property of autopoiesis, as they can generate more of themselves (Morin, Pauwelyn and Hollway 2017).

Evolution

Raustiala and Victor (2004) suggest that regime complexes evolve in ways that are distinct from single regimes. While the development of standard regimes is driven by political contestation over core rules, the evolution of regime complexes is mediated by a process focused on the inconsistencies at the 'joints' between regimes. In addition, arrangements in elemental regimes will constrain the creation of new rules within these elemental regimes. As a result, the regime complex as a whole will evolve in a path-dependent manner. In other words, the evolution of rules within a regime complex will not correspond neatly to changes in the underlying structure of power, interests and ideas.

Building on this, several scholars have set out to examine the drivers and trajectories in the development of a regime complex (Morrison 2017). Regarding the timing and nature of change, Colgan and colleagues (2012) for instance argue that the energy regime complex has evolved according to a pattern of punctuated equilibrium, characterized by both periods of stasis and periods of great innovation, as opposed to a continuous, gradual process of change. Dissatisfaction on the part of powerful actors with the outcomes in the regime complex largely account for this specific pattern.

A more linear model is the 'co-adjustment model' developed by Morin and Orsini (2013). They contend that complexes have a life cycle that consists of four stages. The first stage, *atomization*, precedes the regime complex as the elemental regimes still exist independently from each other. In a second stage, *competition*, the various elemental regimes morph into a wider complex and compete for strategic positions within it. Some regime complexes reach a third stage of *specialization*, whereby elemental regimes coexist in relative harmony and explicitly recognize each other's competence. A fourth stage may eventually emerge, *integration*, when the regime complex becomes unified and reaches internal stability. The quest for greater policy coherence at the domestic level is a major incentive for more integration at the level of the regime complex, although this does not mean that every complex will necessary reach the fourth stage.

Some scholars go a step further and argue that there is a natural tendency in regime complexes to move towards greater synergies and even integration. Normative conflicts and regulatory competition between elemental institutions are frequently assumed to 'drive the institutions towards an accommodation even in the absence of a coordinating institution' (Oberthür and Gehring 2006: 26). Inspired by ideas from institutional ecology, Gehring and Faude (2013) contend that regime complexes evolve in the same way as populations of organizations. Functional overlap between elemental institutions creates competition between institutions over regulatory authority and scarce resources. Over time, they expect selection processes to lead eventually to an internal division of labour in regime complexes, characterized by institutional specialization into specific niches and reduced functional overlap.

Whether regime complexes stay fragmented or develop a division of labour depends partly on the characteristics of the issue area. For example, due to strong interest diversity of major powers in energy policy, Colgan and colleagues (2012) do not expect a coherent energy regime to emerge soon. Likewise, in their study of the climate regime complex, Keohane and Victor (2011) expect fragmentation to persist since they see it as the product of rather stable traits of the issue: strong interest divergence, high uncertainty and the absence of productive linkages between the cooperation problems in climate change. A factor that is likely to foster regime integration might be 'negative spill-overs' between international regimes (Johnson and Urpelainen 2012).

Most scholars, however, do not assume that regime complexes evolve naturally, because of generic forces of competition or the traits of an issue. Instead, the evolution of regime complexes is shaped by the interests and power of the actors who create and operate these regimes (Orsini, Morin and Young 2013). States and non-state actors can employ a host of cross-institutional strategies when faced with a fragmented governance architecture. In forum shopping, the shopper strategically selects the venue to gain a favourable decision for a specific problem (Busch 2007). Actors may also deliberately create strategic inconsistency or strategic ambiguity between parallel venues (Raustiala and Victor 2004; Alter and Meunier 2009).

Yet, not all strategies of state actors result in creating overlap and potential inconsistencies between parallel venues. Aggarwal (1998) distinguished several strategies aimed at 'institutional reconciliation', including nesting broader or narrower regimes hierarchically; establishing a division of labour between parallel regimes; or modifying existing organizations with a view to securing institutional compatibility with other regimes. Governments and non-state actors might also try to link and integrate different forums, by proposing a common normative frame applicable to all forums, a strategy named forum linking (Orsini 2013). Finally, Rabitz (2016) argues that, under certain conditions, actors' cross-institutional strategies are constrained and change in a regime complex takes the form of 'institutional layering'.

Consequences and Effects

There is little doubt that regime complexes have consequences for global governance and international cooperation. A higher institutional density in an issue area is hypothesized to lead to a greater role for implementation in determining outcomes, a greater reliance on bounded rationality in decision-making, more social interaction among key actors, more forum shopping, more institutional competition and more feedback among institutions (Alter and Meunier 2009). Whether these are good outcomes in normative terms continues to be hotly debated, particularly among lawyers (Raustiala 2012; Papa 2015).

On the one hand, regime overlaps could have negative effects, such as introducing confusion over authority, unclear organizational boundaries and rule uncertainty. This may lead in turn to reduced accountability and lower levels of compliance with international commitments (Raustiala 2012). Regime complexes could also lead to duplication of efforts (Orsini 2016), confused vision (Gallemore 2017), turf wars between bureaucracies (Chapter 3) and inefficiencies (Alter and Meunier 2009; Biermann et al. 2009b; Kelley 2009). The proliferation of international treaties may result in 'treaty congestion' (Brown Weiss 1993; Hicks 1998), a term that alludes to conflicts in objectives, obligations or procedures. Finally, the presence of multiple overlapping institutions can strengthen powerful actors who may be able to navigate complex settings while placing a heavy burden on weaker actors, thus exacerbating power imbalances (Benvenisti and Downs 2007; Drezner 2009; Orsini 2016).

On the other hand, regime complexes also bring benefits and opportunities for cooperation that would not occur if a single regime enjoyed a monopoly of governance in an area. The redundant overlap in competences between different institutions makes it less likely that blame avoidance will result in issues being overlooked (Kellow 2012). Regime complexes are also thought to have greater flexibility across issues and adaptability across time over single, legally integrated regimes (Keohane and Victor 2011; Kellow 2012; Stokke 2013). In that sense, institutional diversity and competition should not be seen as a design failure, but regime separation can be wilfully pursued as a design strategy by international negotiators (Johnson and Urpelainen 2012).

Insights from organizational ecology theory (Gehring and Faude 2014; Abbott, Green and Keohane 2016) and complex system theory (Kim and Mackey 2014) further suggest that institutional competition may foster beneficial adaptation. Competing institutions are under pressure to specialize in a specific niche. This can be done thematically, but niche selection could also take place with regard to governance tasks, such as generating knowledge, strengthening norms, enhancing problem-solving capacity or enforcing rule compliance (Stokke 2013). Through competition and niche selection, the population of institutions continually adapts to exogenous shifts and those institutions best suited to their environment thrive.

In theory, polycentric instead of monocentric governance systems also provide more opportunities for experimentation to improve policies over time (Alter and Meunier 2009; Hoffmann 2011; De Búrca, Keohane and Sabel 2014). They also open possibilities for 'inter-institutional learning' (Oberthür 2009; Young 2010) and increase communications and interaction across parties to international institutions, thus helping to foment the mutual trust needed for international cooperation (Ostrom 2010; Cole 2015). Finally, several features of regime complexity – inter-institutional competition, decentralized authority and opportunities for forum shopping – can provide conditions for global democratization (Kuyper 2014, 2015).

Given these mixed effects on outcomes in earth system governance, several analysts have evaluated how regime complexes and interactions could be actively managed (Chapter 10; see also Victor 2011; Young 2012; Abbott 2014; Morin, Pauwelyn and Hollway 2017). One way to achieve greater coherence and coordination is by a more centralizing authority, for instance a world environment organization, as it has been advocated by some (Biermann 2005, 2014) and opposed by others (for example, Whalley and Zissimos 2001; Najam 2003; Oberthür and Gehring 2004). Of course, centralization entails that it is no longer possible to talk about a regime complex, since the non-hierarchical relationship between regimes is a defining feature of a regime complex (Raustiala and Victor 2004; see also Chapter 13).

Since centralization might undo some of the purported benefits of having multiple institutions, other scholars have proposed alternative coordination mechanisms, such as *interplay management*, that is, the 'conscious efforts by any relevant actor or group of actors, in whatever form or forum, to address and improve institutional interaction and its effects' (Chapter 10; see also Oberthür and Stokke 2011: 6). *Orchestration* involves efforts by state actors and intergovernmental organizations to mobilize and work with private actors and institutions to achieve regulatory goals (Chapter 11; see also Abbott and Snidal 2010; Abbott 2012). Under the right conditions, such acts of orchestration can improve transparency and accountability (Bäckstrand and Kuyper 2017). International organizations can also rely on *institutional deference*, recognizing and ceding regulatory authority and jurisdiction to other organizations. These attempts at inter-institutional coordination tend to be based on a division of labour among organizations, ultimately aimed at reducing overlaps and conflicts within regime complexes (Pratt 2018).

In some cases, collaboration between international regimes is achieved through legal means, for example through saving clauses (Raustiala 2012), cooperative agreements (Scott 2011) or clustering international agreements (Von Moltke 2001; Oberthür 2002). In other cases, interaction management boils down to knowledge management, which can be achieved by bringing together stakeholders from different regimes in so-called boundary organizations to facilitate information sharing, joint knowledge production and the improvement of institutional interactions (Morin et al. 2016).

Regime complexity does not only have consequences at the systemic level – the governance architecture – but also at other levels. Institutional proliferation can lead to a restructuring of the mandates of international organizations, for instance. Incumbent organizations confronted with new entrants can become 'challenged organizations' and face pressure to adapt formally (Betts 2013) or informally (Colgan and Van de Graaf 2015) to remain the focal point in their area.

Other studies focus on the influence of regime complexity on actors' strategies. Institutional overlaps provide state and non-state actors with opportunities for forum shopping and regime shifting, as noted above. In general, this increases the menu of institutional options when actors are confronted with a problem that calls for international coordination or collaboration: governments and other actors can decide to use an existing organization, modify one so that it is fit for purpose, select between different institutional venues or create an entirely new organization (Jupille, Mattli and Snidal 2013). The latter option – institutional creation – could further increase the fragmentation and density of governance (Morse and Keohane 2014; Urpelainen and Van de Graaf 2015).

These kinds of cross-regime strategies could in turn contribute to the effectiveness and success of cooperation in regime complexes (Ward 2006). To the extent that these strategies are cooperative rather than opportunistic, state and non-state actors can take advantage of institutional overlaps to exchange information, create and reframe issues, diffuse norms and even develop complementary legal instruments. Cross-regime strategies may in turn offset the negative spill-overs of overlapping institutions (Gómez-Mera 2016).

Methodological Approaches

Over the last fifteen years, empirical studies on regime complexes have grown significantly, not only in quantity but also in methodological scope and sophistication. The early work, which was primarily conceptual, used qualitative and historical cases of different regimes to illustrate taxonomies and typologies of institutional linkages and interactions. As noted above, environmental regimes have been central in this literature. Young (1996, 2002, 2008) used examples from environmental governance to illustrate the distinction between embedded, nested, overlapping and clustering regimes, as well as the emergence of horizontal and vertical institutional interplay. Important conceptual contributions by Stokke (2001), Rosendal (2001), Raustiala and Victor (2004) and Oberthür and Gehring (2006) also focused empirically on different issue areas in global environmental governance.

Building on these insights, numerous cases of regime complexes have been documented in diverse fields, among others on maritime piracy (Struett, Nance and Armstrong 2013), international security (Hofmann 2009), human trafficking (Gómez-Mera 2016, 2017), refugees (Betts 2013), energy (Colgan, Keohane and Van de Graaf 2012), food security (Margulis 2013), human rights (Hafner-Burton 2009), public health (Holzscheiter, Bahr and Pantzerhielm 2016), shipping (Stokke 2013), trade (Gómez-Mera 2015) and intellectual property (Muzaka 2011), to cite but a few. The environment – including fisheries, climate change and biodiversity – has continued to occupy pride of place in the regime complex literature (for example, Paavola, Gouldson and Kluvánková-Orvaská 2009; Michonski and Levi 2010; Gomar et al. 2014; Zelli and Pattberg 2016; Young 2017). To this rich body of empirical work by political scientists, international lawyers have added their own comprehensive studies of fragmentation and conflictive overlaps in various areas of international law (e.g., Koskenniemi and Leino 2002;

International Law Commission 2006; Borgen 2012; M. Young 2012; Pauwelyn and Alschner 2015).

Many of these studies provide detailed maps of regime complexes and identify the main actors and institutions involved in each issue area. While illuminating, most of these studies focus only on single cases of regime complexes or dyads of overlapping regimes, with limited scope for generalizations of insights beyond the specific cases (Koops and Biermann 2017). Moreover, much of this work has relied too heavily on desk research and would benefit greatly from more active field work, including interviews with key players within international governmental and non-governmental organizations and even participant observation (Koops and Biermann 2017). Such in-depth qualitative research is crucial to assess empirically the causal effects of regime overlaps and interactions through, for example, counterfactual analysis (Alter and Meunier 2009).

To overcome some of these methodological problems and constraints to generalizability, scholars have begun to engage in collaborative projects to systematically collect data on larger numbers of cases. The collection of original large-n data has allowed for a quantitative turn in the study of overlapping institutions and regime complexes, particularly regarding environmental and trade governance. The International Environmental Agreements Database Project, for example, includes 1,287 multilateral agreements, 2,170 bilateral agreements and almost 250 other environmental agreements, as well as specific information on when agreements were signed, ratified and entered into force (Chapter 2). In addition, scholars have made significant progress in collecting data on transnational private initiatives, through the Climate Initiative Platform and the Non-state Actor Zone for Climate Action, both of which focus on arrangements driven by non-state actors.[1] Apart from detailed descriptions and mapping of governance structures in each area, these data have been used to test hypotheses on the determinants of fragmentation and cooperation within regime complexes, using other methods that include regression and social network analysis (Widerberg and Pattberg 2017).

Indeed, studies that apply social network analysis to global environmental governance have proliferated in recent years (e.g., Ward 2006; Green 2013; Kim 2013; Hollway and Koskinen 2015; Böhmelt and Spilker 2016; Morin et al. 2016). Social network analysis permits analyzing the multiple interconnections among legal instruments, organizations and public and private actors in regime complexes. It is useful for studying degrees of centrality, clustering and positioning of elements and actors within the complex. Moreover, the community detection methods used in social network analysis are especially relevant for the study of regime

---

[1] See http://climateinitiativesplatform.org/ and http://climateaction.unfccc.int/. See also Widerberg and Stripple (2016) for an overview of another five databases on cooperative initiatives for decarbonization.

complexes, since they may help identify their emergent and dynamic components but also their boundaries. Along these lines, Kim (2013) argues that a network-based approach is necessary to obtain a macroscopic view of international environmental law, which captures the basic patterns of connections among its components. Using a dataset of 1,001 cross-references among 747 multilateral environmental agreements, his study documents the increasing fragmentation in earth system governance. Morin and colleagues (2016) in turn use social network analysis to study the density of relationships among individuals working at the Intergovernmental Science-Policy Platform on Biodiversity and Ecosystem Services to assess the latter's 'social representativeness' and its ability to contribute to the governance of the biodiversity regime complex.[2]

Despite the growing focus on network dynamics and effects, some scholars have continued to focus on how regime complexity and overlapping institutions influence the incentives and choices of actors, particularly states and intergovernmental organizations. Several studies have relied on formal models to generate clear and testable propositions on states' selection among competing fora (Busch 2007) and their decisions to create new overlapping institutions (Urpelainen and Van de Graaf 2015). Others use standard econometric techniques to examine how states respond to cross-regime influences (Gómez-Mera and Molinari 2014).

Conclusions and Future Directions

Since the 2009 Science and Implementation Plan of the Earth System Governance Project has called for greater examination of regime complexes (Biermann et al. 2009a), research has made remarkable theoretical and methodological progress, providing illuminating insights into the sources, evolution and effects of institutional density and overlaps in world politics. Where does this scholarship turn next? We suggest several avenues for future research, which might contribute to a deeper and theoretically informed understanding of regime complexity and its implications for global governance.

First, for all the progress made in the identification of causal pathways and mechanisms through which regime complexity matters, we still lack a coherent and comprehensive theory of regime complexes and their implications. While facilitating comparison across cases, the proliferation of taxonomies of regime overlaps and institutional linkages has also introduced some terminological confusion and stands in the way of a more coherent programme (Zelli, van Asselt and Gupta 2012; Zelli, Gupta and van Asselt 2013). Lack of consensus in the literature over the

---

[2] Network analysis has also been increasingly used to study the trade regime (e.g., Pauwelyn and Alschner 2015; Milewicz et al. 2017; Morin, Pauwelyn and Hollway 2017).

definition of key concepts, such as 'regime complex', 'overlaps' and 'conflict', hinders the development of general hypotheses that could be tested across cases. Moreover, while recent research on networks is fascinating, more could be done to clarify the theoretical contribution of this work to general debates in international relations research. As Alter and Meunier (2009) suggest, it is crucial that social network analysis be driven by theoretical questions concerning the causal links between regime complexity and networked relationships. In this sense, future research on regime complexes would benefit from greater theoretical discipline, perhaps by drawing upon theories in international relations and comparative politics. In this way, research on regime complexity would also contribute to more central questions and debates in these fields.

Second, another promising research line has begun incorporating analytical tools and metaphors from evolutionary biology into the study of regime complexes. International relations theorists have for years incorporated these ecological metaphors into their writings (Jervis 1998; Rosenau 2003; Axelrod 2006). They have only recently started to blossom in the study of international institutions and regime complexes. Indeed, insights from evolutionary biology are helpful in making sense of change and evolutionary dynamics in regime complexes, an issue that earth system governance literature has largely overlooked.

Along these lines, scholars have approached regime complexes as 'complex adaptive systems', in which 'large networks of components with no central control and simple rules of operation give rise to complex collective behaviour, sophisticated information processing and adaptation through learning or evolution' (Mitchell 2009: 13). In these complex systems, moreover, interactions among constituent parts give rise to adaptation through learning or to co-evolution (Kim and Mackey 2014; Pauwelyn 2014). Alter and Meunier (2009: 15) were among the first to link regime complexes to complexity theory, recognizing the usefulness of the idea that 'understanding units does not sum up to the whole and that dynamics of the whole shape the behaviour of units and sub-parts'.

Organizational ecology is also illuminating in understanding the proliferation and density of specific types or groups of organizations and how their evolution is shaped by their environment (Gehring and Faude 2014; Abbott, Green and Keohane 2016). An organizational ecology perspective places emphasis on the competition for resources among overlapping organizations and the process of natural selection determining success or failure of different organizational forms, and more generally, the decline and survival among populations. Similarly, the idea of punctuated equilibrium can also be used to describe patterns of change and evolution in regime complexes and in international law more generally (Goertz 2003; Diehl and Ku 2010; Colgan, Keohane and Van de Graaf 2012). While international legal change is often incremental, modern international law has been characterized by sudden and dramatic breaks, in which a new treaty or instrument is introduced, followed by long periods of stability. Insights from evolutionary biology have been growing in the study of international institutions. Yet, they are still used sparsely, often unreflexively and almost exclusively in the field of global environmental governance. There is much to be gained from a more systematic and theoretically informed application of these concepts to the study of the evolution of regime complexes. This could be easily done given the progress in data availability and methodological tools in the field.

Finally, despite a broadening of the empirical focus in recent years, the regime complex literature remains strongly concentrated in environmental governance. Yet our understanding of the latter would only benefit from direct and indirect comparisons with other issue areas. Greater collaboration with scholars beyond the Earth System Governance Project, focusing on different issue areas, including trade, security and human rights, would also add to the theoretical coherence of the regime complex literature.

References

Abbott, K. W., & Snidal, D. (2010). International regulation without international government: Improving IO performance through orchestration. *Review of International Organizations*, 5 (3), 315–44.

Abbott, K. W. (2012). The transnational regime complex for climate change. *Environment and Planning C: Government and Policy*, 30 (4), 571–90.

Abbott, K. W. (2014). Strengthening the transnational regime complex for climate change. *Transnational Environmental Law*, 3, 57–88.

Abbott, K. W., Green, J. F., & Keohane, R. O. (2016). Organizational ecology and institutional change in global governance. *International Organization*, 70 (2), 247–77.

Aggarwal, V. K. (1998). Reconciling multiple institution: Bargaining, linkages, and nesting. In B. Aggarwal (ed.), *Institutional designs for a complex world: Bargaining, linkages, and nesting* (pp. 1–31). Ithaca, NY: Cornell University Press.

Alter, K. J., & Meunier, S. (2009). The politics of international regime complexity. *Perspectives on Politics*, 7 (1), 13–24.

Andersen, R. (2002). The time dimension in international regime interplay. *Global Environmental Politics*, 2 (3), 98–117.

Axelrod, R. (2006). *The evolution of cooperation*. New York: Basic Book.

Bäckstrand, K., & Kuyper, J. W. (2017). The democratic legitimacy of orchestration: The UNFCCC, non-state actors, and transnational climate governance. *Environmental Politics*, 26 (4), 764–88.

Benvenisti, E. D., & Downs, G. W. (2007). The empire's new clothes: Political economy and the fragmentation of international law. *Stanford Law Review*, 60 (2), 595–632.

Betts, A. (2013). Regime complexity and international organizations: UNHCR as a challenged institution. *Global Governance*, 19 (1), 69–81.

Biermann, F. (2005). The rationale for a world environment organization. In F. Biermann, & S. Bauer (eds.), *A world environment organization: Solution or threat for effective international environmental governance?* (pp. 117–44). Aldershot: Ashgate.

Biermann, F., Betsill, M. M., Gupta, J. et al. (2009a). *Earth system governance: People, places and the planet. Science and implementation plan of the Earth System Governance Project.* Bonn: Earth System Governance Project.

Biermann, F., Pattberg, P., van Asselt, H., & Zelli, F. (2009b). The fragmentation of global governance architectures: A framework for analysis. *Global Environmental Politics*, 9, 14–40.

Biermann, F. (2014). *Earth system governance: World politics in the Anthropocene.* Cambridge, MA: The MIT Press.

Böhmelt, T., & Spilker, G. (2016). The interaction of international institutions from a social network perspective. *International Environmental Agreements*, 16 (1), 67–89.

Borgen, C. J. (2012). Treaty conflicts and normative fragmentation. In D. B. Hollis (ed.), *The Oxford guide to treaties* (pp. 448–71). Oxford: Oxford University Press.

Brown Weiss, E. (1993). International environmental law: Contemporary issues and the emergence of a new world order. *Georgetown Law Journal*, 81, 675–710.

Busch, M. L. (2007). Overlapping institutions, forum shopping, and dispute settlement in international trade. *International Organization*, 61 (4), 735–61.

Cole, D. H. (2015). Advantages of a polycentric approach to climate change policy. *Nature Climate Change*, 5 (2), 114–18.

Colgan, J. D., Keohane, R. O., & Van de Graaf, T. (2012). Punctuated equilibrium in the energy regime complex. *Review of International Organizations*, 7 (2), 117–143.

Colgan, J. D., & Van de Graaf, T. (2015). Mechanisms of informal governance: Evidence from the IEA. *Journal of International Relations and Development*, 18 (4), 455–81.

De Búrca, G., Keohane, R. O., & Sabel, C. (2014). Global experimentalist governance. *British Journal of Political Science*, 44 (3), 477–86.

Diehl, P., & Ku, C. (2010). *The dynamics of international law.* Cambridge, UK: Cambridge University Press.

Drezner, D. W. (2009). The power and peril of international regime complexity. *Perspectives on Politics*, 7 (1), 65–70.

Galaz, V., Olsson, P., Hahn, T., Folke, C., & Svedin, U. (2008). The problem of fit between governance systems and environmental regimes. In O. R. Young, L. A. King, & H. Schroeder (eds.), *Institutions and environmental change: Principal findings, applications and research frontiers* (pp. 147–86). Cambridge, MA: The MIT Press.

Gallemore, C. (2017). Transaction costs in the evolution of transnational polycentric governance. *International Environmental Agreements*, 17, 639–54.

Gehring, T., & Oberthür, S. (2009). The causal mechanisms of interaction between international institutions. *European Journal of International Relations*, 15 (1), 125–56.

Gehring, T., & Faude, B. (2013). The dynamics of regime complexes: Microfoundations and systemic effects. *Global Governance*, 19 (1), 119–30.

Gehring, T., & Faude, B. (2014). A theory of emerging order within institutional complexes: How competition among regulatory international institutions leads to institutional adaptation and division of labor. *Review of International Organizations*, 9 (4), 471–98.

Goertz, G. (2003). *International norms and decision making: A punctuated equilibrium model.* New York: Rowman and Littlefield.

Gomar, J. O. V., Stringer, L. C., & Paavola, J. (2014). Regime complexes and national policy coherence: Experiences in the biodiversity cluster. *Global Governance*, 20 (1), 119–45.

Gómez-Mera, L., & Molinari, A. (2014). Overlapping institutions, learning, and dispute initiation in regional trade agreements: Evidence from South America. *International Studies Quarterly*, 58 (2), 269–81.

Gómez-Mera, L. (2015). International regime complexity and regional governance: Evidence from the Americas. *Global Governance*, 21 (1), 19–42.

Gómez-Mera, L. (2016). Regime complexity and global governance: The case of trafficking in persons. *European Journal of International Relations*, 22 (3), 566–95.

Gómez-Mera, L. (2017). The emerging transnational regime complex for trafficking in persons. *Journal of Human Trafficking*, 3 (4), 303–26.

Green, J. F. (2013). Order out of chaos: Public and private rules for managing carbon. *Global Environmental Politics*, 13 (2), 1–25.

Green, J. F., & Auld, G. (2017). Unbundling the regime complex: The effects of private authority. *Transnational Environmental Law*, 6 (2), 259–84.

Hafner-Burton, E. M. (2009). The power politics of regime complexity: Human rights trade conditionality in Europe. *Perspectives on Politics*, 7 (1), 33–37.

Helfer, L. R. (2004). Regime shifting: The TRIPs agreement and new dynamics of international intellectual property lawmaking. *Yale Journal of International Law*, 29 (1), 1–83.

Hickmann, T. (2015). *Rethinking authority in global climate governance: How transnational climate initiatives relate to the international climate regime*. London: Routledge.

Hicks, B. L. (1998). Treaty congestion in international environmental law: The need for greater international coordination. *University of Richmond Law Review*, 32, 1643.

Hoffmann, M. J. (2011). *Climate governance at the crossroads: Experimenting with a global response after Kyoto*. Oxford: Oxford University Press.

Hofmann, S. C. (2009). Overlapping institutions in the realm of international security: The case of NATO and ESDP. *Perspectives on Politics*, 7 (1), 45–52.

Hollway, J., & Koskinen, J. (2016). Multilevel embeddedness: The case of the global fisheries governance complex. *Social Networks*, 44, 281–94.

Holzscheiter, A., Bahr, T., & Pantzerhielm, L. (2016). Emerging governance architectures in global health: Do metagovernance norms explain inter-organisational convergence? *Politics and Governance*, 4 (3), 5–19.

International Law Commission (2006). *Fragmentation of international law: Difficulties arising from the diversification and expansion of international law*. UN Doc. A/CN.4/L.682.

Jervis, R. (1998). *System effects*. Princeton, NJ: Princeton University Press.

Jinnah, S. (2011a). Marketing linkages: Secretariat governance of the climate-biodiversity interface. *Global Environmental Politics*, 11 (3), 23–43.

Jinnah, S. (2011b). Strategic linkages: The evolving role of trade agreements in global environmental governance. *The Journal of Environment and Development*, 20 (2), 191–215.

Johnson, T., & Urpelainen, J. (2012). A strategic theory of regime integration and separation. *International Organization* 66 (4), 645–77.

Johnson, T. (2014). Organizational progeny: Why governments are losing control over the proliferating structures of global governance. Oxford: Oxford University Press.

Jupille, J. H., Mattli, W., & Snidal, D. (2013). *Institutional choice and global commerce*. Cambridge, UK: Cambridge University Press.

Kelley, J. (2009). The more the merrier? The effects of having multiple international election monitoring organizations. *Perspectives on Politics*, 7 (1), 59–64.

Kellow, A. (2012). Multi-level and multi-arena governance: The limits of integration and the possibilities of forum shopping. *International Environmental Agreements*, 12 (4), 327–42.

Keohane, R. O., & Victor, D. G. (2011). The regime complex for climate change. *Perspectives on Politics*, 9 (1), 7–23.

Kim, J. A. (2004). Regime interplay: The case of biodiversity and climate change. *Global Environmental Change*, 14 (4), 315–24.

Kim, R. E. (2013). The emergent network structure of the multilateral environmental agreement system. *Global Environmental Change*, 23 (5), 980–91.

Kim, R. E., & Mackey, B. (2014). International environmental law as a complex adaptive System. *International Environmental Agreements*, 14 (1), 5–24.

Koops, J., & Biermann, R. (eds.) (2017). *Palgrave handbook of inter-organizational relations in world politics*. London: Palgrave.

Koskenniemi, M., & Leino, P. (2002). Fragmentation of international law? Postmodern anxieties. *Leiden Journal of International Law*, 15 (3), 553–79.

Krasner, S D. (1982). Structural causes and regime consequences: Regimes as intervening variables. *International Organization*, 36 (2), 185–205.

Kuyper, J. W. (2014). Global democratization and international regime complexity. *European Journal of International Relations*, 20 (3), 620–46.

Kuyper, J. (2015). Deliberative capacity in the intellectual property rights regime complex. *Critical Policy Studies*, 9 (3), 317–38.

Leebron, D. W. (2002). Linkages. *The American Journal of International Law*, 96 (1), 5–27.

Margulis, M. E. (2013). The regime complex for food security: Implications for the global hunger challenge. *Global Governance*, 19 (1), 53–67.

Meunier, S., & Morin, J. F. (2015). No agreement is an island: Negotiating TTIP in a dense regime complex. In J. F. Morin, T. Novotna, F. Ponjaert, & M. Telo (eds.), *The Transatlantic Trade and Investment Partnership in a multipolar world* (pp. 173–86). Farnham: Ashgate.

Michonski, K. E., & Levi, M. A. (2010). *The regime complex for global climate change*. New York: Council on Foreign Relations.

Milewicz, K., Hollway, J., Peacock, C., & Snidal, D. (2017). Beyond trade: The expanding scope of the nontrade agenda in trade agreements. *Journal of Conflict Resolution*, 54 (3), 1–31.

Mitchell, M. (2009). *Complexity: A guided tour*. Oxford: Oxford University Press.

Morin, J. F., & Orsini, A. (2013). Regime complexity and policy coherency: Introducing a co-adjustments model. *Global Governance*, 19 (1), 41–51.

Morin, J. F., & Orsini, A. (2014). Policy coherency and regime complexes: The case of genetic resources. *Review of International Studies*, 40 (2), 303–24.

Morin, J. F., Louafi, S., Orsini, A., & Oubenal, M. (2016). Boundary organizations in regime complexes: A social network profile of IPBES. *Journal of International Relations and Development*, 20 (3), 1–35.

Morin, J. F., Pauwelyn, J., & Hollway, J. (2017). The trade regime as a complex adaptive system: Exploration and exploitation of environmental norms in trade agreements. *Journal of International Economic Law*, 20 (2), 365–90.

Morrison, T. H. (2017). Evolving polycentric governance of the Great Barrier Reef. *Proceedings of the National Academy of Sciences of the United States of America*, 114 (15), E3013–E3021.

Morse, J. C., & Keohane, R. O. (2014). Contested multilateralism. *Review of International Organizations*, 9 (4), 385–412.

Muzaka, V. (2011). Linkages, contests and overlaps in the global intellectual property rights Regime. *European Journal of International Relations*, 17 (4), 755–76.

Najam, A. (2003). The case against a new international environmental organization. *Global Governance*, 9, 367.

Oberthür, S. (2002). Clustering of multilateral environmental agreements: Potentials and limitations. *International Environmental Agreements*, 2 (4), 317–40.

Oberthür, S. (2009). Interplay management: Enhancing environmental policy integration among international institutions. *International Environmental Agreements*, 9 (4), 371–91.

Oberthür, S., & Gehring, T. (2006). Institutional interaction in global environmental governance: The case of the Cartagena Protocol and the World Trade Organization. *Global Environmental Politics*, 6 (2), 1–31.

Oberthür, S., & Gehring, T. (2004). Reforming international environmental governance: An institutionalist critique of the proposal for a world environment organisation. *International Environmental Agreements*, 4, 359–81.

Oberthür, S., & Stokke, O. S. (2011). *Managing institutional complexity: Regime interplay and global environmental change*. Cambridge, MA: The MIT Press.

Oberthür, S., & Pożarowska, J. (2013). Managing institutional complexity and fragmentation: The Nagoya Protocol and the global governance of genetic resources. *Global Environmental Politics*, 13 (3), 100–18.

Oh, C., & Matsuoka, S. (2017). The genesis and end of institutional fragmentation in global governance on climate change from a constructivist perspective. *International Environmental Agreements*, 17 (2), 143–59.

Orsini, A., Morin, J. F., & Young, O. R. (2013). Regime complexes: A buzz, a boom, or a boost for global governance? *Global Governance*, 19 (1), 27–39.

Orsini, A. (2013). Multi-forum non-state actors: Navigating the regime complexes for forestry and genetic resources. *Global Environmental Politics*, 13 (3), 34–55.

Orsini, A. (2016). The negotiation burden of institutional interactions: Non-state organizations and the international negotiations on forests. *Cambridge Review of International Affairs*, 29 (4), 1421–40.

Ostrom, E. (2010). Polycentric systems for coping with collective action and global environmental change. *Global Environmental Change*, 20 (4), 550–7.

Paavola, J., Gouldson, A., & Kluvánková-Oravská, T. (2009). Interplay of actors, scales, frameworks and regimes in the governance of biodiversity. *Environmental Policy and Governance*, 19 (3), 148–58.

Papa, M. (2015). Sustainable global governance? Reduce, reuse, and recycle institutions. *Global Environmental Politics*, 15 (4), 1–20.

Pauwelyn, J., & Alschner, W. (2015). Forget about the WTO: The network of relations between preferential trade agreements (PTAs) and 'Double PTAs'. In A. Dür, & M. Elsig (eds.), *Trade cooperation: The purpose, design and effects of preferential trade agreements* (pp. 497–532) Cambridge, UK: Cambridge University Press.

Pauwelyn, J. (2014). At the edge of chaos? Foreign investment law as a complex adaptive system, how it emerged and how it can be reformed. *ICSID Review*, 29 (2), 372–418.

Pratt, T. (2018). Deference and hierarchy in international regime complexes. *International Organization*, 72 (3), 561–90.

Rabitz, F. (2016). Regime complexes, critical actors and institutional layering. *Journal of International Relations and Development*, 21 (2), 300–21.

Raustiala, K., & Victor, D. G. (2004). The regime complex for plant genetic resources. *International Organization*, 58 (2), 277–309.

Raustiala, K. (2012). Institutional proliferation and the international legal order. In J. L. Dunoff, & M. A. Pollack (eds.), *Interdisciplinary perspectives on international law and international relations: The state of the art* (pp. 239–320). Cambridge, UK: Cambridge University Press.

Rosenau, J. N. (2003). *Distant proximities: Dynamics beyond globalization*. Princeton, NJ: Princeton University Press.

Rosendal, G. K. (2001). Impacts of overlapping international regimes: The case of biodiversity. *Global Governance*, 7 (1), 95–117.

Scott, K. N. (2011). International environmental governance: Managing fragmentation through institutional connection. *Melbourne Journal of International Law*, 12, 177–216.

Sprinz, D. F. (2000). *Research on the effectiveness of international environmental regimes: A review of the state of the art.* Potsdam: Potsdam Institute for Climate Impact Research and University of Potsdam.

Stokke, O. S. (1997). Regimes as governance systems. In O. Young (ed.), *Global governance: Drawing insights from the environmental experience* (pp. 27–63). Cambridge, MA: The MIT Press.

Stokke, O. S. (2001). *The interplay of international regimes: Putting effectiveness theory to work.* Lysaker: The Fridtjof Nansen Institute.

Stokke, O. S. (2013). Regime interplay in Arctic shipping governance: Explaining regional niche selection. *International Environmental Agreements*, 13 (1), 65–85.

Struett, M. J., Nance, M. T., & Armstrong, D. (2013). Navigating the maritime piracy regime complex. *Global Governance*, 19 (1), 93–104.

Underdal, A., & Young, O. R. (2004). Research strategies for the future. In A. Underdal, & O. R. Young (eds.), *Regime consequences* (pp. 361–80). Dordrecht: Springer.

Urpelainen, J., & Van de Graaf, T. (2015). Your place or mine? Institutional capture and the creation of overlapping international institutions. *British Journal of Political Science*, 4 (4), 799–827.

Van de Graaf, T. and B. Sovacool (2020). Global Energy Politics. Cambridge: Polity Press.

Van de Graaf, T. (2013). Fragmentation in global energy governance: Explaining the creation of IRENA. *Global Environmental Politics*, 13 (3), 14–33.

Van de Graaf, T., & De Ville, F. (2013). Regime complexes and interplay management. *International Studies Review*, 15 (4), 568–71.

Victor, D. G. (2011). *Global warming gridlock: Creating more effective strategies for protecting the planet.* Cambridge, UK: Cambridge University Press.

Von Moltke, K. (2001). *On clustering international environmental agreements.* Winnipeg: International Institute for Sustainable Development.

Ward, H. (2006). International linkages and environmental sustainability: The effectiveness of the regime network. *Journal of Peace Research*, 43 (2), 149–66.

Whalley, J., & Zissimos, B. (2001). What could a world environmental organization do? *Global Environmental Politics*, 1 (1), 29–34.

Widerberg, O., & Stripple, J. (2016). The expanding field of cooperative initiatives for decarbonization: A review of five databases. *Wiley Interdisciplinary Reviews: Climate Change*, 7 (4), 486–500.

Widerberg, O., & Pattberg, P. (2017). Accountability challenges in the transnational regime complex for climate change. *Review of Policy Research*, 34 (1), 68–87.

Young, M. A. (2012). *Regime interaction in international law: Facing fragmentation.* Cambridge, UK: Cambridge University Press.

Young, O. R. (1996). Institutional linkages in international society: Polar perspectives. *Global Governance*, 2 (1), 1–24.

Young, O. R. (2002). *The institutional dimensions of environmental change: Fit, interplay, and scale.* Cambridge, MA: The MIT Press.

Young, O. R. (2008). *Institutional interplay: Biosafety and trade.* Tokyo: United Nations University Press.

Young, O. R. (2010). Institutional dynamics: Resilience, vulnerability and adaptation in environmental and resource regimes. *Global Environmental Change*, 20 (3), 378–385.

Young, O. R. (2012). Building an international regime complex for the Arctic: Status and next steps. *The Polar Journal*, 2 (2), 291–407.

Young, O. R. (2017). *Governing complex systems: Social capital for the Anthropocene.* Cambridge, MA: The MIT Press.

Zelli, F., Gupta, A., & van Asselt, H. (2013). Institutional interactions at the crossroads of trade and environment: The dominance of liberal environmentalism? *Global Governance*, 19 (1), 105–18.

Zelli, F., van Asselt, H., & Gupta, A. (2012). Horizontal institutional interlinkages. In F. Biermann, & P. Pattberg (eds.), *Global environmental governance reconsidered.* Cambridge, MA: The MIT Press, 175–98.

Zelli, F., & Pattberg, P. (eds.) (2016). *Environmental politics and governance in the Anthropocene: Institutions and legitimacy in a complex world.* New York: Routledge.

8
Governance Fragmentation

FRANK BIERMANN, MELANIE VAN DRIEL, MARJANNEKE J. VIJGE
AND TOM PEEK

The concept of 'architectures' of global governance is a useful heuristic device to help understand the macro level of institutions and governance mechanisms. With it, one may better grasp the complexity of the myriad treaties and agreements in, for instance, climate and energy governance and compare this with a governance architecture on oceans, biodiversity or chemicals. Such comparisons across institutional architectures and issue areas can reveal, especially, lower or higher degrees of governance fragmentation, which might influence performance of an architecture.

We find governance fragmentation at all levels of political institutions, from local administrations up to national political systems and global governance. Architectures of global governance, however, fundamentally differ from national architectures. Within countries, the rights and responsibilities of political actors and institutions are defined in a written or unwritten constitution that lays down procedures in cases of institutional conflict and normative contestation. While this ideal-type description is rarely matched in reality – with political systems being often marked by constitutional ambiguity, conflict, overlap and crisis – the difference between national and global architectures is evident. Some observers see the Charter of the United Nations as a functional equivalent to national constitutions. But even then, global governance follows logics that differ from national political systems. At the global level, institutional fragmentation is much deeper, and it is ubiquitous.

This fragmentation of global governance stands at the centre of this chapter. We start with a conceptualization of governance fragmentation and its relation to concepts such as polycentricity and institutional complexity. We then review the origins of governance fragmentation and its problematization; methodological approaches to studying fragmentation; and the impacts and consequences of fragmentation. We conclude by identifying future research directions in this domain.

Our review is based on a comprehensive study of the literature on governance fragmentation over the last decade. We draw on a Scopus search on all articles published in the subject area of social sciences in 2009–2018 with 'fragmentation' in the title, abstract or keywords, which yielded 6,831 articles. To narrow the scope, we qualitatively scanned the abstracts of these articles and excluded all articles not concerned with governance fragmentation at the global and transnational level. This left us with 242 articles. We then further excluded articles with abstracts where fragmentation did not appear to play a key role in the analysis but was rather context-setting or mentioned without being of further influence. The remaining articles were supplemented for this review with additional studies, such as books, book chapters and a few policy briefs and working papers.

Conceptualization

The academic literature on fragmentation and complexity in global governance dates back to the 1960s and 1970s (Visseren-Hamakers 2015, 2018), with a strong empirical focus on the governance of major planetary systems, such as climate or ocean governance. We find in this debate also different but related terms such as polycentric governance (e.g., Ostrom 2010a, 2010b; Gallemore 2017); interlinkages between institutions and regimes (Chapter 6); and institutional complexity and regime complexes (Chapter 7). Whereas many of these concepts focus on relations among international organizations and regimes, the concept of fragmentation looks at an entire governance architecture in which institutions interact (Biermann et al. 2009). As such, compared to institutional interlinkages (Chapter 6) and regime complexes (Chapter 7) – the other two core structural features identified in this volume – fragmentation has a clear focus on *macro-level governance*.

As a concept of political analysis, governance fragmentation is used in a variety of ways; there is no generally agreed definition. However, three key characteristics delineate the concept and place it in the context of this volume.

(1) First, fragmentation describes the *quality* of an entity but not an entity or phenomenon itself. Fragmentation, as a concept, cannot be used without reference to an empirical phenomenon that is fragmented. This makes fragmentation different from concepts that describe empirical phenomena, such as regime complexes or institutional interlinkages, which are discussed in other chapters in this volume. Notions of regime complexes and interlinkages refer to units and the relationship between units. Fragmentation, instead, describes the quality of entities. In global politics, fragmentation hence relates as a quality to the concept of governance architecture, which we can assess as being more or less fragmented.

(2) Second, this makes fragmentation an inherently *comparable* variable. Different governance architectures – for example, in health, trade or climate governance – can be compared as to the degree of fragmentation, which can be higher or lower. The fragmentation of architectures can also be compared over time, allowing for insights into whether architectures became more or less fragmented. This comparability makes fragmentation, as a concept of political analysis, different from regime complexes: a regime complex is an entity that can be described through its parts, such as the various regimes and organizations that comprise it. But regime complexes cannot easily be compared because the comparable quality is missing, unless the analyst wants to assess whether a regime complex became more complex over time or is more complex compared to other regime complexes, which comes close to tautological reasoning. Fragmentation, however, is a *variable* that we can assess in political research.

(3) Third, the comparability of governance fragmentation makes it a *continuous*, *non-binary* concept. Governance architectures are neither fully fragmented nor entirely non-fragmented. Instead, they will always be more, or less, fragmented compared to other architectures, and more or less fragmented than architectures in the past and those in the future. The endpoints of this continuum would be two politically unrealistic ideal-types. Extreme fragmentation, on the one hand, would be anarchy without institutions. This would run counter to the very idea of a governance architecture and negate the existence of governance in the first place. Zero fragmentation, on the other hand, would bring about the complete institutional integration of all treaties, actors and organizations, which is hardly realistic at the international level. Related to this, the notion of governance fragmentation does not necessarily entail an unrealistic assumption of a primordial, 'pre-existing world polity or order' that is becoming increasingly fragmented (as some argued, see Zürn and Faude 2013). Instead, governance architectures are always fragmented to some degree – and it is the relative degree of fragmentation that is of interest to the political analyst.

This notion of governance fragmentation as a continuum rather than a binary requires careful categorizations and typologies of different stages and phases of governance fragmentation. As one example, Biermann and colleagues (2009) have proposed a threefold categorization of cooperative, conflictive and synergistic fragmentation. Furthermore, they use three criteria to differentiate between these three degrees of fragmentation: (a) the degree of institutional nesting and degree of overlaps between decision-making systems; (b) the existence and degree of norm conflicts; and (c) the type of actor constellations (see Table 8.1).

The typology of Biermann and colleagues (2009) has been used in numerous case studies. From 55 case studies that we analyzed in this chapter and that use the above framework, several explicitly include this typology of fragmentation (Orsini

Table 8.1: *Typology of fragmentation of governance architectures*

	Synergistic	Cooperative	Conflictive
Institutional nesting	One core institution, with other institutions being closely integrated	Core institutions with other institutions that are loosely integrated	Different, largely unrelated institutions
Norm conflicts	Core norms of institutions are integrated	Core norms are not conflicting	Core norms conflict
Actor constellations	All relevant actors support the same institutions	Some actors remain outside main institutions, but maintain cooperation	Major actors support different institutions

Source: Biermann et al. 2009.

2013; Van de Graaf 2013; Zürn and Faude 2013; Richerzhagen 2014; Velázques Gomar 2016; Well and Carrapatoso 2017; Rana and Pacheco Pardo 2018; Fernández-Blanco, Burns and Giessen 2019). In these studies, we find numerous examples of both cooperative fragmentation and conflictive fragmentation. For example, the climate regime is often analyzed as a prominent case of cooperative fragmentation (Biermann et al. 2009), while the energy security regime has been classified as conflictive (Fernández Carril, García Arrazola and Rubio 2013). Cases of synergistic fragmentation seem to be rather exceptional. The agreement of the Nagoya Protocol under the Convention on Biological Diversity has been described as an example of synergistic fragmentation, as it increases regulatory fragmentation but is still embedded in the framework of the convention. The addition of the protocol to the convention supports the objectives of the convention; the protocol is administered by the same secretariat; and it is financed through the same channels, all of which makes the architecture 'quite synergistic' (Richerzhagen 2014: 149).

Though the above shows that fragmentation of architectures as a whole can be classified as synergistic, cooperative or conflictive, the degree and type of fragmentation can also vary within one architecture. In the case of forest governance, for example, Fernández-Blanco and colleagues (2019) show that while synergistic fragmentation is observed among vague institutional elements (such as the norm of sustainability), more concrete and substantial elements (such as the role of civil society in reaching sustainability) coincide with more conflictive fragmentation.

The typology by Biermann and colleagues (2009) relies on both *structural fragmentation* – measured by the quantity of relationships between institutions – and *functional fragmentation*, measured by the quality of relationships between institutions, norms or actors. Most literature on fragmentation seeks to assess functional fragmentation, while a minority focuses on structural fragmentation (e.g., Kim 2013). This is closely related to the prevalence of qualitative analyses of fragmentation compared to quantitative analyses (as we discuss below).

Another typology has been brought forward by Zürn and Faude (2013), who differentiate between segmentary fragmentation (between institutions with similar tasks in different regions), stratificatory fragmentation (in a hierarchical sense, like framework convention and protocols) and functional fragmentation (largely a division of labour between, for instance, economic and environmental institutions). The typology by Zürn and Faude (2013) does not contradict but could rather be combined with the typology by Biermann and colleagues (2009). Their notion of stratificatory fragmentation, for instance, comes close to the notion of institutional nesting, which Biermann and colleagues (2009) use to assess degrees of fragmentation. The notion of segmentary and functional fragmentation is similar to discussions about vertical and horizontal fragmentation. Here, horizontal fragmentation refers to fragmentation between different policy domains (Zelli, Gupta and van Asselt 2012), while vertical fragmentation points at fragmentation between different levels of governance (Busch, Gupta and Falkner 2012). While most literature on governance fragmentation focuses on the horizontal dimension, long-standing academic debates in the broader literature on world polity also centre on whether the world becomes more regionally fragmented or more globalized (for overviews, see for example, Beckfield 2010; Gomez and Parigi 2015). Interestingly, though the distinction between horizontal and vertical governance fragmentation is rarely explicitly made in the literature, there seems to be more attention for vertical versus horizontal policy measures to address the negative consequences of fragmentation, for example through policy integration (Chapter 9), institutional interlinkages (Chapter 6) and interplay management (Chapter 10).

Importantly, all authors agree that fragmentation is ubiquitous, that it varies among policy areas and governance areas and that it is a variable that can be assessed in comparative research across policy areas and over time. We now review research findings on what explains fragmentation and what its consequences are.

Research Findings

The concept of institutional fragmentation originates in international law, where the first studies about the fragmentation of international law and the overlap and conflicts between international treaties date back to the mid-nineteenth century (Isailovic, Widerberg and Pattberg 2013). In this chapter, however, we focus on the literature in political science and governance studies and on how fragmentation of governance architectures is discussed there. Within political science and international relations research, governance fragmentation has been studied especially in the wider domain of earth system governance, and here both broadly regarding larger structures and more narrowly with reference to specific issues, such as climate or ocean governance.

Unsurprisingly, by far most case studies analyzed the increasingly fragmented, deterritorialized and hybrid architecture of global climate governance. Many studies focus on international climate governance in general.[1] Others investigate specific sectors (Hackmann 2012) or specific areas, such as climate finance (Pickering, Betzold and Skovgaard 2017), carbon governance (Biermann 2010; De Coninck and Bäckstrand 2011; Smits 2017), policies of Reducing Emissions from Deforestation and Forest Degradation (REDD+) (Gupta, Pistorius and Vijge 2016; Gallemore 2017; Well and Carrapatoso 2017) and short-lived climate pollutants (Zelli, Möller and van Asselt 2017; Yamineva and Kulovesi 2018). Because of this prominence of climate governance in academic debates around fragmentation, many of the examples that we further discuss in this chapter focus on fragmentation in climate governance.

The concept of fragmentation has also been used, however, to describe and analyze governance architectures for other issues, such as forests (Giessen 2013; Orsini 2013; Carleton and Becker 2018; van der Ven, Rothacker and Cashore 2018), biodiversity (Richerzhagen 2014; Velázques Gomar 2016), energy (Ghosh 2011; Fernández Carril, García Arrazola and Rubio 2013; Van de Graaf 2013; Heubaum and Biermann 2015; Guerra 2018), health (Graham 2014; Holzscheiter 2017), oceans (Ekstrom and Crona 2017), international security and finance (Held and Young 2013), the arctic (Humrich 2013; Yamineva and Kulovesi 2018) and counterterrorism, intellectual property and election-monitoring (Pratt 2018). In contrast, we find it surprising that several important issue areas are still understudied despite the high fragmentation of their broader governance architectures, notably agriculture and food (on corporate food governance, see however Clapp 2018; Scott 2018) and fisheries (but see Young 2009; Hollway 2011; Techera and Klein 2011).

While fragmentation is often analyzed as a quality of governance in particular areas (e.g., Pattberg et al. 2014), only few articles use the concept of fragmentation to analyze the interface between different areas, such as between forests, climate and biodiversity (e.g., van Asselt 2012) or between water, energy and food (e.g., Weitz et al. 2017). Similarly, fragmentation is not often used to analyze the interface between entire policy domains such as between trade and the environment (for

---

[1] See, for instance, Biermann et al. 2009, 2010; Zelli 2011; Zelli et al. 2010; Galaz et al. 2012; Karlsson-Vinkhuyzen and McGee 2013; Palmujoki 2013; Dyer 2014; van Asselt 2014; van Asselt and Zelli 2014; Hjerpe and Nasiritousi 2015; Zelli and van Asselt 2015; Aykut 2016; Widerberg, Pattberg and Kristensen 2016; Dorsch and Flachsland 2017; Oh and Matsuoka 2017.

exceptions, see for example, Young 2009). At these higher levels of analysis, concepts such as institutional interplay, interplay management and regime interaction are more frequently used, even though the broader notion of governance fragmentation might in fact provide more explanatory power.

Fragmentation is most often studied as an independent variable, that is, as a possible explanatory factor for the degree of effectiveness of governance architectures or goal-attainment (Jabbour et al. 2012). Fragmentation is also often seen as a contextual factor that is deemed to complicate governance efforts.

However, even though fragmentation is of key importance in many articles, it is not always sufficiently operationalized with distinguishable criteria for analysis. An exception is the study by Pattberg and colleagues (2014, drawing on Biermann et al. 2009), in which institutional constellations, actor constellations, norm constellations and discourse constellations are used as indicators to empirically measure fragmentation in specific areas. In their effort to map and measure fragmentation, they elaborate a two-step process that includes the mapping of governance architectures based on a set of criteria to demarcate the main actors within an architecture.

Regarding methods, the vast majority of the studies on governance fragmentation are qualitative, usually analyzing multiple sources of data such as scientific literature, grey documents and interviews with experts from government, business or civil society. An in-depth, qualitative lens is indeed suitable to identify norm conflicts, certain types of actor constellations and degrees of institutional nesting. There are also some authors who rely on small-*n* comparative case-study methods (e.g., Held and Young 2013).

A more limited set of studies draws on quantitative methods to analyze fragmentation. An example is the research of Pratt (2018), who uses a dataset of over 2,000 policy documents to describe patterns of deference – a strategy to cope with jurisdictional conflicts – in three policy areas. As argued above, fragmentation can occur in different degrees and is a non-binary concept that allows for a comparison between architectures where the relative degree of fragmentation is of interest. This provides an opportunity for quantitative assessments of degrees of fragmentation that might prove useful in comparative analyses. In the current literature, however, only very few attempts have been made to quantify fragmentation or to compare degrees or types of fragmentation across architectures. Among others, the lack of adequate methods and large datasets hampers empirical research that takes such an approach (Kim 2013; see also Chapter 14).

Another methodological challenge in comparative research is that the larger the scale of the governance architectures that are studied, the higher the degree of fragmentation is likely to be (Biermann et al. 2009; Zelli and van Asselt 2013). Fragmentation is evident in more narrowly defined global governance architectures, that is, between parallel policies and regimes in the same issue area such as climate governance or governance of plant genetic resources. On this scale, comparative analyses of different degrees and types of fragmentation are likely to be most fruitful, though fragmentation is also useful at a higher scale of comparative analysis, for example for entire policy domains such as environment or trade.

One promising and recently introduced method to quantify and comparatively analyze fragmentation is network theory, which has so far mainly been used to analyze links between international organizations (Beckfield 2008, 2010; Gomez and Parigi 2015; Greenhill and Lupu 2017). Network theory can be used to study not only fragmentation but also related concepts such as polycentricity and complexity (e.g., Ahlström and Cornell 2018), and it can be applied as well for analyses of the structure and dynamics of global governance architectures. Network theory also allows for introducing temporal and diverse comparative components into analyses, which can help to study the degree of fragmentation over time and between governance areas. Because network theory requires a careful and thorough justification of how the network is composed, it might not be applicable to all aspects of fragmentation.

So far, network analyses have been done for numerous institutions and actor constellations (for a conceptual discussion, see Pattberg et al. 2014). This approach might be less useful, however, to analyze norm conflicts. Furthermore, the binary character of networks (that is, the identification of either absence or presence of a link between two network components) is not well-suited to reflect the complex nature and quality of institutional interactions (Kim 2013). Hence, network analyses are useful to analyze *structural* fragmentation, which looks into degrees of fragmentation, but not necessarily *functional* fragmentation, which focuses on the types of fragmentation. This would thus call for mixed-method approaches with complementary qualitative analyses; but such a combination of approaches has so far been rarely used in the study of governance fragmentation (for an exception, see Orsini 2013).

We now turn to discussing research that studies the emergence and evolution of governance fragmentation, followed by a review of research on the impacts and consequences of fragmentation.

Emergence and Evolution

Empirically, most literature on the fragmentation of global governance addresses issues of earth system governance. This is not surprising, given that earth system governance seems to be much more fragmented than, for example, trade or health governance. Different from these areas, earth system governance is characterized by a multitude of international organizations with related mandates as well as over

1,000 multilateral agreements, many of which have their own independent secretariats. The origins of this broadly fragmented architecture date back to the late 1800s, when the first multilateral agreements on transboundary environmental concerns were signed (Mitchell 2003). It was not until 1972, however, that a specialized international agency was created with environment as its core mandate, the United Nations Environment Programme (UNEP, now also known as UN Environment). By the time that this programme was created, about 200 multilateral environmental agreements were already in place (Mitchell 2003), and many international organizations existed that had, or later acquired, environment-related mandates.

The establishment of UNEP did not end the increase in fragmentation in this field. One reason is that while UNEP was mandated to coordinate and galvanize actions by other institutions and agencies, it was not given the authority to steer or authoritatively coordinate such actions (Ivanova 2007; Vijge 2013). Time and again, academic literature as well as UN reports (e.g., United Nations 1998, 2006) argue that the earth system governance architecture is becoming increasingly fragmented and that efforts to address this, often in the form of new institutions for coordination, have led to an increase rather than a decrease of fragmentation (Vijge 2013; see also Chapter 13).

Yet, why is the governance of global socio-ecological systems, from climate to biodiversity, so much more fragmented than other governance domains? How can we explain these persistently high degrees of governance fragmentation? Unfortunately, the question has not often been analyzed in depth. As we discuss below, most literature on governance fragmentation focuses on its *consequences* (treating fragmentation as an independent variable), rather than its *causes* (treating fragmentation as a dependent variable).

There are some exceptions, though. For example, one long-standing strand in earth system governance research seeks to explain the absence of a core integrating institution such as an international treaty (such as the non-existence of a global forest treaty; see Dimitrov 2005; Dimitrov, Sprinz and DiGiusto 2007) or the absence of a central – or centralizing – organization (such as a world environment organization, which has been called for since the 1970s but is still not in place; see Vijge 2013). In addition, Johnson and Johannes (2012) studied why some environmental regimes integrate – hence reducing fragmentation – while others remain separated. Their main claim is that fear of negative spill-overs – for instance between the climate regime and the ozone regime – provides the strongest incentive for regime integration, while – unexpectedly – possible positive spill-overs do not drive actors to push for regime integration. They drew in their research on cases that differed by the degree of integration, that is, their dependent variable. This gains limited insights on the independent and possibly intermediating variables in these cases, nor does it yield causal explanations.

Especially international legal scholars have engaged for some time with the question of whether fragmentation emerges organically or whether it is rather powerful actors that consciously create a fragmented international legal system to serve their interests (e.g., Koskenniemi and Leino 2002; Benvenisti and Downs 2007; Broude 2013). Regarding the first perspective of organic emergence, Vijge (2013) uses the concept of institutional path-dependency to explain how the global environmental governance architecture has entered a self-reinforcing cycle wherein incremental changes – in the form of the ad hoc and diffused establishment of a set of fragmented institutions – are more likely than actions that would dismantle, change or replace large institutions. Dryzek (2016) has deemed this type of path-dependency as one of the core problems of governance architectures in the Anthropocene. Such an increasingly fragmented architecture coincides with fragmented or even circular policy debates about possible measures that could be taken at the global level to defragment the architecture.

Regarding the second perspective of purposeful fragmentation by interested states, several scholars have argued that neither powerful countries nor major international organizations have a strong interest in substantially transforming the governance architecture in order to defragment it (for an overview, see Vijge 2013). Some authors went a step further by arguing that powerful countries *consciously* design and maintain a fragmented governance architecture because it serves their interests. For example, Benvenisti and Downs (2007: 595) have argued that powerful states 'maintain and even actively promote fragmentation' because it allows them to maintain some of their power that they fear to lose in a time when hierarchy is increasingly considered illegitimate (see also Chapter 13). Higher degrees of fragmentation would give such states the freedom to make or break rules without negatively affecting the entire system and without being held responsible (Benvenisti and Downs 2007). Paris (2015) even sees a long-term shift in power away from the United States towards emerging countries in the Global South as the core explanation of what he labels the pluralization of global governance, or what we would label in the context of this book as increasing fragmentation. In analyzing the fragmentation of the global environmental governance architecture, Ivanova and Roy (2007: 50) argue that 'governments deliberately create weak and underfunded international organizations with overlapping and even conflicting mandates' because they are '[f]earful of infringement upon their national sovereignty'. Regarding the forest regime, Dimitrov (2005: 19 and 4) even argues that states establish 'hollow institutions' that function as '"decoys" deliberately designed to pre-empt governance'. Fragmentation that arises from the establishment of such decoy institutions can isolate policy issues for which there is no political will from more important, higher-level political fora. This allows governments to conform to the widely held norm of 'doing at least something' to address earth system concerns yet without taking substantive actions, something that has been termed 'symbolic policymaking' (Dimitrov 2005; Vijge 2013).

In addition, Zürn and Faude (2013) have brought forward a theoretical approach that views fragmentation of governance architectures as a 'functional response to the swelling tide of problems that can be handled best on the international level', drawing on differentiation theory in social science (Zürn and Faude 2013: 123). In this perspective, fragmentation is inevitable and neither positive nor negative per se; rather, it constitutes a political challenge to ensure the fruitful coordination of (increasingly) functionally differentiated governance units at the global level (Zürn and Faude 2013).

Although global governance of trade is generally considered less fragmented than earth system governance, several authors have sought to explain the growing fragmentation of trade governance as well. Here they focus on the increase of bilateral trade agreements, as opposed to multilateral agreements that include most countries. Interestingly, also in this domain, fragmentation through the setting up of bilateral agreements seems to be in the interest of powerful countries and hence supported by them. While in multilateral agreements, developing countries in the Global South may gain power vis-à-vis industrialized countries and cannot anymore be marginalized in negotiations, it is rather the rapidly spreading bilateral agreements that seem to serve the commercial interests of the United States and the European Union (Aggarwal and Evenett 2013; Trommer 2017).

Yet despite all this work in political science and international legal studies, additional comparative studies that take fragmentation as the dependent variable are needed and are an interesting venue for future research.

Impacts and Consequences

In addition, an extensive line of research has focused on the consequences and impacts of governance fragmentation. This literature, however, is still fundamentally divided regarding the overall benefits and downsides of more or less fragmented governance architectures. In part, these different perspectives on whether consequences of fragmentation are overall positive or negative depend on the actual degree of fragmentation in the area that is studied.

Positive effects of fragmentation. Several studies point to the benefits of fragmentation. In cases where smaller sets of actors seek to cooperate while others stay out, it is argued, the resulting fragmentation facilitates quicker, more innovative and more far-reaching decision-making among this highly collaborative but smaller set of actors. This is often referred to as minilateralism. Here, some authors expect that a patchwork of multiple minilateral fora can allow for experimentation with unconventional governance frameworks; enable tailor-made decisions with more specialized accounting or reporting frameworks; and ensure that inaction or stalemates in one decision-making process do not jeopardize others. Minilateralism may also facilitate concessions, funding agreements and the transfer of technology between a small group of actors that would hesitate to commit resources in large multilateral agreements (Bodansky 2002; Zelli, Gupta and van Asselt 2012; Dyer 2014; Dorsch and Flachsland 2017).

In climate governance, for example, the repeated deadlocks in multilateral climate negotiations under the United Nations Framework Convention on Climate Change (Falkner, Stephen and Vogler 2010) have led to research on the potential of alternative fora and parallel initiatives, including minilateral institutions such as the Major Economies Forum on Energy and Climate, as well as transnational markets, public–private partnerships, and other layers and networks of rule- and decision-making. These climate governance initiatives take place outside of, yet are loosely related to, the climate convention (Okereke, Bulkeley and Schroeder 2009; Victor 2009; Biermann 2010; Biermann et al. 2012; Dyer 2014; Falkner 2016). Similarly, Keohane and Victor (2011) and Abbott (2012) have argued for a transnational regime complex, in which the climate convention as a central negotiating forum would offer substantial degrees of flexibility and diversity to allow for a loose complex of (sub-)regimes. To be successful, such mix-and-match approaches would need to rely on the principle of subsidiarity to increase self-organization; take into account the site-specific conditions by specifying the preferences, competencies, constraints and interactions of actors; include experimentation and learning at subsidiary levels to test innovations that can later be scaled up; and strengthen the trust across all scales and levels (Dorsch and Flachsland 2017; see also Hackmann 2016 for learning in global environmental governance). However, especially for polycentric systems, robust connections are needed to realize governance with decentralized feedback as one of its main components (Gallemore 2017).

Smaller agreements can be negotiated either by like-minded actors with closely aligned interests, or by actors that otherwise depend on one another, for example in the case of regional agreements (Bodansky 2002; Zelli, Gupta and van Asselt 2012). A fragmented configuration of institutions or decision-making processes that are loosely but cooperatively connected may then enable a larger set of actors to access and participate in the multiple co-existing decision-making processes (Zelli, Gupta and van Asselt 2012; Acharya 2016). Fragmentation may thus offer actors – including non-state actors such as business and civil society representatives – the flexibility to freely enter or leave non-confrontational negotiations wherein decisions are made through consensus (Acharya 2016).

Several studies have also investigated the benefits of vertical, as opposed to horizontal, fragmentation. Scholars studying polycentric approaches in climate governance, for example, highlight the potential of multiple bilateral, national and local forums to deliver solutions, as opposed to having one, exclusively global-level, convention being responsible for crafting solutions. Vertical fragmentation, in the form of a combination of top-down and bottom-up approaches, may offer the potential to exploit co-benefits at multiple decision-making fora and thereby provide incentives for climate action at multiple levels (Dai 2010; Rayner 2010; Hoffmann 2011; Araral 2014; Cole 2015; see also Ostrom 2010b; Galaz et al. 2012; Falkner 2016; Dorsch and Flachsland 2017).

Negative effects of fragmentation. In contrast, several studies emphasize the dangers, downsides and further challenges of strongly fragmented governance architectures.

(1) First, many authors emphasize that smaller institutions cannot function effectively without a broader framework, and hence emphasize the continued relevance of broader, overarching frameworks that bind smaller agreements of only a few countries. As Hafner (2003: 856), for instance, argues, fragmentation 'jeopardizes the credibility, reliability, and, consequently, the authority of international law'. Several studies suggest that in a highly fragmented architecture such as in climate governance, a centralized regime such as the climate convention is necessary to create fair and effective outcomes (Hare et al. 2010; Winkler and Beaumont 2010; Dyer 2014; Dorsch and Flachsland 2017). Also, as Eckersley (2012) points out, creative compromises can only be realized if the diversity among involved member states is enhanced, for instance through the creation of a 'Climate Council' to be constituted based on common but differentiated representation (Eckersley 2012). This could be a way in which minilateralism could relegitimize the United Nations Framework Convention on Climate Change, negating the global powershifts that contributed to the slowdown in multilateral negotiations that took place under its purview (Falkner 2016). Pratt (2018) emphasizes here the concept of institutional deference, where deference to other international organizations makes focused rule-making on sub-issues more likely. Although the concept was used to describe the development of a division of labour within regime complexes, comparable developments might also be apparent or possible within broader governance architectures.

(2) Second, many studies associate governance fragmentation with a lack of coherence, inefficiency or ineffectiveness, and overlapping or even conflicting – and thus potentially counterproductive – policies (Jabbour et al. 2012; Held and Young 2013). Especially in the case of the many agreements resulting from

fragmented architectures, Jabbour and colleagues (2012) claim that the multiplicity of obligations can hamper implementation in countries with limited international policy capacity. Looking at different stages of decision-making for the policy mechanism called Reducing Emissions from Deforestation and Forest Degradation (REDD+), for example, Gallemore (2017) has claimed that broadly fragmented systems – what they refer to as polycentric systems – comprising the coalitions necessary to raise issues to the agenda create high transaction costs when diverse interests must be realized simultaneously during implementation. This draws attention to questions surrounding the effectiveness of polycentric systems throughout their evolution.

A related issue is that for those engaging in fragmented architectures, it is not always clear where to draw the line in terms of participants, and this might also not be possible. When a patchwork of smaller institutions is in conflict or produces conflicting norms, decisions might become contradictory, thereby inhibiting their implementation (Bodansky 2002; Biermann et al. 2009; Zelli et al. 2010; Falkner 2016). In addition, a fragmented governance architecture is argued to increase the potential of duplication (Held and Young 2013). Countries and other actors can cherry-pick from a fragmented set of agreements and choose to engage only in those decision-making processes that align well with their individual interests, thereby creating 'coalitions of the willing' (Falkner 2016: 87; see also Biermann et al. 2009). The more fragmented a system becomes, the more likely it is that multiple actors are involved and spend unnecessary resources on comparable issues, policy solutions and activities. Moreover, conflictive norms or decisions in a fragmented architecture can obstruct the formation of a common vision, ambition and action and create confusion among actors about the direction that global governance should take (Biermann et al. 2009).

(3) Third, several studies argue that a fragmented patchwork of small-*n* agreements may not sustain in the long run, either because the smaller agreements do not address – or perhaps even increase – larger-scale institutional barriers, or because they are not accepted by the wider set of actors responsible for their implementation. Held and Young (2013) describe such a case in international finance. They argue that in this area – and also in international security – fragmentation must be seen as the outcome of *mal*-adaptation. Because established institutions such as the International Monetary Fund and the World Bank were unable to produce system change in the face of the financial crisis, existing ad hoc, informal institutions – especially the Group of 20 – were de facto changed into small-*n* platforms wherein major powers devised plans to cope with the issues. The 'agreements' established within the Group of 20, however, did not sustain in the long run, as the venue lacked an administrative structure, enforcement capacity and mandate to execute

its orders. After having reached their agreements, the Group of 20 directed their plans to the traditional institutions of the international finance governance architecture in which all countries participate; but such proposals rarely gained full support. Instead, the final compromises often resulted in watered-down, incremental reform proposals.

(4) Fourth, several authors point out that when conflicts among institutions and actors arise, fragmentation disadvantages smaller or less powerful actors, which need larger coalitions and broader institutions to increase their collective bargaining power vis-à-vis the more powerful actors, such as the United States (Zelli et al. 2010; Biermann 2014). In the case of climate governance, for example, Eckersley (2012) argues against a patchwork of smaller agreements from the angle of substantive and communicative justice, suggesting that such smaller agreements are elitist, procedurally unjust, self-serving and not in line with the justice principles enshrined in the multilateral climate convention. Karlsson-Vinkhuyzen and McGee (2013) argue that minilateral fora have even allowed powerful states to advocate certain discourses around voluntary commitments that have now been taken up by the multilateral climate convention. They further argue that the proliferation of minilateral fora is characterized by a limited participation by state and non-state actors, a fundamental lack of transparency in decision-making and a lack of accountability towards non-participants. If some form of (functional) differentiation within a fragmented architecture with complex interdependencies exists in a policy domain, actors may become vulnerable to crosscutting and intersecting independent variables over which they have little control (Cerny and Prichard 2017).

Managing the negative consequences of fragmentation. Finally, how to manage governance fragmentation and especially its negative consequences has received much policy and scholarly attention. This is well covered in other chapters in this book. Responses to fragmentation include, for instance, the more general embracement of fragmentation (Rayner, Buck and Katila 2010) combined with active policy measures such as orchestration (Chapter 11) and governance through global goals (Chapter 12). Policy measures to reduce fragmentation also include policy integration (Chapter 9), interplay management (Chapter 10) and eventually hierarchization (Chapter 13).

Conclusions and Future Directions

While the literature on governance fragmentation is vast and still growing, key gaps remain. These include explanatory analyses of the relations between different fragmented governance architectures and governance levels (horizontal fragmentation) (Visseren-Hamakers 2018); research on the relation between problem structure and the degree of fragmentation within a specific issue domain; and analyses of the agency of actors in a fragmented architecture, particularly actors from the Global South (Acharya 2016). More research is also needed that draws on quantitative and mixed-methods approaches to studying governance fragmentation. To analyze *structural* fragmentation that focuses on the degree of fragmentation, more efforts are needed to quantify fragmentation. For this purpose, the creation of adequate methods and large datasets that facilitate these approaches can be highly beneficiary (Kim 2013; see also Chapter 1). Mixed-method research could analyze *structural* and *functional* fragmentation – focusing on both degrees and types of fragmentation – in more detail at the same time.

Future research could also invest in the development of a typology of all potential linkages between entities of a governance architecture. This research could rely on insights from network analysis to continue the inductive work of, for instance, Betsill and colleagues (2015). An example of a less-studied but potentially interesting type of linkage is that of catalytic linkages. These centre around the alteration of the actions of one or more actors to allow third parties to improve the performance of their governance tasks, similar to orchestration (Chapter 11). Mapping these and other linkages not only allows scholars to better comprehend existing architectures but could also be used by actors trying to increase cooperation or even synergies within existing structures or trying to reform existing structures.

Additionally, comparative studies that take fragmentation as the dependent variable remain an interesting venue for future research. This would shift research from trying to explain how fragmentation impacts governance effectiveness to *causal* questions surrounding fragmented global governance architectures. This research can build on the research already undertaken in the field of international law about conscious versus organic emergence (including path-dependency) of fragmentation. However, as governance architectures often cover entire policy areas, it remains empirically and practically challenging to compare such large areas based on variation of the dependent variable.

Finally, research on governance fragmentation – and the continued strong emphasis on the negative impacts of strong governance fragmentation – reinforces the necessity of more research on possible policy responses and options for structural transformation. The field of earth system governance is unique in its high degrees of governance fragmentation, for a variety of historical and structural reasons, and it remains a major challenge for political science and policy analysis to sketch powerful solutions and transformative trajectories that could lead our societies to more integrative, more effective and more equitable global governance.

Such policy interventions are discussed in detail in the following chapters of this book.

References

Abbott, K. W. (2012). The transnational regime complex for climate change. *Government and Policy*, 30 (4): 571–90.

Acharya, A. (2016). The future of global governance: Fragmentation may be inevitable and creative. *Global Governance*, 22, 453–60.

Aggarwal, V., & Evenett, S. (2013). A fragmenting global economy: A weakened WTO, mega-FTAs, and murky protectionism. *Swiss Political Science Review*, 19, 550–7.

Ahlström, H., & Cornell, S. E. (2018). Governance, polycentricity and the global nitrogen and phosphorus cycles. *Environmental Science & Policy*, 79, 54–65.

Araral, E. (2014). Ostrom, Hardin and the commons: A critical appreciation and a revisionist view. *Environmental Science & Policy*, 36, 11–23.

Aykut, S. C. (2016). Taking a wider view on climate governance: Moving beyond the 'iceberg', 'the 'elephant', and the 'forest'. *Climate Change*, 7 (3), 318–28.

Beckfield, J. (2008). The dual world polity: Fragmentation and integration in the network of intergovernmental organizations. *Social Problems*, 55 (3), 419–42.

Beckfield, J. (2010). The social structure of the world polity. *American Journal of Sociology*, 115 (4), 1018–68.

Benvenisti, E. D., & Downs, G. W. (2007). The empire's new clothes: Political economy and the fragmentation of international law. *Stanford Law Review*, 60 (2), 595–632.

Betsill, M., Dubash, N. K., Paterson, M., van Asselt, H., Vihma, A., & Winkler, H. (2015). Building productive links between the UNFCCC and the broader global climate governance landscape. *Global Environmental Politics*, 15 (2), 1–10.

Biermann, F. (2010). Beyond the intergovernmental regime: Recent trends in global carbon governance. *Current Opinion in Environmental Sustainability*, 2 (4), 284–8.

Biermann, F. (2014). *Earth system governance: World politics in the Anthropocene*. Cambridge, MA: The MIT Press.

Biermann, F., Pattberg, P., van Asselt, H., & Zelli, F. (2009). The fragmentation of global governance architectures: A framework for analysis. *Global Environmental Politics*, 9 (4), 14–40.

Biermann, F., Pattberg, P., van Asselt, H., & Zelli, F. (eds.) (2010). *Global climate governance beyond 2012: Architecture, agency and adaptation*. Cambridge, UK: Cambridge University Press.

Biermann, F., Abbott, K. W., Andresen, S. et al. (2012). Transforming governance and institutions for global sustainability: Key insights from the Earth System Governance Project. *Current Opinion in Environmental Sustainability*, 4 (1), 51–60.

Bodansky, D. (2002). *U.S. Climate Policy after Kyoto: Elements for success*. Washington DC: Carnegie Endowment for International Peace.

Broude, T. (2013). Keep calm and carry on: Martti Koskenniemi and the fragmentation of international law. *Temple International and Comparative Law Journal*, 27, 279–92.

Busch, P.-O., Gupta, A., & Falkner, R. (2012). International-domestic linkages and policy convergence. In F. Biermann, & P. Pattberg (eds.), *Global environmental governance reconsidered* (pp. 199–218). Cambridge, MA: The MIT Press.

Carleton, L., & Becker, D. (2018). Forest biomass policy in Minnesota: Supply chain perspectives on barriers to bioenergy development. *Forests*, 9 (5), 254.

Cerny, P. G., & Prichard, A. (2017). The new anarchy: Globalisation and fragmentation in world politics. *Journal of International Political Theory*, 13 (3), 378–94.

Clapp, J. (2018). Mega-mergers on the menu: corporate concentration and the politics of sustainability in the global food system. *Global Environmental Politics*, 18 (2), 12–33.

Cole, D. H. (2015). Advantages of a polycentric approach to climate change policy. *Nature Climate Change*, 5, 114–18.

Dai, X. (2010). Global regime and national change. *Climate Policy*, 10, 622–37.

De Coninck, H., & Bäckstrand, K. (2011). An international relations perspective on the global politics of carbon dioxide capture and storage. *Global Environmental Change*, 21 (2), 368–78.

Dimitrov, R. S. (2005). Hostage to norms: States, institutions and global forest politics. *Global Environmental Politics*, 5 (4), 1–24.

Dimitrov, R. S., Sprinz, D. F., & DiGiusto, G. M. (2007). International nonregimes: A research agenda. *International Studies Review*, 9, 230–58.

Dorsch, M. J., Flachsland, C. A. (2017). Polycentric approach to global climate governance. *Global Environmental Politics*, 17 (2), 45–64.

Dryzek, J. S. (2016). Institutions for the Anthropocene: Governance in a changing earth system. *British Journal of Political Science*, 46 (4), 937–56.

Dyer, H. C. (2014). Climate anarchy: Creative disorder in world politics. *International Political Sociology*, 8 (2), 182–200.

Eckersley, R. (2012). Moving forward in the climate negotiations: Multilateralism or minilateralism? *Global Environmental Politics*, 12, 24–42.

Ekstrom, J. A., & Crona, B. I. (2017). Institutional misfit and environmental change: A systems approach to address ocean acidification. *Science of the Total Environment*, 576, 599–608.

Falkner, R. (2016). A minilateral solution for global climate change? On bargaining efficiency, club benefits, and international legitimacy. *Perspectives on Politics*, 14 (1), 87–101.

Falkner, R., Stephan, H., & Vogler, J. (2010). International climate policy after Copenhagen: Towards a 'building blocks' approach. *Global Policy*, 1, 252–62.

Fernández-Blanco, C. R., Burns, S. L., & Giessen, L. (2019). Mapping the fragmentation of the international forest regime complex: Institutional elements, conflicts and synergies. *International Environmental Agreements*, 19 (2), 187–205.

Fernández Carril, L., García Arrazola, R., & Rubio, J. (2013). Discursive overlap and conflictive fragmentation of risk and security in the Geopolitics of energy. *Sustainability*, 5 (3), 1095–1113.

Galaz, V., Crona, B., Österblom, H., Olsson, P., & Folke, C. (2012). Polycentric systems and interacting planetary boundaries: Emerging governance of climate change–ocean acidification–marine biodiversity. *Ecological Economics*, 81, 21–32.

Gallemore, C. (2017). Transaction costs in the evolution of transnational polycentric governance. *International Environmental Agreements*, 17 (5), 639–54.

Ghosh, A. (2011). Seeking coherence in complexity? The governance of energy by trade and investment institutions. *Global Policy*, 2, 106–19.

Giessen, L. (2013). Reviewing the main characteristics of the international forest regime complex and partial explanations for its fragmentation. *International Forestry Review*, 15 (1), 60–70.

Gomez, C. J., & Parigi, P. (2015). The regionalization of intergovernmental organization networks: A non-linear process. *Social Networks*, 43, 192–203.

Graham, E. R. (2014). International organizations as collective agents: Fragmentation and the limits of principal control at the World Health Organization. *European Journal of International Relations*, 20 (2), 366–90.

Greenhill, B., & Lupu, Y. (2017). Clubs of clubs: Fragmentation in the network of Intergovernmental Organizations. *International Studies Quarterly*, 61, 181–95.

Guerra, F. (2018). Mapping offshore renewable energy governance. *Marine Policy*, 89, 21–33.

Gupta, A., Pistorius, T., & Vijge, M. J. (2016). Managing fragmentation in global environmental governance: The REDD+ Partnership as bridge organization. *International Environmental Agreements*, 16, 355–74.

Hackmann, B. (2012). Analysis of the governance architecture to regulate GHG emissions from international shipping. *International Environmental Agreements*, 12 (1), 85–103.

Hackmann, B. (2016). Regime learning in global environmental governance. *Environmental Values*, 25, 663–86.

Hafner, G. (2003). Pros and cons ensuing from fragmentation of international law. *Michigan Journal of International Law*, 25, 849–1349.

Hare, W., Stockwell, C., Flachsland, C., & Oberthür, S. (2010). The architecture of the global climate regime: A top-down perspective. *Climate Policy*, 10, 600–14.

Held, D., & Young, K. (2013). Global governance in crisis? Fragmentation, risk and world order. *International Politics*, 50 (3), 309–22.

Heubaum, H., & Biermann, F. (2015). Integrating global energy and climate governance: The changing role of the International Energy Agency. *Energy Policy*, 87, 229–39.

Hjerpe, M., & Nasiritousi, N. (2015). Views on alternative forums for effectively tackling climate change. *Nature Climate Change*, 5, 864–7.

Hoffmann, M. J. (2011). *Climate governance at the crossroads: Experimenting with a global response after Kyoto*. Oxford: Oxford University Press.

Hollway, J. (2011). *Taking stock of the fragmentation of the global fisheries governance architecture*. Bonn: German Development Institute.

Holzscheiter, A. (2017). Coping with institutional fragmentation? Competition and convergence between boundary organizations in the global response to polio. *Review of Policy Research*, 34 (6), 767–89.

Humrich, C. (2013). Fragmented international governance of arctic offshore oil: Governance challenges and institutional improvement. *Global Environmental Politics*, 13 (3), 79–99.

Isailovic, M., Widerberg, O., & Pattberg, P. (2013). Fragmentation of global environmental governance architectures: A literature review. Amsterdam: Institute for Environmental Studies.

Ivanova, M. (2007). Designing the United Nations Environment Programme: A story of compromise and confrontation. *International Environmental Agreements*, 7 (4), 337–61.

Ivanova, M. H., & Roy, J. (2007). The architecture of global environmental governance: Pros and cons of multiplicity. In L. Swart, & E. Perry (eds.), *Global environmental governance: Perspectives on the current debate* (pp. 48–66). New York: Center for UN Reform Education.

Jabbour, J., Keita-Ouane, F. Hunsberger, C. et al. (2012). Internationally agreed environmental goals: A critical evaluation of progress. *Environmental Development*, 3, 5–24.

Johnson, T., & Johannes, U. (2012). A strategic theory of regime integration and separation. *International Organization*, 66, 645–77.

Karlsson-Vinkhuyzen, S. I., & McGee, J. (2013). Legitimacy in an era of fragmentation: The case of global climate governance. *Global Environmental Politics*, 13 (3), 56–78.

Keohane, R. O., & Victor, D. G. (2011). The regime complex for climate change. *Perspectives on Politics*, 9 (1), 7–23.

Kim, R. E. (2013). The emergent network structure of the multilateral environmental agreement system. *Global Environmental Change*, 23 (5), 980–91.

Koskenniemi, M., & Leino, P. (2002). Fragmentation of international law? Postmodern anxieties. *Leiden Journal of International Law*, 15 (3), 553–79.

Mitchell, R. B. (2003). International environmental agreements: A survey of their features, formation and effects. *Annual Review of Environment and Resources*, 28, 429–61.

Oh, C., & Matsuoka, S. (2017). The genesis and end of institutional fragmentation in global governance on climate change from a constructivist perspective. *International Environmental Agreements*, 17 (2), 143–59.

Okereke, C., Bulkeley, H., & Schroeder, H. (2009). Conceptualizing climate governance beyond the international regime. *Global Environmental Politics*, 9 (1), 58–78.

Orsini, A. (2013). Multi-forum non-state actors: Navigating the regime complexes for forestry and genetic resources. *Global Environmental Politics*, 13 (3), 34–55.

Ostrom, E. (2010a). Beyond markets states: Polycentric governance of complex economic systems. *American Economic Review*, 100, 641–72.

Ostrom, E. (2010b). Polycentric systems for coping with collective action and global environmental change. *Global Environmental Change*, 20, 550–7.

Palmujoki, E. (2013). Fragmentation and diversification of climate change governance in international society. *International Relations*, 27 (2), 180–201.

Paris, R. (2015). Global governance and power politics: Back to basics. *Ethics and International Affairs*, 29 (4), 407–18.

Pattberg, P., Widerberg, O., Isailovic, M., & Dias Guerra, F. (2014). *Mapping and measuring fragmentation in global governance architectures*. Amsterdam: Institute for Environmental Studies.

Pickering, J., Betzold, C., & Skovgaard, J. (2017). Managing fragmentation and complexity in the emerging system of international climate finance. *International Environmental Agreements*, 17 (1), 1–16.

Pratt, T. (2018). Deference and hierarchy in international regime complexes. *International Organization*, 72 (3), 561–90.

Rana, P. B., & Pacheco Pardo, R. (2018). Rise of complementarity between global and regional financial institutions: perspectives from Asia. *Global Policy*, 9 (2), 231–43.

Rayner, J., Buck, A., & Katila, P. (eds.) (2010). *Embracing complexity: Meeting the challenges of international forest governance. A Global Assessment Report: Prepared by the Global Forest Expert Panel on the International Forest Regime*. Vienna: International Union of Forest Research Organizations.

Rayner, S. (2010). How to eat an elephant: A bottom-up approach to climate policy. *Climate Policy*, 10, 615–21.

Richerzhagen, C. (2014). The Nagoya Protocol: Fragmentation or consolidation? *Resources*, 3 (1), 135–51.

Scott, C. (2018). Sustainably sourced junk food? Big food and the challenge of sustainable diets. *Global Environmental Politics*, 18 (2), 93–113.

Smits, M. (2017). The new (fragmented) geography of carbon market mechanisms: Governance challenges from Thailand and Vietnam. *Global Environmental Politics*, 17 (3), 69–90.

Techera, E. J., & Klein, N. (2011). Fragmented governance: reconciling legal strategies for shark conservation and management. *Marine Policy*, 35 (1), 73–8.

Trommer, S. (2017). The WTO in an era of preferential trade agreements: Thick and thin institutions in global trade governance. *World Trade Review*, 16 (3), 501–26.

United Nations (2006). *Delivering as one. Report of the Secretary-General's High-level Panel on UN System-wide Coherence in the Areas of Development, Humanitarian Assistance and the Environment*. New York: United Nations.

United Nations General Assembly (1998). *Report of the United Nations Task Force on Environment and Human Settlements*. UN Doc. A/53/463.

Van Asselt, H. (2012). Managing the fragmentation of international environmental law: Forests at the intersection of the climate and biodiversity regimes. *New York University Journal of International Law and Politics*, 44 (4), 1205–78.

Van Asselt, H. (2014). *The fragmentation of global climate governance: Consequences and management of regime interactions*. Cheltenham: Edward Elgar.

Van Asselt, H., & Zelli, F. (2014). Connecting the dots: Managing the fragmentation of global climate governance. *Environmental Economics and Policy Studies*, 16 (2), 137–55.

Van de Graaf, T. (2013). Fragmentation in global energy governance: explaining the creation of IRENA. *Global Environmental Politics*, 13 (3), 14–33.

Van der Ven, H., Rothacker, C., & Cashore, B. (2018). Do eco-labels prevent deforestation? Lessons from non-state market driven governance in the soy, palm oil, and cocoa sectors. *Global environmental change*, 52, 141–51.

Velázquez Gomar, J. O. V. (2016). Environmental policy integration among multilateral environmental agreements: the case of biodiversity. *International Environmental Agreements*, 16 (4), 525–41.

Victor, D. (2009). Plan B for Copenhagen. *Nature*, 461, 342–4.

Vijge, M. J. (2013). The promise of new institutionalism: Explaining the absence of a World or United Nations Environment Organisation. *International Environmental Agreements*, 13, 153–76.

Visseren-Hamakers, I. J. (2015). Integrative environmental governance: Enhancing governance in the era of synergies. *Current Opinion in Environmental Sustainability*, 14, 136–43.

Visseren-Hamakers, I. J. (2018). Integrative governance: The relationships between governance instruments taking center stage. *Politics and Space*, 36 (8), 1341–54.

Weitz, N., Strämbo, C., Kemp-Benedict, E., & Nilsson, M. (2017). Closing the governance gaps in the water-energy-food nexus: Insights from integrative governance. *Global Environmental Change*, 45, 165–73.

Well, M., & Carrapatoso, A. (2017). REDD+ finance: Policy making in the context of fragmented institutions, Climate Policy, 17 (6), 687–707.

Widerberg, O., Pattberg, P., & Kristensen, K. (2016). *Mapping the institutional architecture of global climate change governance*. Amsterdam: Institute for Environmental Studies.

Winkler, H., & Beaumont, J. (2010). Fair and effective multilateralism in the post-Copenhagen climate negotiations. *Climate Policy*, 10, 638–54.

Yamineva, Y., & Kulovesi, K. (2018). Keeping the Arctic white: The legal and governance landscape for reducing short-lived climate pollutants in the Arctic region. *Transnational Environmental Law*, 7 (2), 201–27.

Young, M. A. (2009). Fragmentation or interaction: The WTO, fisheries subsidies, and international law. *World Trade Review*, 8 (4), 477–515.

Zelli, F. (2011). The fragmentation of the global climate governance architecture. *Wiley Interdisciplinary Reviews: Climate Change*, 2 (2), 255–70.

Zelli, F., & van Asselt, H. (2015). Fragmentation. In K. Bäckstrand, & E. Lövbrand (eds.), *Research Handbook on climate governance* (pp. 121–31). Cheltenham: Edward Elgar.

Zelli, F., & van Asselt, H. (2013). The institutional fragmentation of global environmental governance: Causes, consequences and responses. *Global Environmental Politics*, 13 (3), 1–13.

Zelli, F., Biermann, F., Pattberg, P., & van Asselt, H. (2010). The consequences of a fragmented climate governance architecture: A policy appraisal. In F. Biermann, P. Pattberg, & F. Zelli (eds.), *Global climate governance beyond 2012. Architecture, agency and adaptation* (pp. 25–34). Cambridge, UK: Cambridge University Press.

Zelli, F., Gupta, A., & van Asselt, H. (2012). Horizontal institutional interlinkages. In F. Biermann, & P. Pattberg (eds.), *Global environmental governance reconsidered* (pp. 175–98). Cambridge, MA: The MIT Press.

Zelli, F., Möller, I., & van Asselt, H. (2017). Institutional complexity and private authority in global climate governance: The cases of climate engineering, REDD+ and short-lived climate pollutants. *Environmental Politics*, 26 (4), 669–93.

Zürn, M., & Faude, B. (2013). Commentary: On fragmentation, differentiation and coordination. *Global Environmental Politics*, 13 (3), 119–30.

Part III

Policy Responses

9

Policy Integration

HENS RUNHAAR, BETTINA WILK, PETER DRIESSEN, NIALL DUNPHY,
ÅSA PERSSON, JAMES MEADOWCROFT AND GERARD MULLALLY

Environmental policy integration is the incorporation of environmental concerns and objectives into non-environmental policy areas, such as energy, transport and agriculture, as opposed to pursuing such objectives through purely environmental policy practices (Persson et al. 2018). Environmental policy integration has been firmly embedded in policy practice. Since 1997, for example, environmental policy integration is a requirement under the Treaty of the European Union, in which Article 6 states that 'environmental protection requirements must be integrated into the definition and implementation of the Community policies . . . in particular with a view to promoting sustainable development' (European Commission 2016). From a more instrumental perspective, environmental policy integration is promoted to overcome policy incoherence and institutional fragmentation (Chapter 8), address driving forces of environmental degradation and promote innovation and synergy (Runhaar, Driessen and Soer 2009).

In the last three decades, environmental policy integration has attracted much scholarly interest (Persson et al. 2018). This has led to a better conceptualization of environmental policy integration (e.g., Persson 2004), a better understanding of its normative foundations (e.g., Lafferty and Hovden 2003) and empirical studies of processes of environmental policy integration, instruments to promote it, barriers and enablers and the outcomes of environmental policy integration (e.g., Runhaar 2016). In this chapter we analyze recent research on environmental policy integration and assess how environmental policy integration has contributed to environmental protection in earth system governance. We first conceptualize environmental policy integration and then provide a meta-analysis of research findings.

Conceptualization

One of the first scientific publications about environmental policy integration is from Lafferty and Hovden (2003). They succinctly define it as the 'integration of environmental concerns into other policy areas' (Lafferty and Hovden 2003: 1). However, Persson and colleagues (2018) observe that both practitioners and scholars have differently interpreted environmental policy integration, what it comprises and what it represents. For many, environmental policy integration is a principle, standard or norm to which the policy process should adhere. This is illustrated, for example, by Lafferty and Hovden's (2003) normative interpretation, in which environmental objectives have a 'principled priority' over other societal objectives. For others, environmental policy integration has a more positivist meaning and relates to the practice of integration in discourse and in everyday political and policy settings (Mullally and Dunphy 2015).

A distinction is often drawn between *weak* environmental policy integration as a procedural input and *strong* environmental policy integration, as it is reflected in policy outputs. Runhaar (2016) refers to these as reflecting 'procedural' and 'substantive' purposes of integration. Following Oberthür (2009), weak environmental policy integration implies that environmental objectives must be considered and weighed against other objectives in decision-making, but not necessarily that decisions reflect environmental objectives. On the other hand, strong environmental policy integration insists on 'principled priority' for environmental objectives in decision-making, which must be reflected in policy outputs. However, Knudsen and Lafferty (2016) clarify that the concept of principled priority, in the context of democratic structures, cannot imply absolute priority for environmental objectives. They argue that it implies a guarantee that 'every effort is made to assess the impacts of policies' and to minimize impacts that represent unacceptable risks.

Lafferty has advanced in several studies (e.g., Lafferty 2004) to look at the horizontal and vertical dimensions of integration to characterize environmental policy integration. Horizontal environmental policy integration involves the question of integrating environmental concerns across sectoral areas of responsibility within governments. Vertical environmental policy integration refers to the extent to which governmental sectors have taken on board and implemented environmental objectives in sectoral objectives.

Often environmental policy integration is also characterized by the *degree* of integration, ranging from small adjustments in sectoral policies to significant changes in policies and approaches to policymaking. Storbjörk and Isaksson (2014) and Persson and colleagues (2018), drawing on Lafferty and Hovden (2003), identify three levels of integration: (a) mere coordination of policies to prevent contradictions; (b) harmonization, which involves a larger degree of integration that gives environmental objectives equal consideration to sectoral objectives to promote synergies; and (c) prioritization, whereby sustainability is considered the overarching guiding principle and environmental objectives take precedence over others and are integrated in all stages of sectoral policymaking.

Furthermore, environmental policy integration is viewed by many as a process of governing, following Meadowcroft, Langhelle and Ruud (2012: 8), who suggest that 'sustainable development is above all about governance'. This perspective sees environmental policy integration as a process 'anchored in the political system' (Jordan and Lenschow 2010: 150). The process-centric perspective considers how the policy process is reorganized to integrate environmental objectives into sectoral policies. Mullally and Dunphy (2015) note that structural societal change emerges from the interplay of many socio-political and socio-technical factors, with many actors and resources all involved in shaping societal transformations. They note the need to coordinate the strategies of actors but also the importance of the concept of interlinkages to explain the relationship between autonomous international institutions (e.g., Oberthür 2009; see also Chapter 6) and that between functional domains or sectors (e.g., Hogl et al. 2012).

Other scholars, however, emphasize not the process but rather the *output* of integration. As Jordan and Lenschow (2010: 154) note, many environmentalists would 'argue that principles are only principles and process is only process; policy outcomes (that is, the influence of any environmental policy integration-related activity on the state of the environment) are what really matter'. This output-centric view is concerned with the degree to which the formal products of the policy process reflect environmental objectives, and how successful policy instruments are in improving the environment. As Adelle and Russel (2013) point out, however, measuring the effectiveness of environmental policy integration processes in terms of outcomes is no easy task.

As for the history of the concept, the 1992 United Nations Conference on Environment and Development (also known as the Rio 'Earth Summit') is often seen as the origin of environmental policy integration. Yet attempts at policy integration have a much longer history. Integration has been a core concern of national and urban planners for over a century. When the environment became a policy focus in the 1960s, the idea of integrating environmental considerations into mainstream decision-making was expressed in varied ways and in different institutional contexts. One of the first formal mechanisms to promote environmental policy integration was the adoption of 'environmental impact assessments', first in the US National Environmental Policy Act in 1969, which was followed by many other countries (Ortolano and Shepherd 1995). The 1987 Report of the Brundtland Commission, *Our Common Future*, not only elaborated the concept of 'sustainable development' – a high-level integrative norm that binds concerns of human welfare, environmental protection and equity between and within generations – but also focused on the concrete 'institutional challenge' that the major central economic and sectoral agencies of governments should be made directly responsible and fully accountable for ensuring that their policies, programmes and budgets support development that is both ecologically and economically sustainable (World Commission on Environment and Development 1987: 13–14).

The principle of environmental policy integration has since diffused rather quickly. It was picked up early and systematically by the European Union, with the so-called Cardiff process in the mid-1990s that aimed at 'greening' sectoral policies and at gradually strengthening its legal foundation (Lenschow 2002). For example, Article 2 of the 2007 Lisbon Treaty states that the internal market of the European Union 'shall work for the sustainable development of Europe ... and a high level of protection and improvement of the quality of the environment'. A number of countries in the European Union and beyond have adopted their own instruments and procedures to promote environmental policy integration (Jacob, Volkery and Lenschow 2008; Jordan and Lenschow 2008). Internationally, an Environmental Management Group was established in the United Nations system to integrate environmental objectives across agencies and programmes, and the principle informed the discussion on upgrading the United Nations Environment Programme to a world environment organization (Biermann, Davies and van der Grijp 2009). Research shows that, out of 50 non-environment intergovernmental organizations, by 2017 around half had incorporated the principle of environmental policy integration in their basic legal documents (Tosun and Peters 2018).

The *effect* of measures and procedures for environmental policy integration, however, has proven to be difficult to assess given a lack of robust comparative research. The literature suggests that while such measures can institutionalize concern for environmental objectives, priority-setting about other societal objectives shifts over time (Jordan, Schout and Unfried 2008; Dunlop et al. 2012). In a review of 30 OECD countries, Jacob and colleagues (2008) found an emphasis on soft and symbolic action for environmental policy integration, rather than operational structures that significantly alter administrative routines and power distribution among interests. In addition, studies emphasize the complexity of pursuing environmental policy integration in multilevel governance, when priorities, resources and capacity for promoting environmental objectives differ across levels (Goria, Sgobbi and Von Homeyer 2010; Söderberg 2011; Storbjörk and Isaksson 2014).

Furthermore, over the last 10–15 years alternative concepts have come up that relate to environmental policy integration in some way, both in public administration more generally (Tosun and Lang 2017) and in environmental policy more specifically. Such concepts often represent a more sector-specific type of environmental policy integration or have a particular coalition of actors behind them. In the world of development cooperation, 'environmental mainstreaming' has been used widely (Persson 2009). The OECD Development Assistance Committee launched in 2016 its Policy Coherence for Sustainable Development programme, which seeks to align development, foreign and trade policies of donors and institutions. The concept of climate policy integration has received increasing attention since the 2000s (Mickwitz et al. 2009; Dupont and Oberthür 2012; Adelle and Russel 2013; van Asselt, Rayner and Persson 2015). Moreover, policy discourses around 'green growth', 'green economy', 'blue economy' and 'low-carbon development' have emerged with a view to promote synergies between environmental and economic objectives through win-win narratives. More examples are found in specific communities of practice, such as nature-based solutions in spatial planning, ecosystem-based adaptation in climate adaptation and eco-engineering in industrial policy.

More recently this integrative notion has been embodied in the 17 Sustainable Development Goals that the United Nations General Assembly adopted in 2015. Here, policy integration is seen as essential by not only governments but also civil society and academia (Stakeholder Forum 2016; International Council for Science 2017; Kanie and Biermann 2017). And, given their expressed focus on sustainable development, the goals represent an integration challenge par excellence. Here, it is not just a traditionally marginalized objective, such as the environment, that is to be integrated into the mainstream and sector objectives, but rather a comprehensive agenda that is to be matched with domestic development priorities and at different time-scales (Nilsson and Persson 2017).

Research Findings

How and to what extent has environmental policy integration contributed to environmental protection? Despite the scientific interest in environmental policy integration that we just reported, the literature still does not provide a compelling answer (Runhaar, Driessen and Uittenbroek 2014). Therefore, we now take stock of empirical research that has been published in peer-reviewed journals and that reports on specific cases of environmental policy integration and its achievements in different countries and policy sectors and at different policy levels.

We used the Scopus database as our main source for data collection and broadly follow the approach used by Runhaar and colleagues (2018) in a similar literature review on climate adaptation mainstreaming. We first identified keywords based on scientific concepts, namely environmental policy integration, environmental mainstreaming, environmental policy coherence and environmental sector integration. Without duplicates our search yielded 154 articles and articles in press (as of 18 August 2017) (Table 9.1). In a second step, we filtered this pool of papers by their empirical value based on an abstract search. We then filtered based on a full text search. We found that some papers were not empirical, or lacked relevance or substantial implications concerning environmental policy integration despite

Table 9.1: *Overview of papers on environmental policy integration in Scopus (as of 18 August 2017)*

Key words search	Number of hits	Empirical	Conceptual	Normative	Irrelevant	Unavailable
'Environmental policy integration'	100	54	16	6	15	9
'Environmental mainstreaming'	11	3	2	1	4	1
'Policy coherence' AND 'Environmental'	34	11	3	4	12	4
'Sector integration' AND 'Environmental'	9	2	0	0	7	0
Total	**154**	**70**	**21**	**11**	**38**	**14**

mentioning keywords in the title or abstract (allocated to the category 'Irrelevant'). Eventually, 70 papers qualified for further analysis, out of which 46 focus on European countries, two on other industrialized countries and 15 on developing countries. Six papers studied countries across these categories and one paper did not specify a geographic location.

Subsequently we coded the papers based on a scheme developed by Runhaar and colleagues (2018). We then organized the variables into four categories: (1) *outputs* of environmental policy integration processes, such as explicit environmental objectives in sectoral policies; (2) *outcomes* or *impacts*, that is, the implementation of these objectives into concrete measures or their effects; (3) *instruments* employed to promote the incorporation of environmental objectives in sectoral policies; and (4) *factors* that stimulated or inhibited incorporating environmental objectives in sectoral policies.

To assess the degree to which outputs and outcomes of environmental policy integration effectively contributed to environmental protection, we translated findings from the analyzed papers into the following categories that each represent a certain degree of environmental policy integration (as discussed above): (1) *coordination*, that is, avoiding contradictory sectoral policies or compensating for adverse environmental consequences of sectoral policies; (2) *harmonization*, that is, an attempt to bring environmental objectives on equal terms with sectoral objectives; and (3) *prioritization*, that is, favouring environmental objectives over sectoral objectives. We employ these degrees for analytical purposes only and do not a priori assume that prioritization is always the ideal outcome of practices of environmental policy integration.

In order to characterize cases of environmental policy integration reported in the papers we selected, we used the coding scheme that we lay out in Table 9.2.

Patterns of Policy Integration

Figure 9.1 shows the number of scientific publications on environmental policy integration per year and their geographical focus. The data shows a growing research interest in environmental policy integration as well as a broadening of the geographical focus of practices from merely European studies to an increasing inclusion of developing countries. Interestingly, research on non-European industrialized countries, such as Australia, Canada, Japan and the United States, is comparably scarce. This does not mean, however, that in these countries environmental considerations are not integrated into decision-making by non-environmental ministries. For instance, in Canada this sometimes happens under the heading of sustainable development with no formal recognition of environmental policy integration as a distinct concept. In contrast, in the European Union the concept of environmental policy integration is institutionally fully embedded and thus more solidly established and used (Meadowcroft and Fiorino 2017).

We identified 97 cases of environmental policy integration from the 70 papers, referring to distinct practices (Runhaar et al. 2018). Some cases had a clear

Table 9.2: *Coding scheme*

Variable	Explanation
Policy sector involved	Transport, energy and so forth
Policy level involved	International or national
Environmental objectives at issue	Biodiversity, climate change and so forth
Direction of integration	Horizontal: integrating environmental protection across policy sectors (e.g., a comprehensive cross-sectoral strategy for policy integration); vertical: interactions 'up and down' between government tiers centred around environmental policy integration.
Strategies for promoting environmental policy integration (adapted from Runhaar 2016)	Supportive constitutional and legal provisions; regulatory tools; bottom-up voluntary policy instruments; economic instruments and incentives; communicative or informational tools; organizational tools and resources; managerial and procedural instruments.

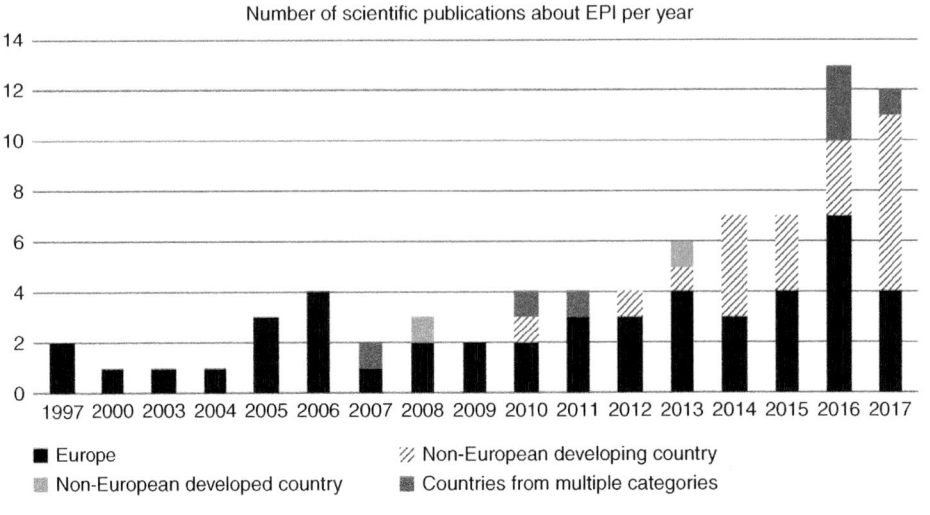

Figure 9.1: Number of scientific publications about environmental policy integration per year (n=70)
Note: whereas for our analysis we used papers that were published until 18 August 2017, this Figure shows all papers published until the end of 2017 to avoid the possible misinterpretation of a substantial drop in papers after 2016.

geographical delineation (European Union-wide, national, regional or local practices), while others represented distinct policies (for example, European Union policies such as the Common Agricultural Policy or the Renewable Energy Directive and its implementation by member states).

That the main share of our sample of empirical scientific studies related to Europe might be partly explained by the explicit incorporation of environmental policy integration as a principle in the European Union. European Union member countries most often reported about are Sweden (11 cases), the Netherlands (8 cases) and the United Kingdom (7 cases) – countries thus that are often considered frontrunners in environmental policy integration (European Environment Agency 2005). However, Norway also appears relatively often in our sample with 11 cases, which may be explained by its ambitious sustainable development policies to combat environmental pressures exerted by its strong oil and energy industry (Lafferty, Knudsen and Larsen 2007). The same applies to Sweden, which experienced an exceptional bioenergy expansion before joining the European Union (Söderberg 2011, 2014).

Many cases (37) relate to the energy sector, followed by transport and telecommunications (31) and agriculture and food (31), as shown in Figure 9.2. This strong representation of these sectors can be explained by the main environmental hazards in industrialized countries, which are chemical and physical in nature: air pollution from industry and transportation, water pollution, chemicals in consumer products and the use of pesticides (Negev 2016). Furthermore, due to their high pollution levels and urgency of taking action towards environmental protection, industry, energy, transport and agriculture have been declared target sectors for environmental policy integration by the European Commission (Dyrhauge 2014).

In terms of what environmental objectives or concerns are integrated most often, depletion, contamination or degradation of natural resources (48 cases), together with ecosystems, biodiversity and nature conservation (47 cases), rank highest, followed by climate mitigation, including renewable energy (37 cases), energy efficiency (33 cases) and reduction of greenhouse gas emissions (20 cases). Because nature conservation and air and water pollution are traditional environmental topics that have been on the policy agenda for much longer than climate mitigation, these results are not surprising. This also explains why climate adaptation is the least represented issue since political awareness around this topic came even later than climate mitigation.

The specific environmental themes addressed in each of the sectoral policies from Figure 9.2 differ. In the energy sector the emphasis is on reducing greenhouse gases. In the transport sector, depletion or contamination of natural resources (probably due to air pollution) scores on equal terms with renewable energy and energy efficiency (17 cases), closely followed by ecosystems (15 cases). The agriculture and food sector more closely resembles the picture presented in other sectors where the integration of nature conservation and the depletion or contamination of natural resources dominate, while the attention for climate mitigation is lower.

Looking at the policy level at which practices of environmental policy integration occur, the papers that we analyzed suggest that the national level is by far the most represented with 57 cases, as shown in Figure 9.3. In terms of the actors involved in processes of environmental policy integration, the level at which environmental policy integration is pursued seems to matter as well. Governments seem to have a major role in driving and facilitating environmental policy integration, for instance by administering supportive constitutional provisions and legal enforcements, although the government's role seems to be less prominent at the local and regional levels than at the international and national levels (Figure 9.3).

Disjuncture between Policies and Practices

In 97 per cent of all cases, efforts for environmental policy integration had led to policy outputs, whereas in only 21 cases did such efforts translate into a policy outcome. This suggests a gap between integrated sectoral policies and their

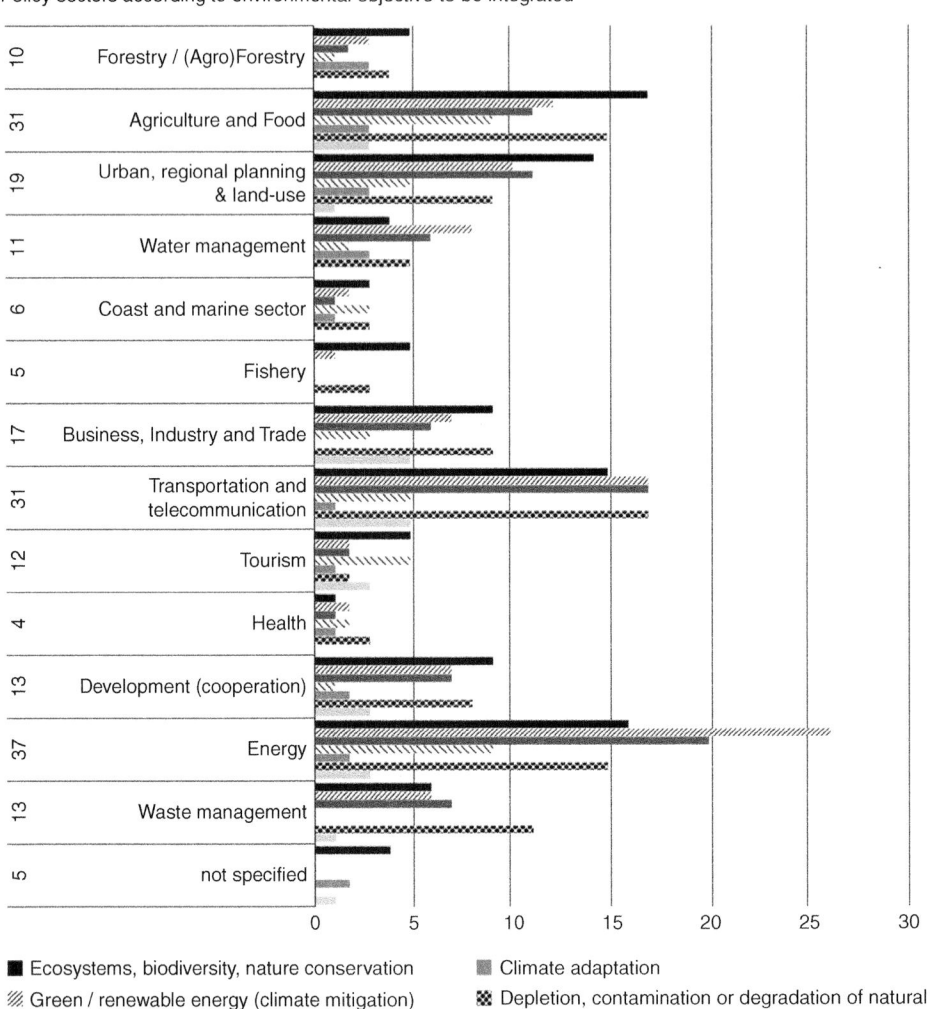

Policy sectors according to environmental objective to be integrated

Figure 9.2: Policy sectors subject to integrating environmental objectives (n=97) Note: the number of cases (n=97) is exceeded, because a case may involve environmental policy integration in multiple policy sectors.

implementation 'downstream' (Runhaar et al. 2018). In some reviewed papers, the implementation gap is explained by a disjuncture between discourses in high-level policy documents and sceptical positions and practices at the operational local level (e.g., Hertin and Berkhout 2003; Nilsson, Eklund and Tyskeng 2009; Huttunen 2015). Several other case studies mention the lack of operationalization of goals and policies into concrete measures for implementation as a cause for the

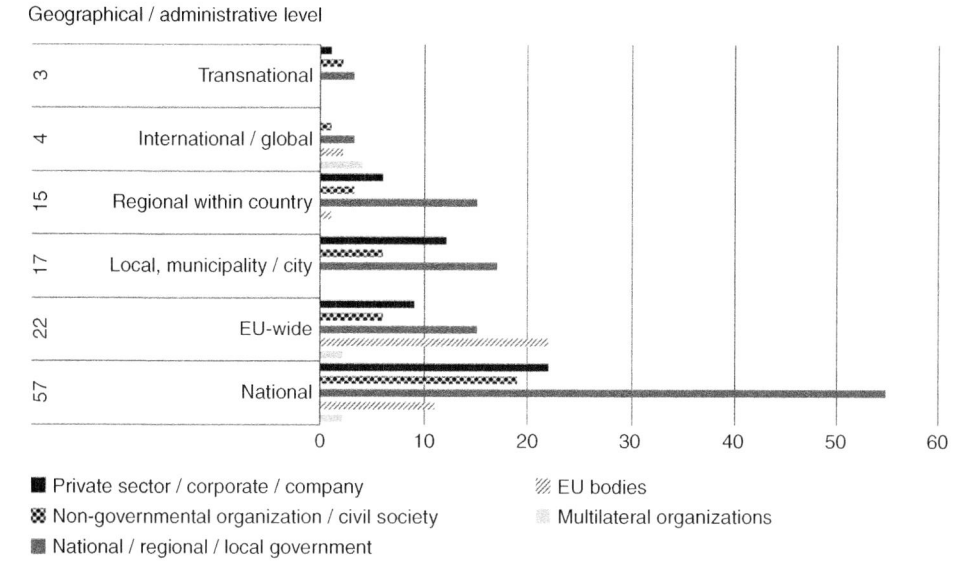

Figure 9.3: Levels at which environmental policy integration is studied and actors involved (n=97)

implementation gap. According to Storbjörk and colleagues (2009: 11), 'those who are aware of the benefits of [sustainable development] do not necessarily know how to make the ideals of "win-win" concrete and operational, thus making the move from words to action difficult'. In addition, 'no readily available definitions currently exist of what constitutes an environmentally positive regional development project. Often it is left to businesses themselves to define their own green profile by putting a mark in a square in their application form that says the project 'benefits environmental sustainability' (Storbjörk, Lähteenmäki-Smith and Hilding-Rydevik 2009: 13). Simeonova and van der Valk (2010) come to a similar conclusion when investigating integrated spatial planning in the Netherlands. Despite the government's efforts to embed environmental concerns in all sectoral policies at the national, regional and local level, this approach has not yet manifested in an operational, integrated policy. Revell (2005: 355) observes the same issue in the transport policy in the United Kingdom where 'the integration of economic and environmental goals has been progressively watered down from policy formulation to implementation, making it increasingly unclear as to what environmental management means in operational terms'.

Looking at the degree of policy integration that can be distilled from the cases, we focus on the 94 cases in which outputs were reported. Outcomes of environmental policy integration were reported in only 21 cases, out of which only 7 cases indicated the degree of environmental policy integration that was achieved.

This does not point to a failure of environmental policy integration as such. We simply could not derive this information from all studies we included in our analysis, given that in 43 cases no outcomes were reported whereas in the other 31 cases policy outcomes did not seem to be part of the analysis. We conclude that despite a large number of policy outputs achieved, 40 per cent of these cases report that not even the lowest degree of integration was achieved, namely policy coordination. In 36 per cent of all cases the degree of policy coordination was achieved, in 12 per cent harmonization was realized and in only 1 per cent were environmental objectives prioritized over sectoral objectives. In 10 per cent of the 94 cases the degree of integration in policy outputs was not mentioned. These findings suggest that environmental policy integration is only partly effective in integrating environmental objectives into sectoral policies. Difficulties in integrating environmental objectives and in overcoming sectoral objectives that conflict with environmental objectives are reported with regard to the European Union's Water Framework Directive, Common Agriculture Policy and Structural Funds and between national and local agri-environmental, waste and water policies (for example, Schout and Jordan 2005; Watson, Bulkeley and Hudson 2008; Rosendal 2012; Huttunen 2015; Regina et al. 2015; Abazaj, Moen and Ruud 2016). Policy incoherence can take various forms, such as contradicting texts within a policy document, the absence of discussion of potential conflicts of different objectives in strategies and plans or non-aligned decisions resulting in inconsistent governance frameworks which are reported to pose prioritization problems on the ground (for example, Kivimaa and Mickwitz 2006; Lafferty, Knudsen and Larsen 2007; Nilsson, Eklund and Tyskeng 2009; Bizikova, Metternicht and Yarde 2018).

As shown in Table 9.3, in most cases with outputs (41 cases) a combination of horizontal and vertical integration efforts or integration processes involving non-state actors were reported. This approach seems most successful for the translation of outputs into outcomes (in 13 of 41 cases) and effective environmental protection (in half of all cases an effect was noted, ranging from coordination up to harmonization). In another 36 cases with outputs, vertical integration processes were reported; yet this strategy seems not very effective for implementation (in 7 out of 36 cases policy outputs were translated into outcomes) or for having environmental effects on sectoral policies and plans (in 16 of the 36 cases). Horizontal integration processes are most often associated with effective environmental protection (8 out of the 13 cases with outcomes) but also face an implementation gap (only 1 out of 13 cases). It is unclear whether the *direction* of the integration efforts (that is, horizontal or vertical) or the *intensity* of efforts (that is, their combination) is important for promoting environmental protection

Table 9.3: *Horizontal and vertical environmental policy integration processes versus number of achievements in terms of environmental protection (n=94)*

Policy integration processes	Outputs (Outcomes)	Effects (Not specified)	Coordination	Harmonization	Prioritization
Horizontal integration	13 (1)	4 (1)	5	3	0
Vertical integration	36 (7)	15 (5)	13	3	0
Horizontal and vertical integration and/or non-state actor involvement	41 (13)	17 (3)	14	6	1
Not specified or unclear	4 (0)	2 (0)	2	0	0
Total	**94 (21)**	**38 (9)**	**34**	**12**	**1**

through environmental policy integration. The latter seems most plausible but this needs further research.

The Effects of Strategies for Policy Integration

The majority of the cases reported a combination of different instruments employed to achieve environmental policy integration. With top-down regulatory instruments ranking highest and supportive constitutional and legal provisions ranking third, top-down instruments are dominant in the reviewed cases (Figure 9.4). Voluntary bottom-up instruments are the least reported strategies. They include voluntary incentive schemes for the implementation of waste-to-energy plants in the private sector in France (McCauley 2015) or self-directed, bottom-up initiatives by farmers across Europe to integrate environmental and agricultural practices (Buizer, Arts and Westerink 2016). However, this picture is not that clear-cut, given that the remainder of the instruments can contain both top-down as well as bottom-up elements (Runhaar 2016). For instance, economic tools comprise top-down financial compensation for abolishing slash-and-burn practices of small-holders (Park and Youn 2017) and agricultural subsidies to foster ecological farming in the European Union Common Agricultural Policy (Rosendal 2012; Huttunen 2015; Regina et al. 2015), but also voluntary green budgeting (Geeraert 2016) or tradable emissions permits and certificates (Nilsson 2005).

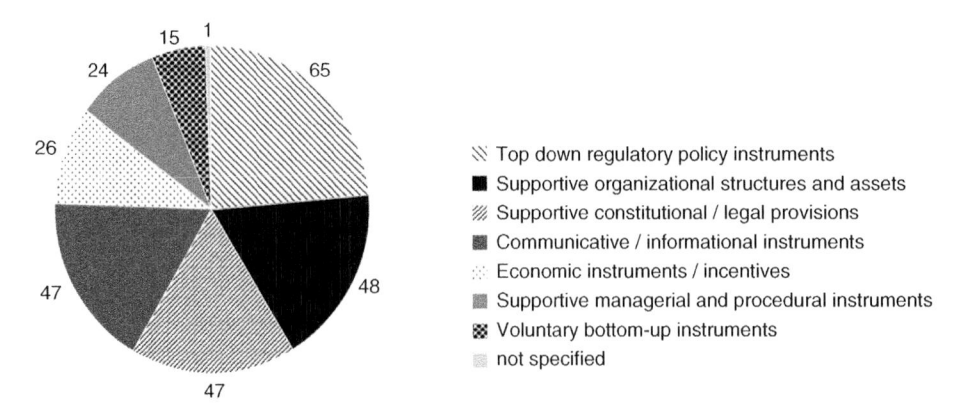

Figure 9.4: Frequency of environmental policy integration instruments reported in the 97 cases

We observed communicative and informational strategies in 17 per cent of all cases. They include capacity-building workshops (Velázquez Gomar 2014), higher level directions regarding classification criteria for environmental quality standards (Söderberg 2016), technical support and guidance in environmental management, such as reviewing environmental impact assessments and monitoring compliance with regulations (Nunan, Campbell and Foster 2012) or multi-stakeholder platforms for managing conflicts, and achieving coherence in sectoral and structural policies and accompanying guidance documents (Abazaj, Moen and Ruud 2016). The presence of these communicative and informational strategies next to top-down strategies is in line with reported tendencies to increasingly supplement top-down legislation with network structures and bottom-up policy instruments to more effectively contribute to incorporating environmental objectives into sectoral policies (Schout and Jordan 2005; Persson, Eckerberg and Nilsson 2016).

It is striking that managerial and procedural tools lag behind, which suggests that the uptake of procedures that facilitate integration (such as reporting, monitoring, ex-post or ex-ante impact assessments for evaluation or checklists) poses challenges. A possible explanation is that the tools provided (for example, by higher policy levels, such as the national government, the European Union or the United Nations) are not instructive or practical enough to translate into daily practices, routines or workflows. However, previous studies suggest these tools are important for the successful implementation of environmental policy integration (Kivimaa and Mickwitz 2006; Runhaar, Driessen and Uittenbroek 2014; Mullally and Dunphy 2015; Persson et al. 2018). Entrenched organizational routines may also inhibit the uptake of new procedures or processes (Uittenbroek et al. 2014). Such explanations resonate with the previously mentioned mismatch between (higher

Table 9.4: *Number of achievements of environmental policy integration strategies in terms of environmental protection (n=94)*

Strategy	Outputs (Outcomes)	No effect (Not specified)	Coordination	Harmonization	Prioritization
Supportive constitutional/legal provisions (47)	46 (8)	19 (4)	18	5	0
Top-down regulatory policy instruments (65)	64 (10)	26 (6)	26	5	1
Voluntary bottom-up instruments (15)	14 (5)	5 (1)	4	4	0
Economic instruments and incentives (26)	25 (6)	13 (1)	9	2	1
Communicative and informational instruments (47)	47 (11)	20 (3)	16	7	1
Supportive organizational structures and assets (48)	47 (14)	14 (3)	22	7	1
Supportive managerial and procedural instruments (24)	24 (10)	7 (3)	8	6	1

level) policy objectives and local practices and the lack of operationalization and concretization of policy measures inhibiting policy implementation.

Table 9.4 indicates the relative effectiveness of the various environmental policy integration strategies. The figures need to be interpreted with some caution. First, the majority of the cases reported a combination of different strategies, which makes an isolation of one particular success category impossible. Second, the relatively low number of strategies that were identified and that could be associated with effects on environmental integration also means we have to interpret the figures as indicative only.

Table 9.4 also suggests that supportive organizational structures and assets and supportive managerial and procedural instruments are the most effective strategies for realizing environmental integration. Interestingly, while the latter are among the least represented in the all cases that we reviewed, they also yielded the highest outcome rate. This implies their relative importance as an ingredient to environmental policy integration that permeates downstream local practices. In contrast, overall we found top-down instruments not to be very effective, neither in terms of translating outputs into outcomes nor in having a strong effect on sectoral policies (except for regulatory instruments). This suggests that they are not a prerequisite for effective practices of environmental policy integration, which contradicts a recent evaluation of strategies for climate adaptation mainstreaming, a specific form of environmental policy integration (Runhaar et al. 2018). We hypothesize that the effectiveness of strategies depends at least in part on the specific environmental objective to be integrated.

Enabling Factors and Barriers

Among the 21 cases in which policy outputs had been translated into outcomes, we found that in 13 cases the main enabling factors include cooperation with private actors, in 11 cases political commitment, in 10 cases the framing of the environmental problem at issue and linking to sectoral objectives and in 9 cases learning. If we only look at cases in which policy *outputs* were achieved with an effect ranging from coordination up to harmonization, we find a similar pattern. The three highest ranking enablers resonate with previous findings for climate adaptation mainstreaming and climate policy integration (Runhaar et al. 2018), while the results differ with regard to the importance of policy entrepreneurs, focusing events and subsidies from higher levels of government. This can be explained by climate adaptation being relatively new on the political agenda and therefore greatly relying on the mechanisms of agenda-setting and external funding. The most often mentioned enabler, cooperation with private actors, illustrates the importance of mixed forms of horizontal and vertical integration with non-state actor involvement for successful environmental policy integration, which had not only the highest representation among the reviewed cases but also the highest success rate in generating outcomes. The apparent importance of political commitment may correspond with our findings of the dominance of top-down instruments. Yet, formal requirements and supportive legal provisions seem to be more salient for higher-level-integration outputs than for creating outcomes, suggesting their relative weight in the early stages of environmental policy integration. The framing and successful linking to sectoral objectives, as indicated in almost half of the cases, is closely intertwined with solving conflicts of interest and thus, achieving higher levels of integration such as policy coordination or harmonization.

Looking at barriers and enablers among cases in which outputs did not translate into outcomes (n=74) is helpful for understanding the implementation gap in policy integration. Conflicting interests (also because of the specific framing of the environmental objectives at issue; 40 cases), organizational structures, routines and practices (44 cases) and a lack of access to knowledge and guidance (38 cases) appear to be important barriers.

The importance of barriers that we found in organizational structures, routines and practices corresponds with the low presence of managerial and procedural strategies for environmental policy integration, which seem very effective for overcoming implementation deficits. It also is in line with a reported disjuncture between policy instruments and objectives and local practices, decision-making procedures and outcomes (Hertin and Berkhout 2003; Nilsson, Eklund and Tyskeng 2009; Huttunen 2015). This disjuncture is clearest in the absence of long-term environmental targets and requirements for reporting about the implementation of sectoral action plans (Lafferty, Knudsen and Larsen 2007; Giordano 2014); a high level of institutionalization of inadequate organizational structures or rigid procedures leaving no room for environmental concerns; ministerial resistance to adopt environmental policy integration into procedures (Jordan and Lenschow 2000; Rosendal 2012; Simeonova and van der Valk 2016); as well as a mismatch between administrative boundaries and those required for efficient environmental or ecosystem management (Ansong, Gissi and Calado 2017).

Conflicting interests are reported in sectoral policies for agriculture, forestry and water management (Revell 2005; Nilsson, Eklund and Tyskeng 2009; Rosendal 2012; Fertel et al. 2013; Kalaba, Quinn and Dougill 2014; Huttunen 2015; Regina et al. 2015; Selianko and Lenschow 2015; Söderberg 2016; Bizikova, Metternicht and Yarde 2018). In trade-offs between environmental and economic objectives, the latter are often prioritized (Ruddy and Hilty 2008; van Stigt, Driessen and Spit 2013; Dyrhauge 2014; Abazaj, Moen and Ruud 2016; Brendehaug, Aall and Dodds 2016; Geeraert 2016). This occurs across all sectors and scales, such as in urban development where Simeonova and van der Valk (2016) observe the undermining of nature policy priorities due to clientelism-oriented municipal practices satisfying the economic needs of landowners, or in the European Commission where sectoral directorates-general keep focusing on their own objectives instead of integrating environmental concerns despite legal obligations to do so (Schout and Jordan 2005).

Another frequently mentioned barrier is the lack of access to, and availability of, knowledge and guidance. Notably, knowledge about the environmental impacts of sectoral policies and developments seems to be lacking (Kivimaa and Mickwitz 2006). Several case studies mention issues with systematic, consistent monitoring, reporting and environmental impact assessment instruments that hamper the development of informed, integrated policies and practices (Lenschow 1997; Scobie 2016; Simeonova and van der Valk 2016; Alons 2017). The call for improved top-down guidance through the European Commission or the climate convention and managerial tools requires improvements of communicative and informational tools, as we have shown above, but also addresses the need for intensifying cross-sectoral collaboration to better disseminate knowledge and best practices.

Conclusions and Future Directions

We have provided here an overview of how environmental policy integration has been conceptualized in scientific research and in policy practice, and we explored its origins and diffusion over time. We then provided a meta-analysis of research on environmental policy integration to assess how environmental policy integration has contributed to environmental protection.

One key finding is that many scientific publications on environmental policy integration report on practices in Europe. Given the institutional embeddedness of the concept as we discussed at the beginning of this paper, this is no surprise. It does not mean that integrating environmental objectives into non-environmental policies does not occur elsewhere. It simply might happen under different labels, such as environmental mainstreaming, green growth and so forth.

Another important finding is the discrepancy between the adoption of environmental policy integration in terms of objectives and commitments and its actual implementation, that is, translation into concrete measures. Overall, we found relatively few cases in which environmental objectives were given a substantial status in non-environmental policies (that is, up to the levels of harmonization and prioritization).

These might raise questions about the effectiveness of pursuing environmental policy integration, also in comparison to traditional environmental policy. We have two remarks here.

First, the low number of cases in which harmonization or prioritization was found does not mean that environmental policy integration is ineffective. That is rather a normative issue: environmental policy integration implies considering environmental objectives versus sectoral objectives and does not imply that environmental objectives should be prioritized always. Yet this also does not mean that the implementation gap – and our observation that in a substantial number of cases no integration at all was found – is not problematic.

Second, one could ask whether the environment is better protected by relying on a specialized environmental administration or by distributing responsibilities more widely across government and trying to integrate environment into mainstream decision-making. In our view, however, posing that question in such general terms is probably a mistake. Both modes of environmental policy delivery appear essential. Modern environmental policy is inconceivable without specialized institutions (agencies, personnel, budgets and knowledge development) dedicated to environmental issues. Yet environmental impacts can never be managed successfully unless they are also integrated into the design of sector-based policies. Climate mitigation is perhaps the clearest example here: specific institutions are required, such as the United Kingdom Climate Committee, and yet an effective approach is unthinkable without integration of climate mitigation into energy policy, agricultural policy, industrial policy, urban design and so on.

The barriers we identified suggest that the *actual detailed design* or *architecture* of the particular institutions and processes, whether specialized or integrative, matters. A specialized environmental administration can be well connected to centres of governmental power or be isolated in a silo, just as policy integration can provide meaningful engagement or amount to little more than a rhetorical flourish. We suspect that it is not so much specialized versus integrative approaches that are at issue, but how each is applied in practice – hence the significance for future research on environmental policy integration to determine from experience the more or less successful approaches.

In general, more comparative research and evaluations are needed to answer the question, 'what works, where and why?' and to see what opportunities are and how barriers might be overcome. More specifically, we see five future research directions, recommending not to restrict the analysis to the concept of environmental policy integration but also to include similar concepts such as environmental mainstreaming.

(1) First, more systematic research is needed to identify the strategies by which environmental policy integration is promoted as well as their specific design (Runhaar, Driessen and Uittenbroek 2014). These strategies should not only include public policies and interactive modes of governance, but also non-state forms of governance that promote the structural integration of environmental protection into sectoral policies (that is, eco-labels, community-based initiatives, philanthropic funding and so forth).

(2) Second, more research is needed into the processes of institutionalization of environmental policy integration both internationally and nationally. For instance, what processes and institutions are developed to strengthen the role of environmental departments and agencies as against other departments and to what effects?

(3) Third, comparative research is needed to increase our understanding of how strategies for environmental policy integration are employed in distinct policy sectors, at different policy levels and in different institutional contexts. As was shown in the previous sections, a majority of the empirical studies are related to Europe. Systematic investigations in countries outside Europe are rare but growing. Also, investigations into the environmental policy integration practices in other than the energy sector, transport sector and agricultural sector are needed. Moreover, our review shows that most papers analyze national strategies for environmental policy integration, leaving the regional and local but also international levels underexplored.

(4) Fourth, we found limited evidence of the achievements of strategies for environmental policy integration in practice in terms of enhancing environmental

protection and environmental quality. The incorporation of environmental concerns in non-environmental policy sectors has been analyzed mostly on output and outcome levels but not in terms of implementation and impact. We should increase our understanding not only of what degrees of integration have been achieved in practice, but also of whether this integration leads to environmental protection. In this respect, it is also interesting to investigate how strategies for environmental policy integration will contribute to societal transformations that are needed to achieve the Sustainable Development Goals and targets from the 2015 Paris Agreement under the climate convention.

(5) Fifth, a last question concerns the transferability of successful practices of environmental policy integration. To what extent can one transfer promising and successful strategies for environmental policy integration to other countries, levels and sectors, and under which conditions? This not only requires a better view on enabling and constraining factors for successful environmental policy integration, but also more research into translation, transformation and policy learning.

References

Abazaj, J., Moen, Ø., & Ruud, A. (2016). Striking the balance between renewable energy generation and water status protection: hydropower in the context of the European Renewable Energy Directive and Water Framework Directive. *Environmental Policy and Governance*, 26 (5), 409–21.

Adelle, C., & Russel, D. (2013). Climate policy integration: A case of déjà vu? *Environmental Policy and Governance*, 23 (1), 1–12.

Alons, G. (2017), Environmental policy integration in the EU's common agricultural policy: Greening or greenwashing? *Journal of European Public Policy*, 24 (11), 1–19.

Ansong, J., Gissi, E., & Calado, H. (2017). An approach to ecosystem-based management in maritime spatial planning process. *Ocean and Coastal Management*, 141, 65–81.

Biermann, F., Davies, O., & van der Grijp, N. (2009). Environmental policy integration and the architecture of global environmental governance. *International Environmental Agreements*, 9 (4), 351–69.

Bizikova, L., Metternicht, G., & Yarde, T. (2018). Environmental mainstreaming and policy coherence: Essential policy tools to link international agreements with national development. A case study of the Caribbean region. *Environment, Development and Sustainability*, 20 (3), 975–95.

Brendehaug, E., Aall, C., & Dodds, R. (2016). Environmental policy integration as a strategy for sustainable tourism planning: Issues in implementation. *Journal of Sustainable Tourism*, 25 (9), 1257–74.

Buizer, M., Arts, B., & Westerink, J. (2016). Landscape governance as policy integration 'from below': A case of displaced and contained political conflict in the Netherlands. *Environment and Planning C: Government and Policy*, 34 (3), 448–62.

Dyrhauge, H. (2014). The road to environmental policy integration is paved with obstacles: Intra- and inter-organizational conflicts in EU transport decision-making. *Journal of Common Market Studies*, 52 (5), 985–1001.

Dunlop, C. A., Maggetti, M., Radaelli, C. M., & Russel, D. (2012). The many uses of regulatory impact assessment: A meta-analysis of EU and UK cases. *Regulation and Governance*, 6, 23–45.

Dupont, C., & Oberthür, S. (2012). Insufficient Climate Policy Integration in EU energy policy: The importance of the long-term perspective. *Journal of Contemporary European Research*, 8 (2), 228–47.

European Commission (2016). *Environmental integration*. Available at: http://ec.europa.eu/environment/integration/integration.htm. Accessed: 17 June 2019.

European Environment Agency (2005). *Environmental policy integration in Europe: Administrative culture and practices*. Available at: www.eea.europa.eu/highlights/Ann1120649962. Accessed: 17 June 2019.

Fertel, C., Bahn, O., Vaillancourt, K., & Waaub, J.-P. (2013). Canadian energy and climate policies: A SWOT analysis in search of federal/provincial coherence. *Energy Policy*, 63, 1139–50.

Geeraert, A. (2016). It's not that easy being green: The environmental dimension of the European Union's sports policy. *Journal of Sport and Social Issues*, 40 (1), 62–81.

Giordano, T. (2014). Multi-level integrated planning and greening of public infrastructure in South Africa. *Planning Theory and Practice*, 15 (4), 480–504.

Goria, A., Sgobbi, A., & Von Homeyer, I. (2010) (eds.). *Governance for the environment: A comparative analysis of environmental policy integration*. Cheltenham: Edward Elgar.

Hertin, J., & Berkhout, F. (2003). Analysing institutional strategies for environmental policy integration: The case of EU enterprise policy. *Journal of Environmental Policy and Planning*, 5 (1), 39–56.

Hogl, K., Kvarda, E., Nordbeck, R., & Pregernig, M. (2012). Effectiveness and legitimacy of environmental governance synopsis of key insights. In K. Hogl, E. Kvarda, R. Nordbeck, & M. Pregernig (eds.), *Environmental governance: The challenge of legitimacy and effectiveness* (pp. 280–304). Cheltenham: Edward Elgar.

Huttunen, S. (2015). Farming practices and experienced policy coherence in agri-environmental policies: The case of land clearing in Finland. *Journal of Environmental Policy and Planning*, 17 (6), 573–92.

International Council for Science (2017). *A guide to SDG interactions: from science to implementation*. Paris: International Council for Science.

Jacob, K., Volkery, A., & Lenschow, A. (2008). Instruments for environmental policy integration in 30 OECD countries. In A. Jordan & A. Lenschow (eds.), *Innovation in environmental policy? Integrating the environment for sustainability* (pp. 24–45). Cheltenham: Edwar Elgar.

Jordan, A. J., & Lenschow, A. (2008). *Innovation in environmental policy? Integrating the environment for sustainability*. Cheltenham: Edward Elgar.

Jordan, A., & Lenschow, A. (2000), 'Greening' the European Union: What can be learned from the 'leaders' of EU environmental policy? *European Environment*, 10 (3), 109–210.

Jordan, A., Schout, A., & Unfried, M. (2008). The European Union. In A. Jordan, & A. Lenschow (eds.), *Innovation in environmental policy? Integrating the environment for sustainability* (pp. 159–79). Cheltenham: Edward Elgar.

Jordan, A., & Lenschow, A. (2010). Environmental policy integration: a state of the art review. *Environmental Policy and Governance*, 20, 147–58.

Kalaba, F. K., Quinn, C. H., & Dougill, A. J. (2014). Policy coherence and interplay between Zambia's forest, energy, agricultural and climate change policies and multilateral environmental agreements. *International Environmental Agreements*, 14 (2), 181–98.

Kanie, N., & Biermann, F. (2017) (eds.). *Governing through goals: Sustainable Development Goals as governance innovation*. Cambridge, MA: The MIT Press.

Kivimaa, P., & Mickwitz, P. (2006). The challenge of greening technologies: Environmental policy integration in Finnish technology policies. *Research Policy*, 35 (5), 729–44.

Knudsen, J. K., & Lafferty, W. M. (2016). Environmental policy integration: The importance of balance and trade-offs. In D. Fisher (ed.), *Research handbook on fundamental concepts in environmental law* (pp. 337–68). Cheltenham: Edward Elgar.

Lafferty, W. M., & Hovden, E. (2003). Environmental policy integration: Towards an analytical framework. *Environmental Politics*, 12 (3), 1–22.

Lafferty, W. M. (ed.) (2004). *Governance for sustainable development: The challenge of adapting form to function*. Cheltenham: Edward Elgar.

Lafferty, W. M., Knudsen, J., & Larsen, O. M. (2007). Pursuing sustainable development in Norway: The challenge of living up to Brundtland at home. *European Environment*, 17 (3), 177–88.

Lenschow, A. (1997). Variation in EC environmental policy integration: Agency push within complex institutional structures. *Journal of European Public Policy*, 4 (1), 109–27.

Lenschow, A. (2002). Greening the European Union: An introduction. In A. Lenschow (ed.), *Environmental policy integration: Greening sectoral policies in Europe* (pp. 3–21). London: Earthscan.

McCauley, D. (2015). Sustainable development in energy policy: A governance assessment of environmental stakeholder inclusion in waste-to-energy. *Sustainable Development*, 23 (5), 273–84.

Meadowcroft J., Langhelle, O., & Ruud, A. (eds.) (2012). *Governance, democracy and sustainable development: Moving beyond the impasse*. Cheltenham: Edward Elgar.

Meadowcroft, J., & Fiorino, D. J. (eds.) (2017). *Conceptual innovation in environmental policy*. Cambridge, MA: The MIT Press.

Mickwitz, P., Aix, F., Beck, S. et al. (2009). *Climate policy integration, coherence and governance*. Helsinki: Partnership for European Environmental Research.

Mullally, G., & Dunphy, N. P. (2015). *State of play review of environmental policy integration literature*. Dublin: National Economic and Social Council.

Negev, M. (2016). Interagency aspects of environmental policy: The case of environmental health. *Environmental Policy and Governance*, 26 (3), 205–19.

Nilsson, M. (2005). Learning, frames and environmental policy integration: The case of Swedish energy policy. *Environment and Planning C: Government and Policy*, 23 (2), 207–26.

Nilsson, M. Eklund, M., & Tyskeng, S. (2009). Environmental integration and policy implementation: Competing governance modes in waste management decision making. *Environment and Planning C: Government and Policy*, 27 (1), 1–18.

Nilsson, M., & Persson, Å. (2017). Policy note: Lessons from environmental policy integration for the implementation of the 2030 Agenda. *Environmental Science and Policy*, 78, 36–39.

Nunan, F., Campbell, A., & Foster, E. (2012). Environmental mainstreaming: The organisational challenges of policy integration. *Public Administration and Development*, 32 (3), 262–77.

Oberthür, S. (2009). Interplay management: Enhancing environmental policy integration among international institutions. *International Environmental Agreements*, 9 (4), 371–91.

Ortolano, L., & Shepherd, A. (1995). Environmental impact assessment: Challenges and opportunities. *Impact Assessment*, 13 (1), 3–30.

Park, M. S., & Youn, Y.-C. (2017). Reforestation policy integration by the multiple sectors toward forest transition in the Republic of Korea. *Forest Policy and Economics*, 76, 45–55.

Persson, Å., Runhaar, H., Karlsson-Vinkhuyzen, S., Mullally, G., Russel, D., & Widmer, A. (2018). Editorial: Environmental policy integration: Taking stock of policy practice in different contexts. *Environmental Science and Policy*, 85, 113–15.

Persson, Å. (2004). *Environmental policy integration: An introduction*. Stockholm: Stockholm Environment Institute.

Persson, Å., Eckerberg, K., & Nilsson, M. (2016). Institutionalization or wither away? Twenty-five years of environmental policy integration under shifting governance models in Sweden. *Environment and Planning C: Government and Policy*, 34 (3), 478–95.

Persson, Å. (2009). Environmental policy integration and bilateral development assistance: Challenges and opportunities with an evolving governance framework. *International Environmental Agreements*, 9 (4), 409–29.

Regina K., Budiman, A., Greve, M. H. et al. (2015). GHG mitigation of agricultural peatlands requires coherent policies. *Climate Policy*, 16 (4), 522–41.

Revell, A. (2005). Ecological modernization in the UK: Rhetoric or reality? *European Environment*, 15 (6), 344–61.

Rosendal, G. K. (2012). Adjusting Norwegian agricultural policy to the WTO through multifunctionality: Utilizing the environmental potential? *Journal of Environmental Policy and Planning*, 14 (2), 209–27.

Ruddy, T. F., & Hilty, L. M. (2008). Impact assessment and policy learning in the European Commission. *Environmental Impact Assessment*, 28, 90–105.

Runhaar, H. (2016). Tools for integrating environmental objectives into policy and practice: What works where? *Environmental Impact Assessment Review*, 59, 1–9.

Runhaar, H., Wilk, B., Persson, Å., Uittenbroek, C., & Wamsler, C. (2018). Mainstreaming climate adaptation: Taking stock about 'what works' from empirical research worldwide. *Regional Environmental Change*, 18 (4), 1201–10.

Runhaar, H., Driessen, P., & Uittenbroek, C. (2014). Towards a systematic framework for the analysis of environmental policy integration. *Environmental Policy and Governance*, 24 (4), 233–46.

Runhaar, H., Driessen, P., & Soer, L. (2009). Sustainable urban development and the challenge of policy integration. An assessment of planning tools for integrating spatial and environmental planning in the Netherlands. *Environment and Planning B: Urban Analytics and City Science*, 36 (3), 417–31.

Schout, A., & Jordan, A. (2005). Coordinated European governance: Self-organizing or centrally steered? *Public Administration*, 83 (1), 201–20.

Scobie, M. (2016), Policy coherence in climate governance in Caribbean small island developing states. *Environmental Science and Policy*, 58, 16–28.

Selianko I., & Lenschow, A. (2015). Energy policy coherence from an intra-institutional perspective: Energy security and environmental policy coordination within the European Commission. *European Integration Online Papers* 19 (2), 1–29.

Simeonova V., & van der Valk, A. (2010). The role of an area-oriented approach in achieving environmental policy integration in the Netherlands, and its applicability in Bulgaria. *European Planning Studies*, 18 (9), 1411–43.

Simeonova, V., & van der Valk, A. (2016). Environmental policy integration: Towards a communicative approach in integrating nature conservation and urban planning in Bulgaria. *Land Use Policy*, 57, 80–93.

Söderberg, C. (2011). Institutional conditions for multi-sector environmental policy integration in Swedish bioenergy policy. *Environmental Politics*, 20 (4), 528–46.

Söderberg, C. (2014). What drives sub-national bioenergy development? Exploring cross-level implications of environmental policy integration in EU and Swedish bioenergy policy. *European Journal of Government and Economics*, 3 (2), 119–37.

Söderberg, C. (2016). Complex governance structures and incoherent policies: Implementing the EU water framework directive in Sweden. *Journal of Environmental Management*, 183, 90–97.

Stakeholder Forum (2016). *Seeing the whole: Implementing the SDGs in an integrated and coherent way*. London: Stakeholder Forum.

Storbjörk S., Lähteenmäki-Smith, K., & Hilding-Rydevik, T. (2009). Conflict or consensus: The challenge of integrating environmental sustainability into regional development programming. *European Journal of Spatial Development*, 34, 1–22.

Storbjörk, S., & Isaksson, K. (2014). Learning is our Achilles heel: Conditions for long-term environmental policy integration in Swedish regional development programming. *Journal of Environmental Planning and Management*, 57 (7), 1023–42.

Tosun, J., & Lang, A. (2017). Policy integration: Mapping the different concepts. *Policy Studies*, 38 (6), 553–70.

Tosun, J., & Peters, B. G. (2018). Intergovernmental organizations' normative commitments to policy integration: The dominance of environmental goals. *Environmental Science and Policy*, 82, 90–99.

Uittenbroek, C., Janssen-Jansen, L., Spit, T., & Runhaar, H. (2014). Organizational values and the implications for mainstreaming climate adaptation in Dutch municipalities: Using Q methodology. *Journal of Water and Climate Change*, 5 (3), 443–56.

Van Asselt, H., Rayner, T., & Persson, Å. (2015). Climate policy integration. In K. Bäckstrand, & E. Lövbrand (eds.), *Research handbook on climate governance* (pp. 388–99). Cheltenham: Edwar Elgar.

Van Stigt, R., Driessen, P., & Spit, T. (2013). Compact city development and the challenge of environmental policy integration: A multi-level governance perspective. *Environmental Policy and Governance*, 23 (4), 221–33.

Velázquez Gomar, J. O. (2014). Environmental policy integration among multilateral environmental agreements: The case of biodiversity. *International Environmental Agreements*, 16 (4), 525–41.

Watson, M., Bulkeley, H., & Hudson, R. (2008). Unpicking environmental policy integration with tales from waste management. *Environment and Planning C: Government and Policy*, 26 (3), 481–98.

World Commission on Environment and Development (1987). *Our common future*. Oxford: Oxford University Press.

10

Interplay Management

OLAV SCHRAM STOKKE

What is interplay management, and how has it helped to advance earth system governance? These questions are important, because institutional interplay affects domestic and international efforts to guide resource use, consumption and other activities, helping to soften the overall pressure exerted on the environment. Over the past 20 years, scholars have deepened our understanding of how distinct institutional arrangements overlap, complement, or affect each other in various ways. Governance of specific issues often results from interplay involving several public as well as private institutions which typically differ in their objectives and institutional capacities. Interplay management concerns the *awareness* of such interplay, and preparedness to *interfere* with it if necessary, in order to reap synergies or avoid disruptions – at least, those disruptions that do not derive from incompatible objectives.

The institutions targeted for interplay management are separated by one or more of their institutional boundaries – functional, spatial or actor type. Interplay management across functional boundaries is evident, for instance, in efforts to reconcile free-trade rules and trade-restrictive compliance measures (Gehring 2011), fisheries management and preservation of coral reefs (Kvalvik 2012), or forestry-oriented climate measures and biodiversity protection (Ochieng, Visseren-Hamakers and Nketiah 2013; Betsill et al. 2015). Interplay management across spatial boundaries is seen when national environmental measures are harmonized with those taken by a neighbouring state, perhaps by means of a bilateral or regional agreement, or when an advanced regional regime promotes the strengthening of a global institution within the same issue area (Stokke 2001a). Lastly, interplay management across the actor–type boundary is found when international organizations seek to influence the operations of private ones (Abbott and Snidal 2010) – but also when non-governmental organizations are involved in strengthening or implementing international rules.

This chapter first delineates interplay management by distinguishing it from other responses to the complexity of earth system governance. Then I synthesize a string of recent conceptual and empirical findings on the *agency* as well as the *means* of interplay management, notably whether interplay management proceeds by means of coordination or adaptation. The aim of this exercise is to pinpoint conditions that can encourage the search for cross-institutional coherence and to examine how interplay management affects priority-setting among competing policy objectives. The last section summarizes the argument and identifies several research topics regarding interplay management in need of further attention.

Conceptualization

Interplay management is a broad concept that covers any deliberate efforts to improve the interaction of two or more institutions that are distinct in terms of membership and decision-making but that deal with the same issue, usually in a non-hierarchical manner (Stokke 2001b; Oberthür 2009; Oberthür and Stokke 2011; Orsini, Morin and Young 2013; Biermann 2014).

Institutions interact when one institution affects the contents, operations or consequences of another. Whenever the number of interacting institutions exceeds two and their interplay is (or may be) problematic, they are collectively described by terms such as regime or institutional complexes (Chapter 7; on problematic relationships as constitutive of institutional complexes see Orsini, Morin and Young 2013). As intervening in such relationships is costly, interplay managers do so only when they perceive the potential disruption as severe, or the potential synergies as substantial. This section elaborates on the agency and purposes of interplay management, pinpointing some distinctive features of this phenomenon.

Agency

Several major research projects on complexes of international institutions have elaborated on the theme of agency in institutional interaction. In an early taxonomy that also distinguished functional from political linkages, Young (1996) instigated research on institutional complexes in earth system governance by contrasting nested regimes (as defined by Aggarwal 1983) with deliberately clustered ones. Drawing on this approach and linking it to specific pathways of interplay derived from regime-effectiveness research, a project reported by Stokke (2001a) examined efforts by international organizations as well as states and secretariats that operate regional and global fisheries regimes with a view to adapting and advancing certain principles and approaches pioneered in other sectors of environmental governance. Raustiala and Victor (2004) launched the term 'regime complexes', highlighting technology- and industry-driven competition among intergovernmental institutions evident in efforts to create 'strategic inconsistency'. Building on a broad inventory of interactions within 'institutional complexes' compiled by Oberthür and Gehring (2006), Oberthür (2009) differentiated levels of interplay management based on the extent of communication and coordination involved. Further, in drawing implications from a long-term project on the institutional dimensions of global environmental change, Biermann (2008) explored the differential capacities of states, various private organizations and international bureaucracies (also Biermann and Siebenhüner 2009) in influencing the interlocking web of principles, institutions and practices referred to as governance architectures in this volume (Chapter 1). Alter and Meunier (2009) examined 'the politics of regime complexity', highlighting strategic opportunities for various kinds of actors operating or implementing international regimes. Oberthür and Stokke (2011) synthesized numerous studies of interplay management within as well as across institutional complexes, with attention to factors that influence the ability of actors to cope with partly competing objectives.

The 'management' part of interplay management has two potentially misleading connotations – of hierarchy and of harmony. First, management is sometimes associated with 'hierarchical steering, planning and controlling of social relations' (Biermann 2014) – but that is not what is implied in this context. Hierarchy is a mode of coordination that is difficult to achieve in international affairs, as well as being alien to the inclusive, cooperative and deliberative practices that often characterize successful governance of complex systems (Chapter 13). As indicated by terms such as 'co-management' (e.g., Ostrom et al. 1999) and the 'management school of compliance' (e.g., Chayes and Chayes 1993), hierarchical command-and-control is not a necessary component of management as examined in this chapter. Interplay management may be horizontal, provided by consenting agents operating the institutions involved.

Second, interplay management does not necessarily imply harmonious orientation towards synergetic outcomes. Stokke and Oberthür (2011) discuss reserving the term interplay management for efforts aimed at reconciling the objectives of the interacting institutions. A main argument against such delimitation is methodological: actors are likely to have objectives for intervening in institutional interplay that are both complex and difficult to ascertain, and considerable disagreement can be expected on where to locate the operational cut-off point between reconciling and overruling intentions. A better option is to expand the concept of interplay management to include strategically motivated interventions aimed at promoting the objectives of one institution by disrupting those of another (Orsini, Morin and Young 2013). This way, the consequences of interplay management on the prioritization among competing objectives becomes an important part of the research agenda.

Accordingly, neither hierarchy nor harmony are constitutive components of the phenomenon examined here. Rather, interplay management is seen as the subfield of governance dealing specifically with efforts to improve (from the perspective of the agents engaging in it) the interaction or the consequences of two or more institutions.

Coherence

The inclusion of 'improve' in the definition of interplay management calls for a normative component: explicit attention to what would constitute improvement of institutional interplay. As institutions are generally created to promote distinctive objectives, macro-level assessment of complexes must either start out from one of the elemental institutions or seek to aggregate the effects of interplay management on the broader range of core objectives pursued by them. Except for Pareto improvements – where there is a positive impact on at least one core objective without undermining any other objective – such aggregation is bound to be controversial. 'Coherence' is often used to describe a particularly desirable state of institutional interaction (see, e.g., Stokke and Oberthür 2011; Gehring and Faude 2013; Morin and Orsini 2014) in which elemental institutions are well aligned by bringing complementary or synergistic capacities to bear on compatible policy objectives.

Assessing such coherence becomes more concrete if we disaggregate governance by distinguishing among three generic tasks (Stokke 2012), each of which is observable at the level of the elemental institution but also at the macro level of the larger architecture. A first task is cognitional, as earth system governance should reflect the best available knowledge on the scope and severity of the problems faced – and, preferably, awareness of the effects of various options for dealing with them. The litmus test of cognitional coherence in an architecture becomes whether scientific assessment and advice on issues or activities that fall within the domain of several institutions are perceived as credible and legitimate by those with competence to engage in the second governance task: regulating the activities in question. Whenever the rules adopted under separate institutions are compatible or even support each other – for instance, by binding different actors or guiding different aspects of the same activity in the same policy direction – we may speak of regulatory coherence. A third governance task is to induce the behavioural adaptation deemed necessary for problem solving, whether by building capacities or by adding costs to rule violation. Coherence on the behavioural-inducement task within an architecture is visible whenever the elemental institutions provide mutually supportive means for promoting the desired behavioural adaptation: this may be financial or technological muscle that facilitates the desired outcome, or enforcement structures for behavioural monitoring, reviewing and responding to non-compliance. As explained below, however, coherence is not always the aim of those who engage in interplay management.

Distinctiveness

Seeing interplay management as any effort to improve interaction from the viewpoint of one actor implies that several of the earth system governance topics dealt with elsewhere in this volume are variants of this phenomenon. Orchestration, for instance, as examined Abbott and colleagues (Chapter 11), is the variant of interplay management that is conducted by a particular agent, an international organization, and that targets a particular set of other institutions – private actors and institutions deemed capable of contributing to or impeding problem-solving efforts within the domain of that organization (see also Abbott and Snidal 2010).

Conversely, interplay management is considerably narrower than certain other important earth system governance phenomena. The concept of fragmentation of governance architectures (Biermann et al., Chapter 8), for instance, incorporates interplay management by hardwiring it into a more comprehensive typology of governance architectures. Thus, among the dimensions that enable Biermann and colleagues (2009) to differentiate among synergistic, cooperative and conflictive fragmentation is the extent of integration among the decision-making procedures of the institutions that co-govern an issue area. As Zürn and Faude (2013) point out, a certain correspondence exists between this taxonomy and three classical modes of coordination – hierarchy, cooperation and market. However, constitutive of each of these fragmentation categories are also two dimensions that impinge on, but are clearly distinct from, interplay management: the compatibility of the objectives pursued under each institution, and the congruence of the actor constellations backing them (Biermann et al. 2009). Fragmentation therefore characterizes a wide set of interesting relationships among institutions in an architecture, whereas interplay management narrows in on how various agents intervene in those relationships.

Also other phenomena examined in this volume have important similarities with interplay management, but they occur at a different level of governance or with more limited goals. Policy integration, as reviewed by Runhaar and colleagues (Chapter 9), involves domestic-level efforts to incorporate one set of concerns (e.g., environmental ones) in sectoral policies beyond their traditional policy domain (see Runhaar, Driessen and Uittenbroek 2014). As such, it tries to sensitize other domestic sectors to the policy concern in question, akin to what some interplay managers do – those who give clear priority to one among the many national, international and transnational institutions that make up a larger complex.

Agents of Interplay Management

Who conducts interplay management is important, because this affects the legitimacy of the effort as well as the capacities available. In the current literature, states predominate among the interplay managers (Orsini, Morin and Young 2013), but also other agents have distinct properties that can equip them for this role – notably international organizations, industry organizations and civil-society groups.

States

States are particularly well-placed for interplay management because of their legitimacy at all levels of earth system governance. They combine supreme authority within their own territories with a sizable share in the collective authority held by international organizations; moreover, they frequently team up with nongovernmental organizations involved in global governance. Thus, states have more points of access for influencing institutional interplay along the earth system chain of governance tasks, from knowledge building, to regulation and then to behavioural inducement. Unlike other agents of global governance, states also have the competence to regulate citizen behaviour across the full range of issues dealt with by international institutions, including the right to prioritize among competing sector interests.

The centrality of states in interplay management is apparent even in studies of global governance that highlight discourses rather than agency. Zelli and colleagues (2013), for instance, argue that an overarching normative compromise, liberal environmentalism (Bernstein 2000), with its emphasis on the liberal component, predominates in various interactions within regime complexes on climate change and biosafety. As evidence they present records of deliberations, resolutions and decisions within the regimes in question and the emphasis on the desirability of minimizing constraints on market exchange. The agents of interplay in this discursive account are the powerful players who have championed those outcomes – in these cases, the United States, the European Union and coalitions of other states lining up with either (Zelli, Gupta and van Asselt 2013). Dominant discursive themes tend to be those championed by dominant states or the blocs they form.

Concerning the motives that drive states to engage in interplay management, Gehring and Faude (2013) note a rationalist trigger – strategic assessment. According to this argument, states are aware of their existing commitments to various institutions in the complex and will assess any competing proposals in light of whether their own interests are best served by adapting the new commitment to existing ones or by allowing it to undermine them (Gehring and Faude 2013). However, if the institutions in question span different sectors of government, both the awareness and the assessment in question will require strong domestic capacity for aggregation.

International Organizations

Important as states are as interplay managers, other agents too have properties that can support efforts to influence interaction within institutional complexes. International organizations are often mandated to monitor developments in broad policy areas and to initiate programmes or regulatory processes when needed for preventing or mitigating collective problems. Highly consequential for ocean governance has been the role played by the Food and Agriculture Organization (FAO), a UN specialized agency, in developing and diffusing among regional fisheries regimes a series of management approaches and measures for improving the sustainability of fisheries within and outside the 200-mile Exclusive Economic Zones codified in the 1982 UN Convention on the Law of the Sea. The 1984 Strategy adopted by the FAO World Conference on Fisheries Management and Development provided early formulations of the precautionary as well as ecosystem approaches to fisheries management (FAO 1984) and formed an important part of the basis for the oceans segment of the 1992 United Nations Conference on Environment and Development in Rio de Janeiro and its Agenda 21 (Garcia and Newton 1996). One outcome of that conference, the Inter-Agency Committee on Sustainable Development, made the FAO the lead agency for developing approaches to implementation of Agenda 21 with respect to marine living resources (Juda 2002). A flurry of FAO activities ensued – notably, the negotiation of a legally binding 1993 FAO Compliance Agreement, which strengthened the responsibilities of flag states to control the harvesting activities of their vessels. Another important FAO output was the 1995 Code of Conduct on Responsible Fisheries, which helped to specify and concretize the precautionary and ecosystem approaches to fisheries management, building on existing practices within the most advanced regional management regimes.

Illustrating the cascading effects of such interplay management by a global organization mandated to advance governance within broad issue areas, the process of developing these various norms on high seas fisheries deeply influenced the parallel negotiation of the legally binding UN Fish Stocks Agreement (Balton 1996). That agreement has become the legal centrepiece for regional high seas fisheries governance as well as the most authoritative specification of the precautionary approach to fisheries (Rothwell and Stephens 2016). The agent-oriented account presented here of one process of normative harmonization is compatible with the macro-development described by Kim (2013): the increasing frequency of citations among environmental treaties, accelerating with the Rio Conference on Environment and Development in 1992.

Institutional properties that equip an international organization for such interplay management are broad membership, relevant expertise and the mandate to initiate negotiations of binding international commitments within its issue domain. Such efforts are more likely to succeed when the same sector of government operates the global institution and the regional one. That advantage was lacking in the case of the regime bureaucracies that van Asselt (2014) examined when studying interplay management in the complex comprising the climate and biodiversity regimes (typically operated primarily by environmental officials), but also global trade (operated primarily by trade and foreign-affairs officials). Even in this more demanding setting, communication among secretariats helped to raise awareness about cross-regime interactions (van Asselt 2014). Secretariats have limited mandates, however; also the decision-making bodies of regimes with their greater leeway may have reason to move cautiously on matters of cross-regime coordination – especially if the regimes in question differ in membership (van Asselt 2014), or if the coordination risks interfering with the regulatory domain of domestic bureaucracies not represented internationally.

That said, Jinnah (2014) argues that mainstream scholarship has underestimated the ability of determined treaty secretariats, like that of the Convention on Biological Diversity, to frame issues of overlap with other regimes – notably, those for climate and trade – in ways conducive to their prioritized policy agenda. Thus, the secretariat of the Convention on Biological Diversity raised the saliency of biodiversity conservation issues by emphasizing the climate and human-security dimensions (Jinnah 2014). Such attempts to filter information in ways that serve to reframe policy issues rely upon leadership within the secretariat. They are more likely to succeed if the international bureaucracy in question has recognized and (ideally) privileged access to key decision-making forums, if there is limited competition from other interplay managers and if state preferences on the issue have not solidified (Jinnah 2014). As noted by O'Neill and colleagues (2013), the rising interest in interplay management has served to re-energize work on international bureaucracies in earth system governance, often drawing on intensive methods such as interviews and archival research.

Private Institutions

A third category of agents engaging in interplay management are private actors, including industry organizations and civil-society groups, that often seek legitimacy by teaming up in partnerships that may include governmental organizations as well (Newell, Pattberg and Schroeder 2012; Hickmann 2017). Certification schemes in sectors such as forestry and capture fisheries exemplify the increasingly widespread practice of partnerships co-founded by transnational business and transnational environmental groups (Pattberg 2005). For instance, the scheme operated by the Marine Stewardship Council is open to applications from any industry grouping that is engaged in the fishery of a specific stock and is prepared to set up a chain-of-custody system that separates production and distribution chains based on a certified fishery from those that are not (Gulbrandsen 2009). By 2017, the Marine Stewardship Council had certified some 300 fisheries in 34 states – mostly in the Northern Hemisphere but recently including China as well – taking some 9.5 million tonnes, or more than one tenth of the global seafood catch (Marine Stewardship Council 2017). Private governance institutions such as the Marine Stewardship Council acquire legitimacy as interplay managers by representing, from the outset, both industry and environmental actors; by anchoring certification criteria in international norms and principles and mobilizing third-party expertise in assessment of applications; and, when successful, by penetrating seafood markets. The specific capacity that this type of organization adds to the earth system governance architecture is the ability to transform corporate environmental responsibility – whether idealistically or opportunistically derived – among dominant wholesale or retail chains, into tangible incentives for more sustainable industry practices (Stokke 2019).

Institutional Conditions Favouring Agency

To sum up this brief review of the three types of agents that engage in interplay management, certain institutional requirements appear relevant to all of them. First, legitimacy is a crucial ingredient in successful interplay management, and one which goes a long way towards explaining the empirical prominence of states in such efforts. Their supreme legitimacy within their own territories allows them to aggregate competing concerns at the preparatory as well as the domestic-implementation stages of the governance chain. However, concerning the many governance problems that transcend national boundaries, even powerful states may face a range of other legitimate interplay managers – including other states, international organizations and transnational private organizations or partnerships. In these more complex settings, international organizations can be well-placed for interplay management if their legitimacy is buttressed by broad membership and a strong mandate. Private organizations may obtain legitimacy for interplay management provision in various ways – for instance, by a broad membership base, by linking up to governmental norms, or by controlling eco-labels that promote market access.

A second institutional property that stands out as crucial for interplay management, besides representative legitimacy, is expertise that other agents in the architecture recognize as credible and useful.

Third, an important asset for states as well as international or transnational organizations is the ability to compensate other actors engaged in or governed by the institutional complex for taking their prioritized concern into account – whether by flexing sizable financial muscles or, as in the case of private governance, by controlling an eco-label that affects market access in certain sectors.

Means of Interplay Management

Whether interplay management proceeds by means of coordination or adaptation is important, because the two differ in intrusiveness and therefore in the extent of resistance they are likely to trigger. Coordination involves cross-institutional communication and adjustment of governance activities, whereas adaptation requires only awareness of the rules or programmes of other institutions and preparedness to take those into account in one's own decisions. Resistance to interplay management may come from states or other participants in the regimes involved. From a governance perspective, a central question is whether better alignment across institutions can be achieved without jeopardizing the willingness of states to endow those international regimes with regulatory powers or other problem-solving assets. Such willingness may suffer if those operating the institutions, typically representing specific sectors of government, fear that closer alignment with other international institutions will weaken their own regulatory competence over activities currently falling within their domain.

Other things being equal, centralized coordination by an overarching institution that enjoys superior authority holds the greatest potential for effectively aligning problem-solving activities across institutions. However, as other things are rarely equal, this section reviews various earth system governance studies to identify political and institutional conditions that appear to affect whether coordination or adaptation is best-suited for improving institutional interplay – and how they affect prioritization among competing objectives.

Coordination

Centralized coordination is found when those operating interacting institutions acknowledge that one among them has the authority to steer their decisions. The UN Security Council has such authority in some issue areas, and member states and other institutional components of the European Union have agreed to accept centralized coordination in a range of issues relating to the internal market. In international affairs, however, such generalized deference to another institution is rare. Coordination tends to take the decentralized and more modest form of communication among those operating the institutions, either in their secretariats or their political decision-making bodies, with a view to producing joint or at least harmonized decisions on one or more tasks of governance (Oberthür and Stokke 2011).

Calls for centralized coordination have often been voiced in studies of environmental governance (Van de Graaf and De Ville 2013). Biermann, for instance, writing on the architecture of global governance, argues that 'an international institutional core with a clear strategy to ensure effective earth system governance is thus the need of the hour' (Biermann 2014). In his view, the main reason why international bureaucracies fall short of their potential in the international domain is that they are fragmented within separate institutions; they lack 'one major international bureaucracy that is solely devoted to supporting governance processes' in the environmental domain, as a world environment organization would (Biermann 2014). In evaluating such calls for centralized coordination, it is instructive to examine the performance of the United Nations Environment Programme (UNEP), the most prominent among international institutions with an explicit mandate for such coordination.

UNEP was created in the aftermath of the 1972 Stockholm Conference on the Human Environment, to 'promote international cooperation in the field of the environment' and to 'provide general policy guidance for the direction and coordination of environmental programmes within the United Nations system' (United Nations General Assembly 1972). Analysts agree that UNEP has been considerably more successful with the first part of that mandate, as evident in its large-scale and authoritative environmental monitoring and assessment programmes as well as its many international regulatory initiatives, and less successful regarding the demanding task of coordinating others (see, e.g., Andresen 2007; Ivanova 2010; Biermann 2014). While UNEP has co-initiated several international regimes in areas such as ozone, biodiversity, desertification and environmental toxics, it has generally failed to become their institutional home: the new agreements have provided for distinctive decision-making procedures and typically also separate secretarial support (Ivanova 2010). Even treaty secretariats with formal links to UNEP, like those in the biodiversity and chemicals clusters, have been obliged and also inclined to seek guidance primarily from the respective conference of the parties rather than from UNEP (Andresen 2007). Perhaps most importantly, UNEP has remained peripheral in the global climate regime – the foremost environmental process of the past quarter-century and the one with the greatest potential for interacting with other environmental institutions. Thus, it is reasonable to conclude that UNEP has performed very well in knowledge-building, acceptably in catalyzing international regulatory processes, but no more than modestly in the interplay-management part of its mandate: the subsequent operation and coordination of those processes.

Explanations for UNEP's mixed performance highlight certain institutional conditions that appear necessary for putting into practice a mandate for centralized coordination – most importantly, material resources that can motivate other institutions to allow UNEP to coordinate them. UNEP's universal membership and formal mandate for agenda-setting, rule-making and coordination are certainly strong assets for an interplay manager, yet its status has remained that of a UN Programme rather than a UN Specialized Agency. That means less leeway for independent initiatives: for UN programmes, member-state core funding is voluntary and therefore less predictable. While the UNEP budget is quite sizable, some USD 780 million in 2018 (United Nations Environment Programme 2016), or nearly four times that of the World Trade Organization (WTO), it is nevertheless dwarfed by the environmental problem-solving funds available under mechanisms other than UNEP, such as the World Bank or the Global Environment Facility. The ability to offer unique technical and policy expertise is another institutional property that can reinforce coordination efforts, but here UNEP has a modest score. Secretariat staffs and policymakers in other environmental regimes reportedly find the organization excessively bureaucratic and often inefficient (Andresen 2007). Partly because its headquarters is in Nairobi, far from the hubs of world diplomacy, UNEP has had difficulties in recruiting and retaining the competent personnel (Ivanova 2010) necessary for cross-regime coordination.

Thus, the UNEP experience indicates that interplay management by means of centralized coordination requires not only the legitimacy deriving from broad membership and a clear mandate for initiating and coordinating regulatory action but also other institutional assets – notably, relevant technical and policy expertise and enough financial muscle to motivate others to heed the directions of the central coordinator.

Even decentralized coordination among institutions, involving communication and adjustment but not necessarily hierarchy, is rare in earth system governance. It is typically limited to other governance tasks than regulation or segments of the institutional complex that pursue highly compatible policy objectives (Oberthür and Stokke 2011). Consider for instance the coordinated knowledge-building undertaken by technical bodies under the climate and ozone regimes during the early 2000s. Those joint efforts generated a shared understanding concerning options for dealing with emissions of fluorinated greenhouse gases, which are major substitutes for ozone-depleting substances – but the belated and limited regulatory and programmatic responses were conducted by each of the regimes unilaterally (Oberthür, Dupont and Matsumoto 2011). The segment of earth system governance in which cross-institutional coordination has advanced the most is the chemicals cluster of agreements initiated and supported by UNEP, as evident in back-to-back meetings of the parties, shared or jointly headed secretariat functions, and harmonized reporting procedures (see van Asselt 2014). Coordination across the chemicals treaties is clearly facilitated by the shared overarching objective of combating persistent, toxic and bio-accumulative pollutants, as well as the fact that those representing the member states in decision-making bodies tend to be the same persons.

Unfortunately, when conditions become only slightly less benign than in the chemicals case, decentralized coordination appears far more difficult to advance and to implement. Consider the set of proposals put forward by a Joint Liaison Group tasked with identifying synergies across the three Rio conventions (on climate change, biological diversity, and desertification) – a relatively 'easy' case, in the methodological sense that conditions for successful coordination are favoured by compatible objectives and joint efficiency gains. However, practical results have been meagre, due to a disconnect among the respective decision-making bodies as to which proposals to implement (Scott 2011) – in turn deriving from rigid procedures and practices, non-overlapping membership, and resistance (especially from those operating the climate regime) to involve biodiversity and conservation segments of government and their associated stakeholders in decisions on how to achieve rapid and cost-effective reductions in greenhouse gas emissions (van Asselt 2014). Not surprisingly, this pattern of resistance to coordination becomes even clearer when the institutions in question are more heterogeneous in terms of their primary objectives. Casting the net more widely by examining not only biodiversity and climate regimes but also the global trade regime, van Asselt was forced to conclude that variants not involving any coordination among the interacting regimes 'form the primary means of interaction management in practice' (van Asselt 2014).

In summary, considerable impediments exist with regard to applying the most ambitious kinds of interplay management which rely on joint decision-making across institutional boundaries. The centralized variant requires a preparedness – rare among those in charge of international institutions – to cede generalized authority to another international body. Achieving even decentralized coordination has proven difficult, especially with respect to regulatory governance, if the institutions in question pursue only slightly differing objectives. When objectives are partly competitive, as in the interface of environmental and trade regimes operated by different sectors of government, there have been very few cases of coordination beyond exchange of information and joint knowledge-building. That makes it particularly pertinent to explore the conditions for effectively employing less ambitious means for interplay management, means requiring awareness of regulatory and other activities under other institutions but only a one-sided response – adaptation of one's own activities to those of others.

Adaptation

Summarizing a multi-year project on institutional interplay in earth system governance, Oberthür and Stokke (2011) observed: 'Practically all cases reported here of deliberate attempts to influence the nature and impacts of interplay concern either unilateral management within one institution, or autonomous management by states or others.' Such one-sided adaptation on the part of elemental institutions in an architecture enables bypassing the impediments to cross-institutional coordination noted above. However, it also foregoes the processes of communication that can support an integrative search for solutions that maximize the set of priorities pursued by the institutions involved. What conditions and strategies can promote efforts by interplay managers unable or unwilling to engage in joint decision-making, to avoid normative conflict or duplication by searching for and realizing potential synergies across institutions?

The compatibility of the primary objectives pursued under each regime is certainly an important condition that influences the likelihood of integrative solutions. Raustiala and Victor (2004) stressed this point in their analysis of the international politics of plant genetic resources. In that policy area, states favouring more stringent protection of intellectual property championed regulation under the global trade regime, whereas many developing countries and others have pushed for substantively competing rules that would put into practice either the benefit-sharing principle of the Convention on Biological Diversity or the common-heritage principle of the FAO-related network of gene banks (see Rosendal 2001; Jungcurt 2011). It would be mistaken, however, to ascribe the normative tensions that ensue from such 'forum shopping' to institutional complexity. Such tensions result from the underlying political disagreement that would exist also under a more unified institutional framework. As argued by Keohane and Victor (2011), institutional complexity typically results from conditions that constrain the negotiation of an integrated institution – notably, interest heterogeneity, uncertainty and various impediments to issue linkage. Sometimes other aspects like problem diversity reinforce these generic impediments, as in the case of climate governance, which Keohane and Victor use to illustrate their approach. Close institutional coordination is not to be expected under such conditions. More likely alternatives to the existing state of cross-institutional tensions managed by means of pragmatic adaptation are deadlocked negotiations, vague commitments that mask actual disagreements or watered-down rules.

Furthermore, adaptation strategies like the pursuit of strategic inconsistency and forum shopping are two-way streets: they allow not only disruptive interference but also the provision of leadership by smaller sets of states keen to promote more ambitious norms (Keohane and Victor 2011). For instance, environmental regulation under the OSPAR Commission for the Protection of the Marine Environment of the North-East Atlantic was greatly enhanced by the creation of a new institution – a series of North Sea Conferences with narrower membership, higher political saliency and more transparent performance review procedures than those of the OSPAR Commission at the time (Skjærseth 2006). The more stringent targets agreed upon under the North Sea Conferences were subsequently adopted unilaterally by the broader regime. Similarly productive interplay management was seen when 14 European states decided to reinforce the weak compliance structures of the global shipping regime centred on the International Maritime Organization by unilaterally drawing up the 1982 Paris Memorandum of Understanding on port-state controls (Stokke 2013). This instrument, which today has 27 parties and has been emulated in all other maritime regions, commits states to use their jurisdiction over vessels voluntarily in their ports to collectively raise the frequency and quality of vessel inspection and response action regarding sub-standard vessels (Molenaar 2007). Thus, interplay management by means of one-sided yet forward-leaning adaptation can help states to overcome the 'slowest-boat' problem typical of consensus-based international governance, thereby paving the way for more ambitious solutions that may attract a broader following later on.

Similarly conducive processes of adaptation are evident in another OSPAR case of interplay management, involving an institution operated by a different sector of government and thus likely to weigh environmental concerns differently – the North-East Atlantic Fisheries Commission. While initially reluctant to engage with the OSPAR Commission on matters related to fisheries, an issue area it considered as an exclusive regulatory domain, the North-East Atlantic Fisheries Commission gradually warmed to the idea of taking into consideration wider environmental concerns, notably the protection of cold-water coral reefs. Partly because both institutions base their decisions on scientific inputs from the International Council for the Exploration of the Sea, the North-East Atlantic Fisheries Commission adapted its 2009 closure of certain high seas areas for bottom trawling to the spatial boundaries of the OSPAR Commission's emerging network of marine protected areas (Kvalvik 2012). Note, however, that this regulatory alignment did not derive from joint decision-making. On the contrary, the initial reluctance to engage with the OSPAR Commission, on grounds that the interface between environmental and fisheries concerns are national rather than international issues, highlights the inclination among those operating institutions to safeguard their formal role and competence in decision-making (Kvalvik 2012). Viewed from this perspective, exchange of information and cooperation on technical issues, combined with readiness to take into consideration the concerns and the management measures of the other institution, may offer an adequate vehicle for coherence, and one less prone to trigger institutional jealousy. Adaptation – whether one-sided or mutual – is probably more easily achieved than institutional coordination across issue-areas.

The promise of adaptation as a means for improving institutional interplay across sectors of government is also evident in the development of trade-restrictive measures under international environmental regimes. Perhaps 20 out of 200 multilateral environmental agreements include provisions that authorize or oblige members to impose trade restrictions on states that fail to assume or comply with environmental commitments in areas such as stratospheric ozone, endangered species as well as fisheries (World Trade Organization 2018). They have managed to overcome the 'chilling effect' expected from rules set forth in global trade agreements – that is, those administered by the WTO (Axelrod 2011). These agreements generally prohibit discrimination in trade among the 164 WTO members, so coherence depends on designing trade restrictions that fit that organization's 'environmental window', a set of exceptions defined first in Article XX of the 1947 General Agreement on Tariffs and Trade and reproduced in subsequent agreements. Key standards of WTO compatibility, subject to cross-disciplinary dialogue among trade and environmental experts in meetings arranged by the WTO Committee on Trade and Environment (Kent 2014), are whether states have exhausted less-restrictive measures, have minimized and justified any remaining discrimination and have developed criteria for avoiding trade restrictions – criteria that are transparent, non-discriminatory and not excessively intrusive on the jurisdictional autonomy of the target state (see Stokke 2009). Subsequent decisions by dispute settlement bodies have clarified and developed the ramifications of these exceptions, which were accorded even greater prominence by the inclusion of sustainable development in the Preamble to the WTO's constitutive instrument. The standards for WTO-compatible measures, constructed unilaterally by those operating that institution, have guided the equally unilateral adaptation undertaken by interplay managers in a wide range of environmental institutions when hammering out trade-restrictive measures to promote their primary objective (Gehring 2011).

Illustrative of such interplay management by adaptation to an institution promoting a partly competing objective, while aiming to avoid unnecessary tension, are the port-state measures adopted by the North-East Atlantic Fisheries Commission for combating illegal, unreported or unregulated fishing on the high seas (Stokke 2009). Port-state measures are necessary in this combat, due to the near-monopoly that flag states have on regulation and rule enforcement on the high seas and the pervasiveness of fishing vessels flying flags of convenience (DeSombre 2005). Certain features of regional management regimes have proven conducive for development of trade-restrictive compliance measures that meet the WTO compatibility standards. Firstly, a portfolio of other conservation measures provides shelter from claims that less trade-restrictive measures have not been exhausted. Secondly, advanced compliance measures like the catch documentation scheme implemented by the North-East Atlantic Fisheries Commission minimize discrimination, by restricting the trade measure to individual vessels. In order to meet the third compatibility standard – reasonable and transparent criteria for avoiding the trade restriction – advanced port-state measures allow non-members to apply for a status as 'cooperating non-contracting party' to avoid trade restrictions, provided they agree to play by the same rules as the members do. This alignment of regulatory actions on either side of the trade–environment boundary has resulted from learning and adaptation rather than overarching or cross-institutional coordination.

Adaptation also characterizes interplay management across the actor–type boundary, as seen in the orchestration-in-reverse efforts of private governance institutions to support the performance of international regimes. The Marine Stewardship Council, for instance, has no formal status with the North-East Atlantic Fisheries Commission, but it has played a role in that organization's efforts to manage the Northeast Atlantic mackerel fishery. Obtaining and retaining the right to apply the Marine Stewardship Council eco-label on seafood products requires approval from an accredited third-party certifying company, stating that the fishery in question is conducted and managed in accordance with three basic principles derived from international fisheries law – involving the health of the fish stock, the harvesting pressure and the management system (Gulbrandsen and Auld 2016). The legitimacy-seeking practice of pegging private governance standards to globally agreed principles is widespread – for instance, 80 per cent of all private carbon-offset schemes link their criteria to the Clean Development Mechanism under the Kyoto Protocol (Abbott, Green and Keohane 2016). The mackerel fishery in the Northeast Atlantic had lost its Marine Stewardship Council certification in 2012 following the breakdown of coastal-state cooperation on this valuable stock two years earlier, yielding several unilateral quotas and total harvesting pressure well in excess of the advice by the International Council for the Exploration of the Sea (Spijkers and Boonstra 2017). In response, industry groups controlling more than 700 vessels from most regional user-states joined in the Mackerel Industry Northern Sustainability Alliance, promoting reinforced efforts to reach a negotiated solution. In 2016, these companies succeeded in regaining Marine Stewardship Council certification, claiming they had played an important and productive role in negotiations that had brought all user-states but one back into a cooperative arrangement on mackerel (Stokke 2019). In this case, the private governance organization managed to improve the effectiveness of an international resource regime by setting its standard higher than what the governments had been able to agree on or implement, while controlling access to an eco-label with enough market recognition to be perceived as lucrative by the industry.

In summary, one-sided adaptation can be an effective means for identifying and adopting integrative measures also with institutions that pursue partly competing objectives. Compared to coordination based on joint decision-making, adaptation has the advantage of being less threatening to those who operate international bodies endowed with regulatory competence. Therefore, adaptation is less likely to encounter institutional resistance to the pursuit of integrative options. Like the environmental institutions that have achieved greater behavioural compliance among their members by exploiting an existing environmental window in the global trade regime, thereby avoiding unnecessary disruption of legitimate free-trade objectives, private governance institutions typically peg their standards to globally agreed principles, but can unilaterally specify them more stringently than the relevant regional institution has been able to do.

True, interplay management by adaptation is unlikely to overcome institutional fragmentation in cases when normative conflict derives from differences among powerful actors in how they prioritize among competing objectives. However, that is no reason for despair. In such situations, substantive tension among the norms or activities upheld under separate institutions is not necessarily more serious than the bargaining failure or the vague or watered-down compromises that are likely results from efforts to assemble them within a singular, closely integrated institution. That being said, we now ask: how can interplay management affect the formal or informal hierarchy among the institutions that make up an architecture?

Prioritization

Coordination and adaptation are important means for exploring and reaping integrative options for minimizing tension among competing policy objectives. Sometimes, however, interplay managers seek not to achieve peaceful co-existence among institutions but rather to raise the prominence of one over others. Such efforts may assume various forms – encouraging strategic inconsistency among institutions while putting one's weight behind one of them (Raustiala and Victor 2004), ensuring a regulatory initiative in the forum most conducive to one's preferred policy outcome, or forum-shifting efforts aimed at relocating the institutional centre of gravity within a complex of institutions (Alter and Meunier 2009). As rules on normative hierarchy are generally weak in international law, such efforts highlight the power relationships among agents of interplay management but also the legitimacy of agreed principles and certain niche advantages each institution may have in the general tasks of governance.

The few existing, widely accepted rules of precedence in international law are too vague or too easily counteracted to resolve issues of normative conflict among institutions on a purely legal basis (Oberthür and Stokke 2011). Those rules most frequently referred to – lex posterior and lex specialis – can be readily side-lined by invoking legal uncertainties as to exactly when a given treaty provision came into existence, whether two treaties concern the same subject matter or how to differentiate unequivocally the specificity of rules laid out in various treaties (van Asselt 2011). Also contestable is the relative weight of such general rules and the specific savings clauses often included in treaty texts, shielding rights obtained in previous agreements (Axelrod 2011; also Kim 2013). Unless the treaty itself or the decision-making body set up by it explicitly defers to another institution, principles of international law are unreliable instruments for measuring the relative priorities that states assign to the various policy objectives upheld within an institutional complex.

Deference, or acceptance of determinations made by another institution (Leebron 2002: 16; Pratt 2018), is not uncommon. It often reflects a combination of clear policy priorities among powerful actors and the tendency of multilateral negotiations to converge around certain salient solutions legitimized by previous agreements. For instance, clear prioritization of free trade over the climate stabilization objective is evident, for instance, evident in the pledges inscribed into core climate documents that '[m]easures to combat climate change, including unilateral ones, should not constitute a means of arbitrary or unjustifiable discrimination or a disguised restriction on international trade' (United Nations Framework Convention on Climate Change, Article 3.5). That formulation is taken verbatim from the chapeau of Article XX of the General Agreement on Tariffs and Trade, without adding counterbalancing phrases; also echoed in provisions of the Kyoto Protocol, it has been confirmed in the subsequent decision to exclude trade restrictions from the agreed portfolio of compliance measures (Stokke 2004). Conversely, the WTO Agreement on Agriculture deferred to FAO rules on food aid when specifying criteria for distinguishing humanitarian assistance from farm subsidies; that reflected a successful bid by other grain-producing states to counteract US market penetration by means of food aid (Margulis 2013). In both cases, the effort to give one institution precedence over another was facilitated by the high legitimacy of the prevailing rule, based on factors such as determinacy, pedigree and symbolic validation (Franck 1990). However, it is impossible to understand the specific outcomes without paying close attention to the perceived interests of leading states and their relative power, which in turn is closely linked to their ability to pursue their prioritized objective outside the institution in question (Gehring and Faude 2013).

Patterns of deference and precedence among international institutions do not always reflect clashes of consolidated interests mediated by power differentials. As studies of institutional ecology have shown, strategic considerations by those operating international regimes also revolve around how institutional properties such as membership, expertise, regulatory competence or financial resources affect the niche such a regime can realistically opt for in the larger governance complex. In ecology, a 'niche' denotes the placement of a species or population in an ecosystem, notably that segment of a resource domain where it out-competes other local populations. Organizational analysts have used this term to narrow in on the relationship between institutional features and the ability to extract the resources necessary for organizational survival (Hannan and Freeman 1977). In a study of environmental toxics, for instance, Stokke (2011a) explains how the Arctic Council has dismissed proposals to engage in regulation on grounds that the major pollution sources are located outside the region, concentrating instead on the niche advantage inherent in the Council's ability to extract member-state funds for a monitoring programme that provided credible, legitimate and salient scientific advice (Cash et al. 2006) on severe health impacts on Arctic indigenous peoples. In turn, those results helped to catalyze regulatory action under the broader and therefore better-placed Stockholm Convention on Persistent Organic Pollutants (Downie and Fenge 2003). Similar considerations of variation in institutional fitness for specific governance tasks have been used to account for the prominence of national, and to some extent regional, regimes in setting environmental standards for offshore petroleum activities and for subregional institutions in high seas fisheries regulation, as well as environmental capacity enhancement (Stokke 2011b). Considerations of asymmetry in relevant expertise can induce those operating institutions at the highest level of regulatory competence to defer to determinations by a technical body – as the Security Council did when instructing UN member states to implement the legally non-binding standards set by the Financial Action Task Force to counter the financing of terrorism (Pratt 2018).

Recently, the link between organization ecology, niches and institutional competition has been explored also at the population level of regime complexity analysis. Abbott and associates have examined patterns of institutional density, notably the steep numerical rise of private institutions within climate governance, engaging in such activities as climate auditing, certification, and capacity enhancement (Abbott, Green and Keohane 2016). Non-governmental institutions are well-equipped for these governance tasks thanks to institutional features that include their lower costs, compared to international bodies, of entering an issue area and their greater ability to adapt their agenda quickly to specialized and narrow resource domains.

One defining characteristic of an institutional complex is the weakness of normative hierarchies, which helps to explain why decentralized interplay management, conducted unilaterally within the elemental institutions, has been so pervasive. Yet, as shown in this section, those in charge of international regimes sometimes find it both advantageous and within their powers to raise the prominence of one institution over another, either because they assign higher priority to its policy objective or because they acknowledge that the institution is better placed than others for the governance task in need of improvement.

Conclusions and Future Directions

Interplay management involves deliberate efforts by one or more actors to improve the interplay of institutions set up for earth system governance. As this chapter has shown, differing views on what counts as 'improvement' complicate assessment of success and failure. Interplay management actors include states, intergovernmental organizations and industry- or civil-society groups, all seeking to mobilize material resources, expertise or other assets such as legitimacy held by one institution to promote objectives pursued under another. This is typically done in a decentralized manner, operating within elemental institutions rather than through overarching structures. The means employed are frequently some variant of unilateral adaptation to norms and programmes undertaken in other institutions, rather than explicit coordination involving joint decision-making. Despite the typically low levels of cross-institutional integration, interplay management often allows states and other actors to improve the coherence of cognitional, regulatory or behaviour-adaptive activities under separate institutions, even when those institutions have partly competing policy objectives or are operated by different sectors of government.

Major advances have been made in research on interplay management. Attention no longer focuses on the generic question of whether governance is best served by centralized or decentralized approaches or by coordination or adaptation: today, the analysis centres on the conditions that affect the conduciveness of each of these approaches. Such study requires realistic assessment of what international institutions (whether their spatial or functional scopes are narrow or broad) can and cannot achieve in earth system governance. Institutions cannot be expected to overcome deep-rooted, conflicting priorities among the most powerful states. Under such circumstances, a more realistic alternative to normative tension and fragmentation involves broken-off talks and a normative vacuum. Scaling up the most advanced management norms or measures from regional to global levels – as done, for instance, with the precautionary principle in environmental governance and with port-state measures in shipping and fisheries – generally requires both time and political energy, in turn indicating the advantages of a gradualist bottom-up approach. Similarly, the potential benefits of expanding the functional scope of international bodies – as in the proposal to include fisheries among the topics for negotiation of a new instrument for conserving biodiversity beyond national jurisdiction – must be weighed against the risk that states prepared to cede regulatory competence to an international body might think twice if the regime in question were likely be operated by sectors of government and associated stakeholders inclined to prioritize preservation over resource use.

Conditions influencing whether interplay management is conducive to one or more governance tasks revolve around the agency and the means of management as well as the compatibility of the policy objectives pursued. Agents may differ with respect to their legitimacy as interplay managers and the resources they can bring to bear on the process. When such legitimacy is low, as is the case of private institutions, agents typically compensate by staying firmly on the adaptation side of management and by pegging any unilateral standards to global or at least intergovernmentally agreed principles. Even UN-based organizations explicitly mandated to coordinate environmental activities must struggle to put those mandates into practice unless they can deliver unique and respected expertise, institutional means for demonstrating and diffusing best practices, or material resources that can motivate those operating other institutions to accept external leadership. Cross-institutional coordination has obvious advantages that flow from its defining characteristics of explicit communication and joint decision-making. Those advantages are particularly valuable when the institutions govern highly interdependent activities or can bring to bear clearly complementary capacities. As the cases and empirical studies reviewed here have shown, such conditions are often present with respect to the cognitional and behavioural-inducement tasks of governance, and also in regulatory matters involving institutions operated by the same sector of government. Those situations promote functional differentiation within the complex, with each institution exploiting the governance niches it is particularly well equipped to occupy. In contrast, whenever the measures triggering interplay management concern policy objectives with clearly competitive elements – as with trade-restrictive compliance measures under environmental regimes – adaptation has the advantage of triggering less turf-sensitive resistance: unlike coordination, adaptation leaves the original division of competences intact.

Successful interplay management can improve the performance of one or more institutions in a governance architecture without disrupting others, by encouraging the search for integrative options also when policy objectives compete and powerful actors prioritize them differently. Certain trends in the research reported in this chapter should be pursued to make further headway in understanding the conditions for conducive interplay management. Conceptually, although disaggregating the notion of coherence along the cognitional, regulatory and behavioural dimensions has helped to make assessment more concrete, we still struggle with how to weigh progress in the performance of one institution against halts or even setbacks in that of others. Empirically, considerable expansion has occurred regarding the issue areas covered in interplay management studies – but those involving climate, biodiversity and ocean governance with occasional side-glances to trade rules still predominate, leaving room for improvement in how findings from studies of non-environmental institutions are incorporated.

Regarding methodology, recent research on institutional deference and institutional ecology has included quantitative analysis, but the case-study approach remains the most commonly used in the field. Greater employment of complementary techniques such as model-based simulation and formalized qualitative comparative analysis will promote further advances in research on interplay management.

References

Abbott, K. W., Green, J. F., & Keohane, R. O. (2016). Organizational ecology and institutional change in global governance. *International Organization*, 70 (2), 247–77.
Abbott, K. W., & Snidal, D. (2010). International regulation without international government: Improving IO performance through orchestration. *Review of International Organizations*, 5 (3), 315–44.
Aggarwal, V. K. (1983). The unraveling of the multi-fiber arrangement, 1981: An examination of international regime change. *International Organization*, 37 (4), 617–45.
Alter, K. J., & Meunier, S. (2009). The politics of international regime complexity. *Perspectives on Politics*, 7 (1), 13–24.
Andresen, S. (2007). The effectiveness of UN environmental institutions. *International Environmental Agreements*, 7 (4), 317–36.
Axelrod, M. (2011). Savings clauses and the 'Chilling Effect': Regime interplay as constraints on international governance. In S. Oberthür, & O. S. Stokke (eds.), *Managing institutional complexity: Regime interplay and global environmental change* (pp. 87–114). Cambridge, MA: The MIT Press.
Balton, D. A. (1996). Strengthening the law of the sea: The new agreement on straddling fish stocks and highly migratory fish stocks. *Ocean Development & International Law*, 27 (1–2), 125–51.
Bernstein, S. (2000). Ideas, social structure and the compromise of liberal institutionalism. *European Journal of International Relations*, 1, 347–71.
Betsill, M., Dubash, N. K., Paterson, M., van Asselt, H., Vihma, A., & Winkler, H. (2015). Building productive links between the UNFCCC and the broader global climate governance landscape. *Global Environmental Politics*, 15 (2), 1–10.
Biermann, F., Pattberg, P., van Asselt, H., & Zelli, F. (2009). The fragmentation of global governance architectures: A framework for analysis. *Global Environmental Politics*, 9 (4), 14–40.
Biermann, F. (2008). Earth system governance: A research agenda. In O. R. Young, L. King, & H. Schroeder (eds.), *Institutions and environmental change: Principal findings, applications and research frontiers* (pp. 277–301). Cambridge, MA: The MIT Press.
Biermann, F. (2014). *Earth system governance: World politics in the Anthropocene*. Cambridge, MA: The MIT Press.
Biermann, F., & Siebenhüner, B. (2009). *Managers of global change: The influence of international environmental bureaucracies*. Cambridge, MA: The MIT Press.

Cash, D. W., Clark, W. C., Alcock, F. et al. (2006). Knowledge systems for sustainable development. *Proceedings of the National Academy of Sciences of the United States of America*, 100 (14), 8086–91.

Chayes, A., & Chayes, A. H. (1993). On compliance. *International Organization*, 47 (2), 175–205.

DeSombre, E. R. (2005). Fishing under flags of convenience: Using market power to increase participation in international regulation. *Global Environmental Politics*, 5 (4), 73–94.

Downie, D. L., & Fenge, T. (2003). *Northern lights against POPs: Combatting toxic threats in the Arctic*. Montreal: McGill-Queens University Press.

FAO. (1984). *Report of the FAO world conference on fisheries management and development*. Rome: Food and Agriculture Organization of the United Nations.

Franck, T. M. (1990). *The power of legitimacy among nations*. Oxford: Oxford University Press.

Garcia, S. M., & Newton, C. H. (1996). Responsible fisheries: An overview of FAO policy developments (1945–1994). *Marine Pollution Bulletin*, 29 (6–12), 528–36.

Gehring, T. (2011). The institutional complex of trade and environment: Toward an interlocking governance structure and a division of labor. In S. Oberthür, & O. S. Stokke (eds.), *Managing institutional complexity: Regime interplay and global environmental change* (pp. 227–54). Cambridge, MA: The MIT Press.

Gehring, T., & Faude, B. (2013). The dynamics of regime complexes: Microfoundations and systemic effects. *Global Governance*, 19 (1), 119–30.

Gulbrandsen, L. H. (2009). The emergence and effectiveness of the Marine Stewardship Council. *Marine Policy*, 33 (4), 654–60.

Gulbrandsen, L. H., & Auld, G. (2016). Contested accountability logics in evolving nonstate certification for fisheries sustainability. *Global Environmental Politics*, 16 (2), 42–60.

Hannan, M. T., & Freeman, J. (1977). The population ecology of organizations. *American Journal of Sociology*, 82 (5), 929–64.

Hickmann, T. (2017). The reconfiguration of authority in global climate governance. *International Studies Review*, 19 (3), 430–51.

Ivanova, M. (2010). UNEP in global environmental governance: Design, leadership, location. *Global Environmental Politics*, 10 (1), 30–59.

Jinnah, S. (2014). *Post-treaty politics: Secretariat influence in global environmental governance*. Cambridge, MA: The MIT Press.

Juda, L. (2002). Rio plus ten: The evolution of international marine fisheries governance. *Ocean Development & International Law*, 33 (2), 109–44.

Jungcurt, S. (2011). The role of expert networks in reducing regime conflict: Contrasting cases in the management of plant genetic resources. In S. Oberthür, & O. S. Stokke (eds.), *Managing institutional complexity: Regime interplay and global environmental change* (pp. 171–98). Cambridge, MA: The MIT Press.

Kent, A. (2014). Implementing the principle of policy integration: institutional interplay and the role of international organizations. *International Environmental Agreements*, 14 (3), 203–24.

Keohane, R. O., & Victor, D. G. (2011). The regime complex for climate change. *Perspectives on Politics*, 9 (1), 7–23.

Kim, R. E. (2013). The emergent network structure of the multilateral environmental agreement system. *Global Environmental Change*, 23 (5), 980–91.

Kvalvik, I. (2012). Managing institutional overlap in the protection of marine ecosystems on the high seas. The case of the North East Atlantic. *Ocean and Coastal Management*, 56, 35–43.

Leebron, D. W. (2002). Linkages. *American Journal of International Law*, 96 (1), 5–27.

Margulis, M. E. (2013). The regime complex for food security: Implications for the global hunger challenge. *Global Governance*, 19 (1), 53–67.

Molenaar, E. J. (2007). Port state jurisdiction: Towards comprehensive, mandatory and global coverage. *Ocean Development & International Law*, 38 (1–2), 225–57.

Morin, J. F., & Orsini, A. (2014). Policy coherency and regime complexes: The case of genetic resources. *Review of International Studies*, 40 (2), 303–24.

Marine Stewardship Council (2017). *Annual Report 2016–17: The MSC at 20. Wild, certified, sustainable.* London: Marine Stewardship Council.

Newell, P., Pattberg, P., & Schroeder, H. (2012). Multiactor governance and the environment. *Annual Review of Environment and Resources*, 37, 365–87.

O'Neill, K., Weinthal, E., Suiseeya, K. R. M., Bernstein, S., Cohn, A., Stone, M. W., & Cashore, B. (2013). Methods and global environmental governance. *Annual Review of Environment and Resources*, 38, 441–71.

Oberthür, S. (2009). Interplay management: Enhancing environmental policy integration among international institutions. *International Environmental Agreements*, 9 (4), 371–91.

Oberthür, S., Dupont, C., & Matsumoto, Y. (2011). Managing policy contradictions between the Montreal and Kyoto protocols: The case of fluorinated greenhouse gases. In S. Oberthür, & O. S. Stokke (eds.), *Managing institutional complexity: Regime interplay and global environmental change* (pp. 115–41). Cambridge, MA: The MIT Press.

Oberthür, S., & Gehring, T. (eds.) (2006). *Institutional interaction in global environmental governance: Synergy and conflict among international and EU policies.* Cambridge, MA: The MIT Press.

Oberthür, S., & Stokke, O. S. (eds.) (2011). *Managing institutional complexity: Regime interplay and global environmental change.* Cambridge, MA: The MIT Press.

Ochieng, R. M., Visseren-Hamakers, I. J., & Nketiah, K. S. (2013). Interaction between the FLEGT-VPA and REDD+ in Ghana: Recommendations for interaction management. *Forestry Policy and Economics*, 32, 32–9.

Orsini, A., Morin, J. F., & Young, O. R. (2013). Regime complexes: A buzz, a boom, or a boost for global governance? *Global Governance*, 19, 27–39.

Ostrom, E., Burger, J., Field, C. B., Norgaard, R. B., & Policansky, D. (1999). Revisiting the commons: Local lessons, global challenges. *Science*, 284 (5412), 278–82.

Pattberg, P. (2005). The institutionalization of private governance: How business and on-profit organizations agree on transnational rules. *Governance*, 18 (4), 589–610.

Pratt, T. (2018). Deference and hierarchy in international regime complexes. *International Organization*, 72 (3), 561–90.

Raustiala, K., & Victor, D. G. (2004). The regime complex for plant genetic resources. *International Organization*, 58 (2), 277–309.

Rosendal, G. K. (2001). Overlapping international regimes: The case of the Intergovernmental Forum on Forests (IFF) between climate change and biodiversity. *International Environmental Agreements*, 1 (4), 447–68.

Rothwell, D. R., & Stephens, T. (2016). *The international law of the sea.* London: Bloomsbury Publishing.

Runhaar, H., Driessen, P., & Uittenbroek, C. (2014). Towards a systematic framework for the analysis of environmental policy integration. *Environmental Policy and Governance*, 24 (4), 233–46.

Scott, K. N. (2011). International environmental governance: Managing fragmentation through institutional connection. *Melbourne Journal of International Law*, 12 (1), 177–216.

Skjærseth, J. B. (2006). Protecting the North-East Atlantic: Enhancing synergies by institutional interplay. *Marine Policy*, 30 (2), 157–66.

Spijkers, J., & Boonstra, W. J. (2017). Environmental change and social conflict: The Northeast Atlantic mackerel dispute. *Regional Environmental Change*, 17 (6), 1835–51.

Stokke, O. S. (ed.) (2001a). *Governing high seas fisheries: The interplay of global and regional regimes*. Oxford: Oxford University Press.

Stokke, O. S. (2001b). *The interplay of international regimes: Putting effectiveness theory to work*. Lysaker: The Fridtjof Nansen Institute.

Stokke, O. S. (2004). Trade measures and climate compliance: Interplay between WTO and the Marrakesh Accords. *International Environmental Agreements*, 4 (4), 339–57.

Stokke, O. S. (2009). Trade measures and the combat of IUU fishing: Institutional interplay and effective governance in the Northeast Atlantic. *Marine Policy*, 33 (2), 339–49.

Stokke, O. S. (2011a). Environmental security in the Arctic: The case for multilevel governance. *International Journal*, 66 (4), 835–48.

Stokke, O. S. (2011b). Interplay management, niche selection and Arctic environmental governance. In S. Oberthür, & O. S. Stokke (eds.), *Managing institutional complexity: Regime interplay and global environmental change* (pp. 143–70). Cambridge, MA: The MIT Press.

Stokke, O. S. (2012). *Disaggregating international regimes: A new approach to evaluation and comparison*. Cambridge, MA: The MIT Press.

Stokke, O. S. (2013). Regime interplay in Arctic shipping governance: Explaining regional niche selection. *International Environmental Agreements*, 13, 65–85.

Stokke, O. S. (2019). Management options for high seas fisheries: Making regime complexes more effective. In J. R. Caddell, & E. J. Molenaar (eds.), *Strengthening international fisheries law in an era of changing oceans* (pp. 51–79). London: Hart Publishers.

Stokke, O. S., & Oberthür, S. (2011). Introduction: Institutional interaction in global environmental change. In S. Oberthür, & O. S. Stokke (eds.), *Managing institutional complexity: Regime interplay and global environmental change* (pp. 1–24). Cambridge, MA: The MIT Press.

United Nations Environment Programme (2016). *Programme of work and budget for the biennium 2018–2019*. UN Doc. UNEP/EA.2/16.

United Nations General Assembly (1972). *Institutional and financial arrangements for environmental cooperation*. UN Doc. A/RES/2997(XXVII).

Van Asselt, H. (2011). Legal and political approaches in interplay management: Dealing with the fragmentation of global climate governance. In S. Oberthür, & O. S. Stokke (eds.), *Managing institutional complexity: Regime interplay and global environmental change* (pp. 59–85). Cambridge, MA: The MIT Press.

Van Asselt, H. (2014). *The fragmentation of global climate governance: Consequences and management of regime interactions*. Cheltenham: Edward Elgar.

Van de Graaf, T., & De Ville, F. (2013). Regime complexes and interplay management. *International Studies Review*, 15 (4), 568–71.

World Trade Organization (2018). *The environment: A specific concern*. Available at: www.wto.org/english/thewto_e/whatis_e/tif_e/bey2_e.htm. Accessed: 17 June 2019.

Young, O. R. (1996). Institutional linkages in international society: Polar perspectives. *Global Governance*, 2 (1), 1–24.

Zelli, F., Gupta, A., & van Asselt, H. (2013). Institutional interactions at the crossroads of trade and environment: The dominance of liberal environmentalism? *Global Governance*, 19 (1), 105–18.

Zürn, M., & Faude, B. (2013). On fragmentation, differentiation, and coordination. *Global Environmental Politics*, 13 (3), 119–30.

11

Orchestration

KENNETH W. ABBOTT, STEVEN BERNSTEIN
AND AMY JANZWOOD

When Abbott and Snidal (2009a, 2009b) originated the idea of orchestration as a mode of global governance, earth system concerns figured prominently in their work. The limited direct authority of the United Nations Environment Programme (UNEP), on one hand, and the ubiquity of transnational actors 'governing' in these arenas, on the other, were emblematic examples of the 'orchestration deficit' that inspired their early work. The concept quickly caught on in earth system governance research. At the time, many earth system governance scholars were developing new research programmes to capture complex governing arrangements, multiple governance actors and the challenges of not only coordinating responses to environmental problems, but also providing the necessary resources, legitimacy and coherence to respond effectively where direct governing authority was weak or absent. The orchestration research programme addressed many of these challenges.

Abbott and colleagues (2015a) define orchestration as a mode of governance in which one actor (the orchestrator) enlists one or more intermediary actors (intermediaries) to govern a third set of actors (targets) in line with the orchestrator's goals. It is a strategy of *indirect* governance, which contrasts with both mandatory regulation and softer collaborative approaches in which governance actors directly engage their ultimate policy targets. An orchestrator lacks firm control over its intermediaries; it must therefore enlist their voluntary cooperation through leadership, persuasion and incentives. Enlistment is only feasible when the policy goals of (potential) intermediaries are broadly aligned with those of the orchestrator.

Earth system governance scholars have most frequently documented orchestration in climate change governance (e.g., Abbott 2012a, 2018; Chan and Pauw 2014; Hale and Roger 2014; Chan et al. 2015; Chan, Brandi and Bauer 2016; Bäckstrand and Kuyper 2017; Bäckstrand et al. 2017; Gordon and Johnson 2017; Kuyper, Linnér and Schroeder 2018; Hickmann et al. 2019). They have also analyzed its use by diverse intergovernmental organizations, treaty bodies and supranational actors in addressing other environmental issues and in sustainable development

233

governance more broadly (e.g., Abbott and Hale 2014; Abbott and Bernstein 2015; Graham and Thompson 2015).

This empirical work has led scholars to develop and investigate a series of hypotheses as to the conditions under which we can expect to see orchestration (Abbott et al. 2015a). Among the major findings of the last 10 years: states often support and even initiate orchestration by intergovernmental organizations, out of a desire to obtain a modicum of governance without creating strong institutions. Yet at the same time, orchestration can enhance the focality of intergovernmental organizations (their institutional position as leaders in their domains) and their authority, thereby increasing their independence from member states.

As the role of orchestrating institutions in environmental and sustainability governance has become increasingly apparent, earth system governance scholars have also begun to ask more normative questions: how does power manifest in indirect governance relationships, and what challenges does indirect governance pose for accountability, legitimacy and democracy (e.g., Bäckstrand and Kuyper 2017; Dryzek 2017; Kuyper, Linnér and Schroeder 2018)? Ultimately, we argue, a forward-looking research agenda in line with the new science plan of the Earth System Governance Project – focused centrally on the need for transformative earth system governance (Earth System Governance Project 2018; see also Burch et al. 2019) – must take these critiques seriously. Only by doing so can a new wave of scholarship understand how orchestration can support more effective governance in complex domains, in ways that promote the kinds of just transformations demanded by the environmental and sustainability crises we face.

We proceed as follows. After a brief review of the contributions of orchestration scholarship to the goals of the Earth System Governance Project and its first science plan (Biermann et al. 2009), we summarize key empirical findings of orchestration research in relevant governance domains. We then examine the contributions of earth system governance scholars to theoretical development in the orchestration research programme. We engage with critiques and normative concerns, particularly regarding power, accountability and legitimacy. We conclude with thoughts on a forward-looking orchestration research agenda that supports the transformative goals of the new earth system governance science plan.

Conceptualization

Orchestration had not yet emerged in the literature when Biermann and colleagues (2009) drafted the first science plan of the Earth System Governance Project. Yet two of the central policy challenges they identified – complexity and fragmentation – have been major drivers of innovative modes of global governance, and of efforts to study and promote them. Under the heading of architecture, the plan identified the importance of institutional and organizational multiplicity; governance by diverse actors, not just by states and domestic governments; multilevel governance; and vertical as well as horizontal linkages and interactions, especially 'within multi-layered institutional systems' (Biermann et al. 2009: 33). Orchestration touches all these bases.

What the first science plan failed to anticipate, however, is that governance performance or effectiveness may not depend simply on questions of institutional design, fit (with underlying environmental or sustainability problems) and interplay (among institutions, levels of governance or broader institutional architectures) (Young 2002). Rather, new modes of governance would be required to manage an evolving governance architecture and increasingly complex problems. By the early 2000s, policy-oriented research in areas such as forestry and climate change, prompted by weak or failing multilateral institutions, began to emphasize the importance of embracing institutional complexity, as opposed to assuming that architects of global governance could design their way out of the challenges it poses (e.g., Rayner, Buck and Katila 2010; Hoffmann 2011). Orchestration research highlighted the central role intermediaries play as bridges between institutions mandated to address global problems and the wide range of organizations and actors that possess the resources and capabilities to do so.

Orchestration research has focused especially on the conditions required to enlist, support and steer intermediaries. Some of these conditions, such as legitimacy and focality, resonate with broader research on legitimacy in global governance (e.g., Bernstein 2005, 2011; Tallberg, Bäckstrand and Scholte 2018). However, orchestration research has also highlighted the resources and capabilities required to govern through intermediaries.

Absent hard control, support is the key tool by which orchestrators generate the incentives needed to enlist intermediaries and to steer them towards desired goals. Support can come in many forms. While some orchestrators can provide administrative, financial or other material support, support is often ideational: orchestrators provide information, cognitive or normative guidance or endorsement. Orchestrators can also use their convening authority – derived from their institutional position and legitimacy – to catalyze formation of new intermediaries and to coordinate the activities of multiple intermediaries, increasing their impact. Hale and Roger (2014: 68–9) thus identify two types of orchestration: *initiating*, helping transnational actors to overcome collective action problems that impede organization, and *shaping*, supporting and steering existing transnational initiatives (see also Abbott and Hale 2014). Abbott (2017: 738) proposes that orchestrators could promote controlled governance experiments by intermediaries, deploying their expertise, authority and support 'to promote innovation, comparability, analysis and systematic learning', and selectively supporting those that demonstrate better results.

Orchestration also fit with the agency theme in the 2009 Science and Implementation Plan of the Earth System Governance Project (Biermann et al. 2009). The plan noted that involvement of 'local, domestic and transnational nonstate actors' is increasingly common and 'may become imperative'. Such actors 'frequently become [authoritative] agents of earth system governance in that they substantively participate in and/or set their own rules related to the interactions between humans and their natural environment' (Biermann et al. 2009: 37). The plan also noted that '[a]gents may contribute to the purposeful steering of constituents either indirectly (by influencing the decisions of other actors) or directly (by making steering decisions). They are thus a constituent part of the cumulative steering effort towards preventing, mitigating or adapting to earth system transformation' (Biermann et al. 2009: 38).

The orchestration literature has made particular contributions to understanding the governance roles of non-state actors. Those actors are frequently the ultimate targets of orchestrators. For example, an intergovernmental organization may encourage firms to disclose their environmental impacts, or seek to provide vulnerable individuals with benefits (e.g., humanitarian assistance). In such cases, the intermediaries too are frequently non-state actors, including non-governmental organizations, business associations and public–private partnerships (van der Lugt and Dingwerth 2015). These organizations have better access to non-state targets than do intergovernmental organizations; they can also deploy different forms of authority and governance instruments to influence them.

The new science plan of the Earth System Governance Project (2018) explicitly acknowledges the contributions orchestration research has made since 2009 to furthering our understanding of architecture and agency. It also highlights its potential to contribute to understanding new power relations in earth system governance.

Research Findings

Empirical Applications

Orchestration has been studied in areas of environmental governance including biofuels (Schleifer 2013; Henriksen 2015), aviation (Henriksen and Ponte 2017), shipping (Lister, Poulsen and Ponte 2015) and forestry (Abbott and Snidal 2009a). But orchestration research has particularly burgeoned in the field of climate change governance. Much of this research was in response to the expanding governance roles of substate and non-state actors in the post-Kyoto Protocol period, as multilateral negotiations became gridlocked. Orchestration has also been extended to human rights governance (Ruggie 2014; Pegram 2015) and the UN Global Compact (Shoji 2019). This section reviews orchestration research in two principal empirical domains.

First, orchestration has been a strategy of *sustainable development governance* for some years (Abbott and Hale 2014). Most notably, the 2002 World Summit on Sustainable Development initiated a major governance innovation by institutionalizing public–private partnerships as a primary means of implementation, enlisting them as intermediaries and promoting their formation (Pattberg, Biermann and Mert 2012; Bäckstrand and Kylsäter 2014). Non-state and networked forms of governance have proliferated in the wake of perceived weaknesses of the multilateral system in several sustainable development governance (e.g., Rayner, Buck and Katila 2010; Hoffmann 2011).

Intergovernmental organizations with sustainable development mandates – many of which have very limited capacities – have used orchestration creatively to promote norms and launch innovative governance mechanisms. For example, UNEP's role in helping to create, endorsing and supporting the Global Reporting Initiative is the paradigmatic case of orchestration (Abbott and Snidal 2010). Similarly, UNEP and the UN Global Compact orchestrated the formation of the Principles for Responsible Investment, to set standards for private investment firms (van der Lugt and Dingwerth 2015). The World Health Organization also facilitated creation of public–private partnerships such as Roll Back Malaria and Stop TB Partnership to act as intermediaries implementing operational programs (Hanrieder 2015).

In the run-up to the 2012 United Nations Conference on Sustainable Development, Abbott (2012b) identified an 'engagement gap' between state-based institutions such as UNEP and the United Nations Commission on Sustainable Development, and the diverse non-state actors and organizations engaged in sustainability governance, including business firms and associations, non-governmental organizations, multi-stakeholder initiatives and public–private partnerships. He also identified potential benefits of more active intergovernmental engagement with private sustainability governance.

Now, orchestration is arguably the central governance strategy for implementing the UN's Agenda 2030 for Sustainable Development (United Nations General Assembly 2015; Bernstein 2017; see also Paulo and Klingebiel 2016). Kanie and Biermann's (2017) volume on governance of the Sustainable Development Goals suggests that orchestration is the most suitable method for implementing the goals, and that the Sustainable Development Goals themselves help to orchestrate sustainable development governance. Bernstein's (2017) contribution to the volume argues specifically that the High-level Political Forum on Sustainable

Development, the lead UN body for reviewing and implementing the Sustainable Development Goals, must operate essentially as an 'orchestrator of orchestrators' (see also Abbott and Bernstein 2015). Boas and colleagues (2016) likewise show how orchestration by the High-level Political Forum on Sustainable Development can help states and organizations with different mandates to jointly address 'nexus' issues where interactions among issues or sectors, such as water-energy-food, must be addressed coherently.

Second, before the Paris climate agreement of 2015 – with no new multilateral climate agreement emerging from negotiations under the climate convention – transnational climate governance initiatives proliferated, drawing attention to the possibility of orchestrating them. Abbott (2012a) and Hale and Roger (2014) first mapped this expanding institutional universe; Hale and Roger found that about a third of transnational climate initiatives were orchestrated. Abbott (2012a) argued that orchestration could be used to manage transnational institutional complexity, 'to harness the benefits of decentralization while minimizing the costs'. Abbott (2014: 57) envisages orchestration of transnational actors as a way to bypass 'recalcitrant national governments'. Hale and Roger (2014) also propose a set of conditions to explain when states and intergovernmental organizations are likely to orchestrate (see below) by applying this theory to two case studies, namely the World Bank's Global Gas Flaring Reduction Partnership and the United Kingdom's Quality Assurance Scheme for carbon offsets.

Hale and Roger (2014: 61) argued that orchestration 'remains under-explored in the academic literature'. But that situation changed rapidly. For example, that same year, in a discussion paper prepared in consultation with the secretariat of the climate convention, Chan and Pauw (2014) propose a Global Framework for Climate Action to link multilateral climate governance with transnational initiatives. They argue that orchestration could be key in improving accountability and effectiveness in a 'fragmented climate-governance landscape' (Chan and Pauw 2014: 1). They draw lessons from two initiatives that failed to mobilize non-state and subnational initiatives effectively: the 2002 Partnerships for Sustainable Development framework and the 2010 Private Sector Initiative under the Nairobi work programme of the climate convention. In both cases, they argue, low participation requirements and weak screening procedures enabled business-as-usual activities to pass for climate or sustainable development action. They therefore propose a more comprehensive Global Framework for Climate Action, with functions including mobilizing new initiatives, monitoring progress and providing assessments.

Building on this and other frameworks (e.g., Pattberg, Biermann and Mert 2012; Hsu et al. 2015), Chan and colleagues (2015) propose that the 2015 UN Climate Conference in Paris adopt a broader institutional framework to catalyze, support and steer non-state initiatives. Orchestration would be a means to promote 'ambition, experimentation and accountability', while limiting overlap (Chan et al. 2015: 466). Orchestration has also been used as a lens to assess post-Paris climate governance. Chan and colleagues (2016: 245), for example, suggest that, while the Paris conference enhanced the visibility of non-party stakeholders and took steps to institutionalize their role, it failed to create a comprehensive orchestration framework. In particular, they argue, the conference provided few ways to ensure intermediary transparency, or to track and evaluate non-state and subnational climate actions (Chan, Brandi and Bauer 2016: 246).

While much of the literature on climate governance has focused on the ability of orchestrators to govern effectively, recent work, such as that by Hermwille (2018), focuses on feedback mechanisms, or the capacity of orchestrators to learn the impact of their orchestration efforts. Hermwille (2018: 449) finds, similarly to Chan, Brandi and Bauer (2016), that the climate convention has been an institutional one-way street, with few feedback mechanisms.

Other scholars have considered climate governance orchestration outside the climate convention. Graham and Thompson (2015), for example, show how the Global Environment Facility orchestrates financing for climate adaptation by developing countries by working through other intergovernmental organizations, which serve as implementing agencies. Pattberg (2017) uses orchestration of carbon disclosure by 'governance entrepreneurs' (that is, individuals and organizations that aim to change public regulations and rule systems) to illustrate the high level of transnational governance in climate change disclosure. Gordon and Johnson (2017) unpack the practice of orchestration in urban climate change governance. They expand the universe of orchestrators beyond states and intergovernmental organizations, with theoretical implications explored below.

Theoretical Advancement

Orchestration theory has built on and/or contributed to numerous theories of governance, including theories of New Governance (Abbott and Snidal 2009a), the performance of intergovernmental organizations (Abbott and Snidal 2010), responsive regulation (Abbott and Snidal 2013), regulatory intermediaries (Abbott, Levi-Faur and Snidal 2017) and experimentalist governance (Abbott 2017). Earth system governance research has also contributed to orchestration theory. Several studies of climate and sustainability governance have refined our understanding of the modes and methods of orchestration and elaborated the conditions under which actors are likely to adopt those methods and do so successfully. We summarize the key findings below.

Abbott and Snidal (2009a) identify two forms of orchestration, directive and facilitative. Directive orchestration entails stronger exercises of authority. For example, states with regulatory authority could promote transnational standards schemes by relaxing regulatory requirements for firms that adhere to them, or by conditioning public procurement opportunities on adherence. Intergovernmental organizations have much weaker authority, but can still condition procurement, financial assistance and other benefits on adherence to approved standards schemes. Facilitative orchestration involves a range of softer approaches, from catalyzing formation of intermediary organizations to providing material and ideational support. These two forms have been taken up in the literature. For example, Schleifer (2013) identifies both forms in EU biofuel policies, while Henriksen and Ponte (2017) observe them in transnational environmental governance of the aviation sector.

Abbott and colleagues (2019; see also Abbott et al. 2016) contrast orchestration with three other modes of indirect governance: delegation, trusteeship and co-optation. In delegation and trusteeship, governors devolve some of their authority to intermediaries, allowing intermediaries to use their capabilities to perform specified governance tasks. For example, some environmental conventions authorize non-state actors to perform specified governance functions (Green 2017). In co-optation and orchestration, in contrast, governors enlist the pre-existing authority of intermediaries, drawing on both their authority and their capabilities. In co-optation, governors retain some ex post control; orchestration, however, is wholly non-hierarchical.

Even within orchestration, several scholars have suggested refinements and sub-modes, with implications for governance and political control. Gordon and Johnson (2017) identify three modes of orchestration in urban climate governance; each creates new forms of power and political control. Where the orchestrator's objective is to align city actions with multilateral climate norms, 'complementary' orchestration creates a politics of inclusion and conformity. 'Concurrent' orchestration – in which actors attempt to coordinate actions by cities to produce autonomous effects – creates a politics of authority and contestation, alongside the multilateral regime. Finally, 'emergent' orchestration – in which climate governance is a by-product of competition for material resources – generates a politics of structural domination (Gordon and Johnson 2017: 15). In this analysis, orchestration is not simply a governance strategy, but an 'orientation' that reflects 'the dissemination and diffusion of experimental norms, metrics and best practices' (Gordon and Johnson 2017: 15).

Van der Ven and colleagues (2017: 2), in their study of the valuation of decarbonization initiatives, highlight that orchestration can create winners and losers. An orchestration platform such as the Non-state Actor Zone for Climate Action, for example – while ostensibly depoliticizing transnational action by gathering and publicizing information on voluntary emission reduction commitments – can overlook initiatives' 'contributions toward the broader, and often contentious, transformations required to achieve global decarbonization' (van der Ven, Bernstein and Hoffmann 2017: 1). Platforms may miss the ways initiatives scale up and entrench themselves (or fail to do so) if they evaluate performance or potential performance in the short term. The risk is that such evaluations can prematurely create winners and losers, by signaling endorsement that helps to mobilize resources, or conversely by signaling that an initiative is undeserving of further support (see also Gordon 2016; Hsu et al. 2019). For example, van der Ven and colleagues (2017) argue that the domestic failure of the United Kingdom's Carbon Trust carbon labelling initiative could lead observers to overlook its long-term contributions: improving supply chain efficiency and disseminating its standards to non-United Kingdom jurisdictions.

Explaining Orchestration

Abbott and colleagues (2015a) advance four general hypotheses as to the conditions under which orchestration is likely to occur; these are broadly supported by the empirical studies in the volume. Orchestration is more likely where: (a) the orchestrator lacks sufficient authority or capabilities to use harder modes of governance (see also Hale and Roger 2014: 66–7); (b) satisfactory intermediaries are available; (c) the orchestrator is focal in its domain; and (d) the orchestrator has an entrepreneurial culture. Henriksen and Ponte (2017) find that the absence of these conditions explains why the International Commercial Aviation Organization has not regulated the environmental impacts of the aviation industry.[1]

Incentives for orchestration also arise where efforts at intergovernmental cooperation fail. In that case, orchestration is often a second-best or fall-back solution. Yet where a governing institution lacks authority or capacity, where stronger governance approaches have high costs or undesirable consequences, and where intermediaries can more effectively interact with and influence targets, orchestration can be a first-best solution.

For intergovernmental organizations (or other institutions with strong principals), Abbott and colleagues (2015a) initially hypothesized that orchestration is more likely where member states (or other principals) have diverging preferences and weak control mechanisms, so that they cannot block entrepreneurial actions by officials from intergovernmental organizations. An important finding of their

---

[1] Henriksen and Ponte (2017: 4) distinguish their understanding of orchestration; in their usage, orchestrators, 'when using soft instruments, may operate both directly and through intermediaries'.

research, however, is that member states sometimes intend that an intergovernmental organization operate through orchestration (Abbott et al. 2015b). Abbott and Bernstein (2015) refine these findings by suggesting that an intergovernmental organization or other institution may act as an orchestrator *by default*, simply because it lacks sufficient authority and resources for more demanding modes of governance, or *by design*, because states or other principals intended it to be (only) an orchestrator, and thus endowed it with (only) the necessary level of authority and resources. The danger with orchestration by design is that principals may not provide even the modest levels of authority and resources necessary for orchestration: intentionally, because powerful players want the institution to be weak, or unintentionally, e.g., because of bargaining problems.

Several studies have analyzed the characteristics necessary for successful orchestration. Hale and Roger (2014) argue that these include an orchestrator's material, epistemic, moral and relational authority. Hale and Roger (2014: 207) also find that the absence of certain capacities – in particular, a perceived lack of expertise – in addition to a lack of focality explain the failure of orchestration in the case of the United Kingdom's Carbon Offset Quality Assurance Scheme. Relatedly, Abbott and Hale (2014) argue that an orchestrator's focality, resources (e.g., technical, material, administrative; see also Henriksen and Seabrooke 2016), legitimacy (stemming, for example, from moral authority or expertise) and organizational culture are key conditions of successful orchestration.

Henriksen and Ponte (2017) find that orchestrators are more likely to succeed when they use a combination of directive and facilitative instruments, including incentives backed by regulation or threat of regulation, and when they are robustly embedded in transnational networks of experts and stakeholders (Henriksen and Ponte 2017: 15). They also find that an additional set of factors, first developed by Lister and colleagues (2015), explain successful orchestration in the aviation industry: high issue visibility, interest alignment between regulators and targets, narrow issue scope and low regulatory fragmentation and uncertainty. Lister and colleagues (2015) had found that the absence of these factors impeded orchestration by the International Maritime Organization.

Other work has also studied cases of orchestration failure or fragility. Graham and Thompson (2015), for example, show that the Global Environment Facility's use of orchestration is eroding as the institution moves towards more direct modes of governance. Graham and Thompson identify both internal factors and external trends to explain this evolution (Graham and Thompson 2015: 136–8). Hanrieder (2015) finds that the ability of the World Health Organization to orchestrate, at least in certain areas, has eroded as its focality has declined, due to the rise of powerful competitors, notably the Bill and Melinda Gates Foundation. Schleifer (2013: 542) finds that the EU's efforts to orchestrate private sustainability governance for biofuels failed because EU officials set the bar for intermediaries too low; weak standards ended up undercutting good private governance instead of promoting it.

Who Orchestrates?

International institutions, especially intergovernmental organizations, have received the most attention in the literature as actual or potential orchestrators. In climate governance, Hickmann (2016) suggests that the secretariat of the climate convention should be strengthened and opened up so it can more effectively orchestrate sub- and non-state climate initiatives (see also Abbott 2014; Hermwille et al. 2017: 166; Hickmann et al. 2019). Kuyper, Linnér and Schroeder (2018: 7) suggest that the Paris Agreement 'solidifies the UNFCCC as orchestrator of the transnational (nonstate) actors'. Chan and colleagues (2015) also acknowledge orchestration by the secretariat of the climate convention, and call for it to collaborate with other intergovernmental organizations, transnational initiatives and research institutions to orchestrate substate and non-state actors. Abbott (2018) identifies a range of organs of the climate convention as climate orchestrators, including the secretariat, the conference of the parties and its presidencies, and subsidiary organs such as the 'high-level champions' established by the 2015 conference of the parties in Paris.

Intergovernmental organizations, however, are not the only orchestrators of transnational governance. Abbott and Hale (2014) identify a range of 'organizational entrepreneurs' that orchestrate policy networks. Of the 53 orchestrated networks they identify, intergovernmental organizations orchestrated half, national governments a third, and non-governmental organizations 16 per cent (Abbott and Hale 2014: 198). States and national and regional regulators are important orchestrators in both domestic and transnational governance (Hale and Roger 2014; Henriksen and Ponte 2017; Chan, Ellinger and Widerberg 2018). Other orchestrators include city networks, corporations, and philanthropic organizations. Gordon and Johnson (2017) identify several of these in urban climate governance. They suggest that orchestrators can emerge endogenously within transnational initiatives, or from 'the broader firmament of non-nation state actors engaged in the process' (Gordon and Johnson 2017: 701).

Van der Ven and colleagues (2017) introduce the concept of 'orchestration platforms', such as databases and registries of voluntary commitments that not only aggregate information on transnational initiatives, but also try to steer them towards objectives including transparency, accountability and effectiveness. Orchestration platforms primarily use information as the means of orchestration, through techniques such as collecting quantifiable data on the greenhouse gas emission reduction potentials of activities and assessing outputs against intended functions. Platforms also use techniques such as ratings or rankings. Orchestration platforms do not direct activity or make decisions, but the information they provide is a resource that serves to 'coordinate, mobilize and value the contributions of private, hybrid and subnational actors, in effect enlisting them as intermediaries to achieve defined regulatory goals' (van der Ven, Bernstein and Hoffmann 2017: 2; see also Abbott and Snidal 2010). The authors caution, however, that such 'ostensibly apolitical' methods may have political consequences, a finding in line with emerging scholarship on indirect governance through benchmarking, ratings and other forms of measurement (Best 2017; Broome, Homolar and Kranke 2018). The power dynamics in such techniques have been picked up in critical assessments of orchestration, discussed further below.

Similarly to the idea of orchestration platforms, Abbott (2017, 2018) identifies a development in global environmental governance whereby states, intergovernmental organizations and other governance institutions establish 'voluntary commitment systems'. These systems encourage non-state actors to make voluntary commitments alongside agreements or initiatives undertaken by states, gather information on those commitments, track their progress and publicize them, to provide incentives for commitment as well as learning opportunities. Two voluntary commitment systems – on sustainable development and on climate – actively encouraged collective commitments, for example, by partnerships and multi-stakeholder initiatives, as these institutions can act as intermediaries vis-à-vis firms and other target actors. Because they establish mechanisms for enabling, supporting and governing entire classes of organizations and activities, both platforms and voluntary commitment systems can be seen as engaging in systemic orchestration.

This body of work has identified several future directions for research. Gordon and Johnson (2017: 8) call for 'an enlarged conceptualization of *who orchestrates the globe*'. They argue for the importance of understanding which actors orchestrate, and in service of which objectives. Similarly, Abbott (2015: 493) suggests that future research should focus on orchestration as an independent variable, analyzing its effects rather than the conditions that encourage it (see also Widerberg 2017). Gordon and Johnson (2017: 8) also echo Pegram's (2015) call for scholars to study the relationships between orchestrators, intermediaries and targets. Using a wider analytic lens, Lelieveldt (2018) explores how EU competition law constrains domestic sustainability orchestration, calling for scholars to look at interactions among governance modes. Widerberg (2017) encourages scholars to open the 'black box' of orchestration relationships, identifying the theory of change they reflect and tracing their impacts on the governance goals being pursued (e.g., decarbonization in the case of orchestrating climate change initiatives) (see also van der Ven, Bernstein and Hoffmann 2017).

Orchestration and Complex Governance Architectures

As orchestration involves governance through intermediaries, it demands a certain degree of institutional complexity: at least the availability of suitable intermediaries, as well as one or more governors and targets. Moreover, a central finding of orchestration research is that governors often create, or catalyze the creation of, new organizations that can act as intermediaries, as in the climate voluntary commitment system. Thus, orchestration not only demands organizational complexity, but often increases it. When a governor creates purpose-built intermediaries, it has substantial ex ante influence on their form, governance, mandates, decision-making procedures, resources and other features (McCubbins, Noll and Weingast 1987), so it can be sure they are suitable.

Abbott (2018) looks at orchestration in the context of polycentric governance theory, which generally emphasizes decentralized mutual adjustment as the principal mechanism producing ordering in polycentric systems of institutions. Using climate governance as an example, however, he finds that orchestration is pervasive as a tool for the strategic ordering of the institutional complex. The clearest example is the role of multiple actors within the climate regime, as well as the UN Secretary-General and other intergovernmental organizations, in orchestrating non-state and substate commitments and initiatives, both individually and through the voluntary commitment system (see also Abbott 2017).

In terms of institutional complexity, however, these orchestration efforts most often seek to influence the structure and operation of the system by promoting formation of desired types of initiatives (thus increasing multiplicity and complexity), rather than by inducing coordination or behavioural change among existing initiatives. To be sure, in some cases orchestration of individual organizations (e.g., UNEP orchestration of the Global Reporting Initiative and the Principles for Responsible Investment) helps establish them as focal in their areas; this deters other institutions from entering the field and avoids increased complexity. In addition, in both the climate and the sustainable development voluntary commitment systems, orchestrators encouraged multi-stakeholder intermediaries that could elicit additional commitments and coordinate them to some degree.

Other ongoing research considers how the rise of new types of governance institutions that can be created at relatively low cost (e.g., informal intergovernmental organizations, transgovernmental networks, public–private partnerships and private initiatives) has encouraged the strategy of institutional creation as a mode of steering, thereby increasing institutional complexity (Abbott and Faude 2019). At the same time, however, promoting the creation of diverse institutions increases the prevalence of multilevel governance, which facilitates orchestration by institutions that possess greater authority, resources and focality.

Critiques and Normative Issues

Orchestration is generally seen as normatively good and is often suggested as a solution to governance problems where other forms of governance are unavailable or blocked (e.g., Chan and Pauw 2014; Abbott and Bernstein 2015; Abbott 2017, 2018; Bäckstrand and Kuyper 2017; Bernstein 2017; Chan, Ellinger and Widerberg 2018; Chan et al. 2018). Most critiques of orchestration are relatively soft. For example, according to Schleifer (2013), the orchestration literature contains little evidence of the benefits and complications of orchestration in practice (but see the many empirical case studies cited here). He argues, for example, that biofuel governance in the EU was a 'partial failure' because the orchestrator did not set high enough standards for intermediaries.

But several studies do identify important normative concerns that arise due to the nature of orchestration. Kuyper and colleagues (2018) argue that soft steering through orchestration, rather than hard principal-agent relationships, makes accountability difficult, because it is not always clear who is responsible for determining whether non-state actors make good on their commitments. In addition, where there are multiple orchestration efforts or multiple intermediaries, actors can make multiple pledges, raising the issue of double counting and reducing accountability. They argue that non-state actors outside the arrangements must act as watchdogs of orchestration platforms, and more generally argue for balancing orchestration efforts with mechanisms to monitor them (Kuyper, Linnér and Schroeder 2018: 11). Finally, Kuyper and colleagues (2018: 10) suggest that orchestrators should maintain space for 'authentic deliberation and participation by nonstate actors' to ensure legitimacy.

Another challenge to accountability is that orchestration by intergovernmental organizations inherently breaks the 'chains of delegation' that run from citizens to states and then to intergovernmental organizations (Bäckstrand and Kuyper 2017: 769; see also Abbott et al. 2015b). Because an intergovernmental organization engaged in orchestration cannot control its intermediaries, intermediary authority should be independently legitimated. In addition, they argue, the initial decision to engage in orchestration, and the ongoing relationships among orchestrator, intermediaries and targets, should all be legitimated. The burden of doing so falls mainly on the orchestrator.

Bäckstrand and Kuyper emphasize a particular basis of legitimacy: democratic legitimacy. Like all governance arrangements, they argue, orchestration should be assessed according to democratic criteria including participation, deliberation, accountability and transparency. Bäckstrand and Kuyper identify democratic deficits in the orchestration of the Lima-Paris Action Agenda and the Non-state Actor Zone for Climate Action. For example, in the Non-state Actor Zone for Climate

Action, non-state actor participation and leadership come largely from the Global North (on North-South participation gaps, see Chan et al. 2018). Bäckstrand and Kuyper (2017: 781) argue that this is largely due to the bottom-up mode of orchestration under the Non-state Actor Zone for Climate Action, with limited criteria or screening. They also find that both the Lima-Paris Action Agenda and the Non-state Actor Zone for Climate Action lack accountability and transparency (Bäckstrand and Kuyper 2017: 781).

These arguments pose a difficult challenge. Few international governance processes – and even fewer transnational processes – fully meet Bäckstrand and Kuyper's criteria. Democratic procedures raise the cost of orchestration and may thus prevent the emergence of governance initiatives that would produce valuable public goods. Where action at a large scale is critical – as, arguably, with voluntary climate initiatives – this trade-off is especially sharp. Although the responsibility for legitimation may rest with orchestrators, the tools of orchestration available to them may be insufficient to demand fully democratic structures along the orchestration chain. There may, moreover, be alternative bases of legitimacy; in polycentric governance, for example, institutions may play a legitimate public role by acting as a check on other powerful institutions (Ostrom 1991).

Gordon and Johnson (2017: 14) suggest that the orchestration literature has overlooked power dynamics, different 'logics of acquiescence' and different forms of control. Orchestration, they argue, is not an instrumental tool free of politics, but an arena of politics and power. Dryzek (2017) also explores the relationship between power and orchestration but comes to more provocative conclusions. In embracing the role of orchestrated intermediary, non-state actors, with limited capabilities, may become unable to fulfil other roles. Acting as an intermediary may restrict an actor's ability to criticize the status quo, hold power holders to account or represent alternative views (Dryzek 2017: 792–3). Dryzek also invokes Foucault's notion of governmentality to warn that these effects risk reinforcing an unjust order.

How one evaluates this risk depends on one's view of the most important governance problems. In an ecological context, Dryzek (2017: 793) argues, the most important virtue is 'reflexivity', the capacity of institutions to remake themselves after reflecting on their own performance – as the climate convention did when it began to orchestrate non-state actors. Paradoxically, while non-state actors can play a critical role in reflection and deliberation, acting as intermediaries may diminish their ability to contribute (Dryzek 2017: 795).

Conclusions and Future Directions

The 2018 Science and Implementation Plan of the Earth System Governance Project focuses on transformations (Burch et al. 2019; see also Chapter 14).

Governance must contend with transformations: in humans' impact on earth systems, technology, politics and socio-economic systems. It must also encourage transformations in socio-economic and political practices, to achieve more just and sustainable futures. The latter effort requires governance *for* transformation – creating the conditions for beneficial transformations – as well as governance *of* transformations – triggering and steering them (Earth System Governance Project 2018: 26).

What role can orchestration play in addressing these challenges? To date, it is fair to say, orchestration research has not grappled with the nature of the goals this mode of governance can achieve; it has generally assumed simply that orchestration helps institutions with limited capacities to achieve their stated goals. In the next stage of earth system governance orchestration research, then, it would be both important and promising to consider explicitly whether and how this mode of governance can be used to respond to transformations of socio-ecological systems and the earth system as a whole, and to encourage and steer beneficial and just sustainability transitions.

Three concerns about the capacity of orchestration to achieve these goals stand out. First is the scope of orchestration. Contending with human impacts on the earth system, and promoting global social, economic and technological transformations such as decarbonization, require coordinated – or at least mutually supporting – changes around the world, involving many disciplines and affecting many diverse social groups, with potential winners and losers who must be given appropriate voice. The second concern is the role of power in orchestration. Orchestration is a mode of soft, indirect governance; how can it effectively create conditions for such transformations, trigger and steer them, in the face of societal inertia and determined opposition? The third concern is the role of justice in orchestration. The accountability and legitimacy concerns discussed above make it important to consider how orchestration can accomplish just transformations, through just processes.

To be sure, no one should expect orchestration, or any mode of governance, to achieve such results on its own. Orchestration can be a valuable complement to intergovernmental agreements, national regulations, global goals and other mechanisms; and it can help fill gaps left by those mechanisms. But it is no panacea.

Orchestration does, however, have highly relevant strengths. We briefly mention three.

First, orchestration often operates ideationally, given its generally weak coercive force. Orchestrator and intermediaries interact around common and divergent governance goals, and through ideational and normative support and steering. Many intermediaries also lack coercive power and use ideational and normative strategies to influence targets. Thus, orchestration can be useful in triggering the rethinking of problems, normative deliberation, and new social relationships. While specific actions may appear incremental, they can contribute – often unpredictably – to transformative effects (Earth System Governance Project 2018: 26–7).

Second, orchestration engages multiple, diverse actors. Many actors can orchestrate; even more can serve as intermediaries. Some coordination is needed to enhance governance impact, but this need not be tight. The polycentric nature of orchestration helps it to address problems with broad geographical scope or to address the nexus between multiple problems, involving multiple disciplines and diverse societal groups (Boas, Biermann and Kanie 2016: 450). It gives orchestrated governance an experimental quality, promoting investigation and learning, whether experiments are formal or informal (Abbott 2017). It allows many actors to participate in governance, enhancing their voice. And it supports the simultaneous use of multiple, parallel governance approaches, targeting actors ranging from states to business firms to individuals.

Third, systemic orchestration creates supportive frameworks of institutions, ideas and norms, which encourage, support and steer whole classes of actors and initiatives. This approach too provides orchestration with scope and power, based on multiple, reinforcing actions and broad participation. Potentially, then, systemic orchestration can trigger broad social change.

Interactions between top-down and bottom-up techniques will be crucial in the governance of transformations. Broad goals like the Sustainable Development Goals, global institutions like the High-level Political Forum and intergovernmental agreements like the Paris Agreement can create conditions that make orchestration more effective; orchestration in turn can mobilize societal actors to implement the goals they set. In the end, a portfolio of governance approaches will be necessary to move the earth system towards sustainability.

References

Abbott, K. W. (2012a). The transnational regime complex for climate change. *Environment and Planning C: Government and Policy*, 30 (4), 571–90.

Abbott, K. W. (2012b). Engaging the public and the private in global sustainability governance. *International Affairs*, 88 (3), 543–64.

Abbott, K. W. (2014). Strengthening the transnational regime complex for climate change. *Transnational Environmental Law*, 3 (1), 57–88.

Abbott, K. W. (2015). Orchestration. In P. H. Pattberg, & F. Zelli (eds.), *Encyclopedia of global environmental governance and politics* (pp. 487–95). Cheltenham: Edward Elgar.

Abbott, K. W. (2017). Orchestrating experimentation in nonstate environmental commitments. *Environmental Politics*, 26 (4), 738–63.

Abbott, K. W. (2018). Orchestration: Strategic ordering in polycentric governance. In A. Jordan, D. Huitema, H. van Asselt, & J. Forster (eds.), *Governing climate change:*

Polycentricity in action? (pp. 188–209). Cambridge, UK: Cambridge University Press.

Abbott, K. W., & Bernstein, S. (2015). The High-level Political Forum on Sustainable Development: Orchestration by default and design. *Global Policy*, 6 (3), 222–33.

Abbott, K. W., & Faude, B. (2019). *Choosing low-cost institutions in global governance.* Unpublished manuscript.

Abbott, K. W., Genschel, P., Snidal, D., & Zangl, B. (2015a). Orchestration: Global governance through intermediaries. In K. W. Abbott, P. Genschel, D. Snidal, & B. Zangl (eds.), *International organizations as orchestrators* (pp. 3–36). Cambridge, UK: Cambridge University Press.

Abbott, K. W., Genschel, P., Snidal, D., & Zangl, B. (2015b). Orchestrating global governance: From empirical findings to theoretical implications. In K. W. Abbott, P. Genschel, D. Snidal, & B. Zangl (eds.), *International organizations as orchestrators* (pp. 349–79). Cambridge, UK: Cambridge University Press.

Abbott, K. W., Genschel, P., Snidal, D., & Zangl, B. (2016). Two logics of indirect governance: Delegation and orchestration. *British Journal of Political Science*, 46 (4), 719–29.

Abbott, K. W., Genschel, P., Snidal, D., & Zangl, B. (2019). Competence versus control: The governor's dilemma. *Regulation & Governance*, in press.

Abbott, K. W., & Hale, T. (2014). Orchestrating global solutions networks: A guide for organizational entrepreneurs. *Innovations: Technology, Governance, Globalization*, 9 (1–2), 195–212.

Abbott, K. W., Levi-Faur, D., & Snidal, D. (2017). Theorizing regulatory intermediaries. *The Annals of the American Academy of Political and Social Science*, 670 (1), 14–35.

Abbott, K. W., & Snidal, D. (2009a). Strengthening international regulation through transnational new governance: Overcoming the orchestration deficit. *Vanderbilt Journal of Transnational Law*, 42, 501–78.

Abbott, K. W., & Snidal, D. (2009b). The governance triangle: Regulatory standards institutions and the shadow of the state. In W. Mattli, & N. Woods (eds.), *In whose benefit? Explaining regulatory change in global politics* (pp. 49–84). Princeton, NJ: Princeton University Press.

Abbott, K. W., & Snidal, D. (2010). International regulation without international government: Improving IO performance through orchestration. *Review of International Organizations*, 5 (3), 315–44.

Abbott, K. W., & Snidal, D. (2013). Taking responsive regulation transnational: Strategies for international organizations: Taking RR transnational. *Regulation & Governance*, 7 (1), 95–113.

Bäckstrand, K., & Kuyper, J. W. (2017). The democratic legitimacy of orchestration: The UNFCCC, nonstate actors, and transnational climate governance. *Environmental Politics*, 26 (4), 764–88.

Bäckstrand, K., Kuyper, J. W., Linnér, B. O., & Lövbrand, E. (2017). Nonstate actors in global climate governance: From Copenhagen to Paris and beyond. *Environmental Politics*, 26 (4), 561–79.

Bäckstrand, K., & Kylsäter, M. (2014). Old wine in new bottles? The legitimation and delegitimation of UN public–private partnerships for sustainable development from the Johannesburg Summit to the Rio+20 Summit. *Globalizations*, 11 (3), 331–47.

Bernstein, S. (2011). Legitimacy in intergovernmental and nonstate global governance. *Review of International Political Economy*, 18 (1), 17–51.

Bernstein, S. (2005). Legitimacy in global environmental governance. *Journal of International Law and International Relations*, 1 (1–2), 139–66.

Bernstein, S. (2017). The United Nations and the governance of Sustainable Development Goals. In N. Kanie, & F. Biermann (eds.), *Governance through goals: New strategies for global sustainability* (pp. 213–39). Cambridge, MA: The MIT Press.

Best, J. (2017). The rise of measurement-driven governance: The case of international development. *Global Governance*, 23 (2), 163–81.

Biermann, F., Betsill, M. M., Gupta, J. et al. (2009). *Earth system governance: People, places and the planet. Science and implementation plan of the Earth System Governance Project.* Bonn: Earth System Governance Project.

Boas, I., Biermann, F., Kanie, N. (2016). Cross-sectoral strategies in global sustainability governance: Towards a nexus approach. *International Environmental Agreements*, 16 (3), 449–64.

Broome, A., Homolar, A., & Kranke, M. (2018). Bad science: International organizations and the indirect power of global benchmarking. *European Journal of International Relations*, 24 (3), 514–39.

Burch, S., Gupta, A., Inoue, C. Y. A. et al. (2019). New directions in earth system governance research. *Earth System Governance*, 1.

Chan, S., van Asselt, H., Hale, T. et al. (2015). Reinvigorating international climate policy: A comprehensive framework for effective nonstate action. *Global Policy*, 6 (4), 466–73.

Chan, S., Brandi, C., & Bauer, S. (2016). Aligning transnational climate action with international climate governance: The road from Paris. *Review of European, Comparative & International Environmental Law*, 25 (2), 238–47.

Chan, S., Ellinger, P., & Widerberg, O. (2018). Exploring national and regional orchestration of nonstate action for a < 1.5 C world. *International Environmental Agreements*, 18 (1), 135–52.

Chan, S., Falkner, R., Goldberg, M., & van Asselt, H. (2018). Effective and geographically balanced? An output-based assessment of nonstate climate actions. *Climate Policy*, 18 (1), 24–35.

Chan, S., & Pauw, P. (2014). *A global framework for climate action (GFCA): Orchestrating nonstate and subnational initiatives for more effective global climate governance.* Bonn: German Development Institute.

Dryzek, J. S. (2017). The meanings of life for nonstate actors in climate politics. *Environmental Politics*, 26 (4), 789–99.

Earth System Governance (2018). *Earth system governance: Science and implementation plan of the Earth System Governance Project.* Utrecht: Earth System Governance Project.

Gordon, D. J. (2016). The politics of accountability in networked urban climate governance. *Global Environmental Politics*, 16 (2), 82–100.

Gordon, D. J., & Johnson, C. A. (2017). The orchestration of global urban climate governance: Conducting power in the post-Paris climate regime. *Environmental Politics*, 26 (4), 694–714.

Graham, E. R., & Thompson, A. (2015). Efficient orchestration? In K. W. Abbott, P. Genschel, D. Snidal, & B. Zangl (eds.), *International organizations as orchestrators* (pp. 114–38). Cambridge, UK: Cambridge University Press.

Green, J. F. (2017). Transnational delegation in global environmental governance: When do non-state actors govern? *Regulation & Governance*, 12 (2), 263–76.

Hale, T., & Roger, C. (2014). Orchestration and transnational climate governance. *Review of International Organizations*, 9 (1), 59–82.

Hanrieder, T. (2015). WHO orchestrates? Coping with competitors in global health. In K. W. Abbott, P. Genschel, D. Snidal, & B. Zangl (eds.), *International organizations as orchestrators* (pp. 3–36). Cambridge, UK: Cambridge University Press.

Henriksen, La. F. (2015). The global network of biofuel sustainability standards-setters. *Environmental Politics*, 24 (1), 115–37.

Henriksen, L. F., & Ponte, S. (2017). Public orchestration, social networks, and transnational environmental governance: Lessons from the aviation industry: Public orchestration in aviation. *Regulation & Governance*, 12 (1), 23–45.

Henriksen, L. F., & Seabrooke, L. (2016). Transnational organizing: Issue professionals in environmental sustainability networks. *Organization*, 23 (5), 722–41.

Hermwille, L. (2018). Making initiatives resonate: How can nonstate initiatives advance national contributions under the UNFCCC? *International Environmental Agreements*, 18 (3), 447–66.

Hermwille, L., Obergassel, W., Ott, H. E., & Beuermann, C. (2017). UNFCCC before and after Paris: What's necessary for an effective climate regime? *Climate Policy*, 17 (2), 150–70.

Hickmann, T. (2016). *Rethinking authority in global climate governance: How transnational climate initiatives relate to the international climate regime*. Routledge Research in Global Environmental Governance. London: Routledge.

Hickmann, T., Widerberg, O., Lederer, M., & Pattberg, P. (2019). The United Nations Framework Convention on Climate Change Secretariat as an orchestrator in global climate policymaking. International Review of Administrative Sciences, in press.

Hoffmann, M. J. (2011). *Climate governance at the crossroads: Experimenting with a global response after Kyoto*. Oxford: Oxford University Press.

Hsu, A., Höhne, N., Kuramochi, T. et al. (2019). A research roadmap for quantifying nonstate and subnational climate mitigation action. *Nature Climate Change*, 9 (1), 11–17.

Hsu, A., Moffat, A. S., Weinfurter, A. J., & Schwartz, J. D. (2015). Towards a new climate diplomacy. *Nature Climate Change*, 5 (6), 501–3.

Kanie, N., & Biermann, F. (eds.) (2017). *Governing through goals: Sustainable Development Goals as governance innovation*. Cambridge, MA: The MIT Press.

Kuyper, J. W., Linnér, B. O., & Schroeder, H. (2018). Nonstate actors in hybrid global climate governance: Justice, legitimacy, and effectiveness in a post-Paris era: Nonstate actors in hybrid global climate governance. *Wiley Interdisciplinary Reviews: Climate Change*, 9 (1), e497.

Lelieveldt, H. (2018). Out of tune or well tempered? How competition agencies direct the orchestrating state: Directing the orchestrating state. *Regulation & Governance*, in press.

Lister, J., Poulsen, R. T., & Ponte, S. (2015). Orchestrating transnational environmental governance in maritime shipping. *Global Environmental Change*, 34, 185–95.

McCubbins, M. D., Noll, R. G., & Weingast, B. R. (1987). Administrative procedures as instruments of political control. *The Journal of Law, Economics, and Organization*, 3 (2), 243–77.

Ostrom, V. (1991). Polycentricity: The structural basis of self-governing systems. In V. Ostrom (ed.), *The meaning of American federalism: Constituting a self-governing society* (pp. 233–44). San Francisco: ICS Press.

Pattberg, P. H. (2017). The emergence of carbon disclosure: Exploring the role of governance entrepreneurs. *Environment and Planning C: Politics and Space*, 35 (8), 1437–55.

Pattberg, P. H., Biermann, F., & Mert, A. (eds.) (2012). *Public–private partnerships for sustainable development: Emergence, influence and legitimacy*. Cheltenham: Edward Elgar.

Paulo, S., & Klingebiel, S. (2016). *New approaches to development cooperation in middle-income countries: Brokering collective action for global sustainable development*. Bonn: German Development Institute.

Pegram, T. (2015). Global human rights governance and orchestration: National human rights institutions as intermediaries. *European Journal of International Relations*, 21 (3), 595–620.

Rayner, J. M., Buck, A., & Katila, P. (eds.) (2010). *Embracing complexity: Meeting the challenges of international forest governance. A global assessment report*. Vienna: International Union of Forest Research Organizations.

Ruggie, J. G. (2014). Global governance and 'New Governance Theory': Lessons from business and human rights. *Global Governance*, 20 (1), 5–17.

Schleifer, P. (2013). Orchestrating sustainability: The case of European Union biofuel governance: Orchestrating sustainability. *Regulation & Governance*, 7 (4), 533–46.

Shoji, M. (2019). The UN Global compact for transnational business and peace: A need for orchestration? In M. Mahmudur Rahim (ed.), *Code of conduct on transnational corporations: Challenges and opportunities* (pp. 89–110). Cham: Springer International Publishing.

Tallberg, J., Bäckstrand, K., & Scholte, J. A. (eds.) (2018). *Legitimacy in global governance: Sources, processes, and consequences*. Oxford: Oxford University Press.

United Nations General Assembly (2015). *Transforming our world: The 2030 agenda for sustainable development*. UN Doc. A/RES/70/1.

Van der Lugt, C., & Dingwerth, K. (2015). Governing where focality is low. In K. W. Abbott, P. Genschel, D. Snidal, & B. Zangl (eds.), *International organizations as orchestrators* (pp. 237–61). Cambridge, UK: Cambridge University Press.

Van der Ven, H., Bernstein, S., & Hoffmann, M. (2017). Valuing the contributions of nonstate and subnational actors to climate governance. *Global Environmental Politics*, 17 (1), 1–20.

Widerberg, O. (2017). The 'black box' problem of orchestration: How to evaluate the performance of the Lima-Paris action agenda. *Environmental Politics*, 26 (4), 715–37.

Young, O. R. (2002). *The institutional dimensions of environmental change: Fit, interplay and scale*. Cambridge, MA: The MIT Press.

12

Governance through Global Goals

MARJANNEKE J. VIJGE, FRANK BIERMANN, RAKHYUN E. KIM, MAYA BOGERS,
MELANIE VAN DRIEL, FRANCESCO S. MONTESANO, ABBIE YUNITA
AND NORICHIKA KANIE

In recent years, a relatively new mechanism of global governance has gained prominence: the use of broad global policy goals to orchestrate the activities of governments, international organizations, civil society, the private sector, and eventually all citizens of the world. Global governance through goal-setting works through the joint commitment of all governments to collective policy ambitions. These ambitions are then enshrined in the form of multilaterally agreed goals that are not legally binding but come with more specific targets, indicators and time frames, all of which are expected to steer public and private actors collectively into desired trajectories (Kanie and Biermann 2017). While governance through global goal-setting has featured in global governance since the second half of the twentieth century, its role has become much stronger in the last two decades (Fukuda-Parr 2014). The Millennium Development Goals, agreed by the United Nations in 2000, were a first attempt at comprehensive global steering through goals. But global goal-setting has gained much more importance when the United Nations General Assembly agreed, in 2015, on 17 Sustainable Development Goals to be implemented by 2030.

Like other attempts at global governance through goal-setting, the Sustainable Development Goals share four key characteristics (Biermann, Kanie and Kim 2017). First, they are not legally binding and cannot be enforced as law within national or international adjudication. Second, they are marked by weak institutional arrangements that are not supported by international treaty organizations, formal monitoring agencies, strong dispute settlement bodies and the like. Third, they are meant to be highly inclusive, covering all countries and sectors of society. Fourth, they are broadly framed and hence leave much leeway to national implementation and interpretation. While none of these characteristics is specific to this type of governance, the combination of these four characteristics amounts to a unique approach to global governance.

In this chapter, we review recent literature on these four key characteristics of governance through global goals. We first conceptualize governance through goals as a mechanism of global governance. We then delve into key literature around the four main characteristics of governance through goals, with a view to understanding how they affect the performance of governance architectures. We then distil how these characteristics, taken together, can affect governance architectures, for instance by leading to new actor constellations, by galvanizing efforts and by transforming or creating new institutions. Thereafter, we identify future research directions that might help increase understanding of whether and how global goals could effectively deal with the challenges that result from the institutional complexity of global governance architectures.

Conceptualization

We define global goals as internationally agreed non-legally binding policy objectives that are time-bound, measurable and aspirational in nature. Notably, in this definition, we exclude legally binding international legal rules and norms such as those often established through multilateral agreements. We also leave out widely proclaimed aspirations of global civil society and other non-state actors, such as those reflected in transnational private regulations. These goals from non-governmental bodies do not enjoy the formal support of governments and inter-governmental organizations; they are rather part of the realm of non-state, transnational governance (Chapter 4). Furthermore, while we acknowledge that goals have been a feature of global governance since the first United Nations Development Decade in the 1960s, we focus on the more recent, and much more ambitious, global goals, and especially the Sustainable Development Goals from 2015.

The concrete mechanisms through which global goals function are yet to be examined in detail. There is consensus, however, that a key defining feature of governance through goals is that it does not seek to directly change existing institutional architectures, and that it does not seek to regulate existing institutions or actors by demanding or enforcing behavioural change (see contributions in Kanie and Biermann 2017). Rather, governance through goals relies on non-legally binding global public policy goals, generally negotiated under the purview of intergovernmental institutions and organizations, most notably the United Nations. Such goals are hence largely aspirational, but they are typically endorsed by governments and non-state actors around the world, which could enable them to guide actions and policies at global, national and subnational levels.

Although it is unknown to what extent governance through global goals can really lead to immediate and radical governance transformations, many commentators and supporters expect them to have some impacts, for example by triggering incremental but widespread changes when goals are taken up in national and international policies and programmes. Governance through goals can thus have some influence by setting priorities that shape the international and national allocation of scarce resources, as well as by galvanizing action through specific and time-bound targets with which actors track their progress towards goal achievement (Young 2017; see also Chapter 13). As such, governance through goals can trigger and orchestrate, rather than enforce, some of the policy responses to governance fragmentation and institutional complexity that have been analyzed in this volume, such as policy integration (Chapter 9), interplay management (Chapter 10), orchestration (Chapter 11) and hierarchization (Chapter 13).

The effects and effectiveness of governance through goals remain contested, however (see discussion in Kanie et al. 2017). While some observers argue that global goals can have significant impacts (Hajer et al. 2015; Stevens and Kanie 2016), others criticize this governance mechanism for its lack of enforcement and compliance mechanisms. Will the goals be effective in the end? In this chapter, we review the body of social science literature that deals with this question. We are less interested in whether goals are actually implemented but rather in the prior, first step: whether goals have any effects on governance systems and processes, and here in particular on whether goals have the potential to affect entire governance architectures, for example by advancing institutional integration between decision-making systems or reducing norm conflicts. While some observers are optimistic that the Sustainable Development Goals of 2015 will help foster institutional integration at the international level (Le Blanc 2015), others doubt such claims, arguing that the goals themselves simply reflect the fragmented structure of global governance (Kim 2016). So far, however, there has been little, if any, empirically grounded research on the effects of governance through goals on governance architectures. Therefore, our review attempts here to lay the foundation for new inquiries into this research domain.

Research Findings

We now review recent research findings and conceptual contestations on the four key characteristics of governance through goals mentioned above, namely their non-legally binding nature; the underlying weak institutional arrangements; the inclusiveness of the goal-setting process; and the national leeway in the implementation of the goals.

Non-legally Binding Nature

A first key characteristic of governance through goals is that they are not legally binding (Biermann, Kanie and Kim 2017). Both the Millennium Development

Goals of 2000 and the Sustainable Development Goals of 2015 were formally established by a non-binding United Nations General Assembly Resolution as part of a broader development agenda. Although some scholars claim that the United Nations General Assembly has quasi-legal competences (Falk 1966), the United Nations Charter clearly deems its resolutions as being only recommendations, as they are not formally signed and ratified by states. These sets of global goals are hence not part of international law but are essentially political agreements (Kim 2016).

Some scholars have argued, therefore, that goal-setting through non-binding agreements is merely a suboptimal, ineffective or even counterproductive strategy. Some even see it as contributing to increasing institutional complexity and fragmentation, with the potential to complicate international cooperation (Elliot 2017). For those global goals that are grounded in international agreements – as is the case with some targets under the Sustainable Development Goals – legal scholars have emphasized the need to create additional mechanisms to ensure that these goals are not just a reflection of, but reach further than the existing fragmented and compartmentalized system of international law (Kim 2016: 17; see also Kim and Bosselmann 2015; Underdal and Kim 2017).

Others have questioned the ability of non-binding goal-setting to influence a wider political arena and to mobilize societal forces in modern systems of multilevel governance (Bodansky 2016; Young 2017). A non-binding status could potentially limit the compliance-pull and legitimacy of globally agreed goals at the national level, because acceptance can be limited to mere executive approval, without the need for governments to seek domestic legislative approval and formal adoption (Bodansky 2016). For example, domestic courts are not obliged to use the Sustainable Development Goals as a judicial source when resolving disputes.

Furthermore, the non-binding status of global goals might limit the sense of urgency, commitment and acceptance, especially among government officials who are expected to assume key roles in realizing the goals (Young 2017: 43; see also Franck 1990; Raustiala 2005; Bodansky 2016). That governments generally attribute some value to the legal status of agreements is emphasized by the strong disappointment expressed by many governments when the outcome of the 2009 Copenhagen conference of the parties under the climate convention proved to be 'only' a political agreement. Another example are the continued discussions over the legal status of the subsequent 2015 Paris Agreement (Bodansky 2010, 2016).

In addition, given the lack of legal standing, internationally it could be unclear how new global goals, such as the Sustainable Development Goals, relate to all the earlier agendas, agreements and plans. In the case of the Millennium Development Goals, for example, it has been argued that they disrupted ongoing processes for the implementation of the 1990s conference agendas through cherry-picking of issues, the modification of previously agreed targets and the disruption of nascent initiatives (Fukuda-Parr, Yamin and Greenstein 2014; Langford and Winkler 2014; van der Hoeven 2014).

Yet, while it does seem that lack of legal force limits the effectiveness of global goals, the opposite argument is also found in the literature. Serious questions have been raised, for instance, about the effectiveness of international environmental law (Kim and Bosselmann 2013) and the extent to which it affects state behaviour (Goldsmith and Posner 2005). Bodansky (2016) even argued that some merely political agreements – including the 2009 Copenhagen Accord – have had a greater influence on state behaviour than legal agreements. Proponents of goal-setting add here that its underlying premises differ substantially from those of rule-making (Young 2017: 34). Whereas rule-making creates indefinite behavioural prescriptions formulated as requirements and prohibitions for specified actors, goal-setting articulates time-bound aspirations, procedures and targets that need to rely on enthusiastic support among a wide range of actors to induce self-governance (Young 2017). The expectation of behavioural constraints that legally binding documents potentially create can even lead to pick-and-choose strategies among countries, resulting in many narrow agreements with only few parties that leave out important countries. The more flexible instrument of goal-setting, however – especially when it provides possibilities for the adaptation to national and local realities – might motivate all governments to make at least some contributions on sensitive topics (Zelli et al. 2010). For example, although the reduction of inequality between and within states was a bone of contention during the negotiations of the Sustainable Development Goals, all countries have in the end agreed to Goal 10 on inequality, including many highly hesitant parties such as the United States (Kamau, Chasek and O'Connor 2018: 184). This would not have been possible if that goal had been legally binding.

Another dimension of 'bindingness' is the precision with which goals are formulated. Although the Paris Agreement included non-legally binding Nationally Determined Contributions, its provisions are formulated in terms that do not create clear individual obligations (Bodansky 2016: 146). Also its provisions on adaptation and means of implementation lack the precision to create enforceable legal obligations (Bodansky 2016). An increasing number of legal norms and provisions can result in the progressive proliferation of normative ambiguity with little effect, whereas non-legally binding commitments might in some cases be more precise and effective (Victor, Raustalia and Skolnikoff 1998). This is what some argue could be the case with the non-binding but sometimes very precise indicators for the Sustainable Development Goals.

Whether global goals as legally non-binding political agreements can have some effect will, hence, depend more on the detail and on additional elements that add alternative dimensions to bindingness that could enhance compliance (Bodansky 2016: 149). An important example is the extent to which accountability mechanisms are in place to support global goals, for instance through systems of transparency and review. In the case of the Sustainable Development Goals, the Voluntary National Reviews provide such a system. Although it will still take more time for all governments to bring forward their Voluntary National Reviews, in the end these reports may have the potential to serve as a detection mechanism for poor performance. This again could raise the reputational cost of non-compliance. In addition, Voluntary National Reviews could help mobilize and empower domestic supporters and increase a sense of urgency among participants. In sum, with these mechanisms in place, the Sustainable Development Goals could have important effects despite their lack of legal standing.

Weak Institutional Arrangements

A second characteristic of governance through global goal-setting is that it needs to rely on weak institutional arrangements at the international level. By 'weak' arrangements, we mean that global goals do not rely on legal authority or on a formal status within the United Nations hierarchy. This also implies that they lack significant resources to execute their mandate and the capacity to create norms, resolve disputes and enforce compliance with further rules and regulations.

Generally, weak institutional arrangements are often associated with claims about the ineffectiveness of global governance that comes from inefficiency, the lack of an overall vision, duplication and conflicts between the mandates and activities of organizations, lack of implementation and enforcement and lack of adequate and predictable funding (Lodefalk and Whalley 2002; Elliott 2005; Biermann 2014). Such criticisms often coincide with negative views on governance fragmentation (Chapter 8). Many of the discussions regarding the institutional reform of the global architecture for earth system governance, for instance, revolve around an upgrade in authority of existing organizations or the establishment of an authoritative international organization dealing with the environment (Chapter 13).

Several authors, however, have framed weak institutional arrangements also as a possible way to deal with governance fragmentation. One such way is known as orchestration, a strategy closely linked to governance through goals. Orchestration relies not on legal authority and enforcement but rather on 'soft modes of influence' (Abbott et al. 2015: 223). Orchestrators gain influence through intermediary organizations and can steer actors in desired directions, typically through 'bottom-up, non-confrontational, country-driven and stakeholder-oriented' strategies

(Biermann, Kanie and Kim 2017: 27). Despite a lack of formal authority, orchestrators are believed to be able to exercise leadership, provided that they are considered as legitimate by intermediary and target organizations and that they are the key focal point and expert within their areas, which grants them political weight (Chapter 11).

A prime example of orchestration is the High-level Political Forum on Sustainable Development, which is responsible for the institutional oversight in formulating and implementing the Sustainable Development Goals (Persson, Weitz and Nilsson 2016). The High-level Political Forum was established during the 2012 United Nations Conference on Sustainable Development, replacing the United Nations Commission on Sustainable Development that was often seen as a mere 'talk shop' with no authority to make or facilitate formal decisions (Ivanova 2013: 219; see also Bernstein 2017). The High-level Political Forum did not gain much formal authority or resources compared to its predecessor (Abbott and Bernstein 2015).

Yet, despite these shortcomings, some scholars perceive the High-level Political Forum as rather influential. The Forum has been granted legitimacy through a formal resolution on its establishment; it has universal membership, high-level representation and participation of not only United Nations member states but also international organizations and non-state actors. The High-level Political Forum is hence regarded by some as a focal point for implementing the Sustainable Development Goals at the global level. It is a forum within the United Nations General Assembly, which may provide it with some political weight (Abbott and Bernstein 2015; Bernstein 2017). Though this points to a potential for success, the High-level Political Forum is bound to face challenges in exercising leadership within an architecture that is still characterized by fragmentation and partial competition among a plethora of international organizations that all work in the field of sustainable development.

In short, the jury is still out on whether weak institutional arrangements harm or help with the effectiveness of governance. While some see little promise in organizations with weak arrangements, others are more optimistic, provided that the right policy measures – such as purposeful orchestration strategies – are in place.

Inclusiveness

A third characteristic of governance though goals is the inclusion of a plurality of state and non-state actors in both goal formation and goal implementation. We distinguish here between procedural inclusiveness – that is, the openness of the process to a wide range of state and non-state actors – and substantial inclusiveness, which relates to the broad range of targets of a given policy. Both dimensions of inclusiveness are related: procedural inclusiveness can shape substantial inclusiveness, because including a wider range of actors in the setting of goals can favour the establishment of goals with broader objectives.

In global goal-setting, the attention to inclusiveness is linked to the search for greater (input) legitimacy in global governance. This, again, relates to the concern of addressing democratic deficits in global governance that result from insufficient participation and accountability (Bäckstrand 2006a; Biermann and Gupta 2011; Keohane 2011; Gellers 2016). Some even see goal-based governance as a way to pursue what they call stakeholder democracy – a type of hybrid governance that responds to the argument that more deliberative input legitimacy results in greater output legitimacy and hence better governance results (Bäckstrand 2006b). Inclusiveness is generally viewed by proponents as a crucial step to more 'reflexive' forms of governance. Reflexivity is seen as a form of resilience and deliberation that embodies the institutional ability to be something else (as opposed to do something else) to effectively deal with changing circumstances (Voß and Kemp 2006; Dryzek 2014; Feindt and Weiland 2018). Also empirically, we observe since the 1990s a participatory turn in global governance that started with the Agenda 21 of 1992 and later evolved into the 2002 World Summit on Sustainable Development, which led to a shift from 'mere' participation to multi-stakeholdership. New forms of hybrid governance emerged, including dialogues and public–private partnerships. These play important roles in the governance of sustainability issues (e.g., Glasbergen, Biermann and Mol 2007; Bitzer, Francken and Glasbergen 2008), regimes (Gupta and Vegelin 2016) and interactions between regimes (Visseren-Hamakers, Arts and Glasbergen 2011; Visseren-Hamakers and Verkooijen 2013), even though concerns about their actual effectiveness and equity effects remain.

These mechanisms have been criticized, for example, for lack of participation from marginalized groups, insufficient monitoring and reporting and the biased funding that is generated through strong private sector involvement (Bäckstrand 2006a; Biermann et al. 2012; Bäckstrand and Kylsäter 2014). Studies on the failure of some partnerships suggest, for example, the importance of clear links with intergovernmental organizations, as well as the existence of measurable targets, effective leadership and systematic reviews for the reporting and monitoring of targets (Bäckstrand 2006a; Bäckstrand and Kylsäter 2014; Pattberg and Widerberg 2016).

This importance of effectiveness and measurability has informed the adoption of the Millennium Development Goals in 2000: a very concise set of development goals, praised for their clarity and simplicity and hailed as a historic example of global mobilization to achieve important priorities (Sachs 2012; Solberg 2015).

And yet, the Millennium Development Goals have also faced sharp criticism with regard to their inclusiveness.

First, the Millennium Development Goals were aimed only at developing countries, with industrialized countries envisaged almost as tutors, reflecting a unidirectional and not very inclusive understanding of development (Deacon 2016). Procedurally, the earlier stages did reflect some inclusiveness, with the United Nations inviting input from non-state actors and eventually publishing 'We the Peoples: The Role of the United Nations in the 21st Century', which included a list of global values and priorities. However, the actual Millennium Declaration, and the extraction of the Millennium Development Goals from it, were largely based on input from the OECD's Development Assistance Committee, thereby attesting to the scarce inclusiveness of a supposedly global goal-setting process (Honniball and Spijkers 2014; Chasek et al. 2016). In addition, there has also been criticism about the strong emphasis of the Millennium Development Goals on measurability, which has caused a certain reductionism and may have led to the exclusion or marginalization of crucial qualitative elements of comprehensive development (Fukuda-Parr, Yamin and Greenstein 2014: 115). At the same time, the partnerships that were established around the Millennium Development Goals were criticized for their weak review mechanisms and performance measurements (Bäckstrand et al. 2012; Bernstein 2017).

Considering these deficits of the Millennium Development Goals, some have described the strong focus of the Sustainable Development Goals on inclusiveness as a transformative moment in development policy (Stevens and Kanie 2016). Unlike other UN goals, the Sustainable Development Goals emerged from a 'mould-breaking' negotiation process that involved the establishment of an Open Working Group, which, in line with the official aim to conduct an 'inclusive and transparent intergovernmental process on sustainable development goals that is open to all stakeholders' (United Nations General Assembly 2012: 63), strived to be as open and inclusive as possible. Unlike most United Nations General Assembly working groups, whose meetings are generally closed to observers and lack official and publicly available records, the Open Working Group pursued the full involvement of stakeholders and the gathering of expertise from civil society, the scientific community and the United Nations system. It actively reduced delegation rigidity and set up a stocktaking process – including meetings with civil society – aimed at providing all negotiators with the same terms of reference and at fostering a high level of cohesion and a common sense of purpose (Chasek and Wagner 2016). In this light, also noting the role played by UN agencies in the UN task force and by the wide consultations with civil society, some scholars have referred to this goal-setting as 'global social governance' (Deacon 2016: 118). Instrumental in the procedural success of the 'largest development dialogue ever held' (Solberg 2015: 61) has also been the experimental use of new technologies – such as the creation of a global questionnaire – in the consultation phase (Sachs 2012; Gellers 2016), with some scholars arguing that the very future of global participation lies in the application of information technologies (Honniball and Spijkers 2014).

Against this rather optimistic backdrop, however, more critical voices have pointed at some weaknesses in the inclusiveness in global goal-setting, even with the Sustainable Development Goals. First, there is a difference between inclusive invitation and inclusive participation (that is, actual influence on the final outcomes), with the process leading to the Sustainable Development Goals faring better in the former than in the latter (Deacon 2016; Gellers 2016). Second, the combined emphasis on growth (Gupta and Vegelin 2016) and on nationally determined commitments presents the risk of stifling inclusiveness at the later stages of goal implementation, in that it might incentivize a 'sovereignist' policymaking reversal away from the concern to improve global governance along 'social' lines (Deacon 2016: 129). Third, from a discursive standpoint, it has been pointed out that the Sustainable Development Goals do not constitute a major revolution vis-à-vis the overwhelmingly neoliberal narrative of the Millennium Development Goals. While the Sustainable Development Goals do include more references to Keynesian, feminist and 'world society' sustainability elements, they still retain an emphasis on neoliberal tenets such as economic growth. And they do not, as pointed out by critics, include any strong criticism of the existing global trade and financial architecture (Briant Carant 2017).

National Leeway

A fourth characteristic of governance through goal-setting is that it grants much leeway for national choices and preferences. While global goals provide a roadmap of what ought to be done, they remain subject to contestation, negotiation and translation at the national level (Fukuda-Parr 2014).

To start with, this again brings in concerns about the legitimacy, fairness and accountability of national goal implementation. For example, the inclusiveness with which the Sustainable Development Goals have been crafted at the global level would imply that such inclusiveness is also important for the implementation of the goals at national and subnational levels, but this is not always the case. So, the national leeway left in the implementation of the Sustainable Development Goals might result in less inclusiveness in some countries than was originally envisaged.

Second, should global goals be nationally implemented without adaptation to national circumstances, the results could be unfair outcomes (Easterly 2009; Fukuda-Parr 2014) and the omission of important priorities for inclusive and equitable development (Kabeer 2010). If countries with different levels of development are held up to the same measures of performance – as was implicitly the case with the Millennium Development Goals – then the special conditions in the least developed countries would make it very difficult for them to meet the goals (Easterly 2009; Hailu and Tsukada 2011). African countries, for instance, have performed poorly in implementing the Millennium Development Goals despite having made significant progress in that period (Easterly 2009). Furthermore, the translation of the Millennium Declaration into an agenda for action has created a dissonance between the Goals' original intent and their implementation (Fukuda-Parr 2010). The Millennium Development Goals distilled complex development challenges into merely 21 quantitative targets, which affected how development was understood and how decisions were made (Fukuda-Parr 2014). Poverty, for example, was narrowly framed as material deprivation with little attention paid to inequality, and it therefore overlooked the multidimensional, intersectional causes of poverty such as race, gender and ethnicity (Kabeer 2010). Even though the Millennium Development Goals had established a clear and communicable focus, the subsequent measures of progress did not account for whether such progress was equitable or sustainable (Hill, Ghulam and Claudio 2010; Hulme 2010; Kabeer 2010).

Third, however, nationally owned strategies for implementing the Sustainable Development Goals might also foster greater accountability at national and other levels, through the development of appropriate monitoring and evaluation mechanisms. Such mechanisms, including the national Sustainable Development Goals reports and the Voluntary National Reviews, provide important means through which states could reflect, confront and fill institutional gaps towards goal attainment. As a key feature of governance through global goal-setting, national leeway encourages self-regulation or self-steering (Fukuda-Parr 2014), the translation and adoption of goals into national policies and institutions (Galli et al. 2018) and more integrated institutional arrangements fit to address cross-sectoral issues and challenges. All of this can affect governance architectures discursively and materially.

However, while the Sustainable Development Goals somewhat remedy the shortcomings of the Millennium Development Goals, potential pitfalls remain. As Oran Young aptly states, '[i]t is relatively easy to establish a causal connection between the articulation of goals and the establishment of organizational arrangements to promote their attainment. It is another matter to demonstrate such a connection between goal-setting and actual progress toward fulfilling the relevant goals' (Young 2017: 37). Given the politics that animate development policy and practices across scales, there are risks of simplification and selectivity of goals through national implementation (Fukuda-Parr 2016). A recent analysis of Voluntary National Reviews indicates that various efforts are underway to incorporate the Sustainable Development Goals across all levels of governance, from setting up new institutions and engaging with local governments to realigning national plans with the Sustainable Development Goals (Sarwar and Nicolai 2018). Yet very few governments clearly articulate how to execute their respective agendas or how to monitor and evaluate their progress (Sarwar and Nicolai 2018). This may result in 'slippage in ambition and vision' in the processes of moving from goals to targets to indicators, all of which guide the orientation of policies and institutions (Fukuda-Parr and McNeill 2019: 12; see also Merry 2019). A study on Sustainable Development Goal 12 discusses the divergent framings of what sustainable production and consumption means and how to get there, arguing that quantitative indicators are vital to ensure accountability and avoid the continuation of 'green growth' trajectories that overlook planetary boundaries (Gasper, Shah and Tankha 2019). At the same time, quantification may lead to misleading or distorted information with significant policy implications (Merry 2019), leaving the Inter-agency and Expert Group on Sustainable Development Goal Indicators to adopt a pragmatic approach of 'measuring what we know how to measure', while addressing remaining challenges (Elder and Olsen 2019: 80).

Fourth, the national leeway might foster important learning processes within countries. Institutional integration requires much re-learning and must transpire through a multi-actor, multi-sector and multilevel process, providing new possibilities to engage with different types of knowledge (Meuleman and Niestroy 2015). For instance, the tendency to simplify global goals may be because of genuine operational challenges in formulating and implementing policies, which can reflect the multiplicity of linkages and foster integration among goals (Elder and Olsen 2019). Some indicators for the Sustainable Development Goals are still not based on established methodologies and standards, and some lack the required data for measurement (MacFeely 2019). Additionally, moving from sectoral to integrated approaches to goal implementation and measurement at the national level is challenging, given that many institutional structures are still arranged in silos (Elder and Olsen 2019). In sum, all these processes at the national level can facilitate social learning both within and across institutions, all in order to create policies that respond to local, national and global aspirations (Patel et al. 2017).

Governance through Goals and the Performance of Architectures

We now turn to the final question of whether global goals can affect global governance architectures, and under what circumstances. Drawing on the typology of fragmentation offered by Biermann and colleagues (2009; see also Chapter 8), we assess whether global goals can strengthen institutional integration and reduce overlaps between decision-making systems, limit norm conflicts and influence the type of actor constellations, all possibly leading to less conflictive and more cooperative or synergistic fragmentation.

With regard to institutional integration, since goals are not legally binding and operate through weak institutional arrangements, their contribution to normative and institutional integration in global governance might seem limited. In the same vein, global goals do not offer much detail on how to reach the goals through specific policies or procedures. In the case of the Sustainable Development Goals, it is left to states to develop their own strategy to achieve the goals. Self-steering is encouraged, which results in the development or adaptation of institutional arrangements at the national level by each state's own preference. Though there are clear benefits to this approach, it also implies – at the global level – that a variety of institutions emerges that are not necessarily integrated.

Yet global goals may still contribute to institutional integration despite their lack of formal authority. As goals can play an important role in creating overarching and crosscutting norms (Biermann, Kanie and Kim 2017), they may serve as a key soft law instrument to orchestrate international agreements and institutions (Kim and Bosselmann 2013; Bridgewater, Kim and Bosselmann 2014). In the case of the Sustainable Development Goals, it has been argued that goals might spur clustering of the agreements within their own area and serve as an overarching set of principles, eventually modifying the application of other norms (Kim 2016). Indeed, it has been observed that the Sustainable Development Goals are already influencing international and national law, for example European trade and investment law (Huck and Kirkin 2018). In this respect, it seems that goals can indeed provide a tool for orchestration through normative guidance; their soft power can lead to more institutional integration in a fragmented system.

A second defining criterion of governance fragmentation is substantial norm conflicts between institutions. Again, global goals may be instrumental here in the creation of overarching norms, as long as broad support for the goals is present. In the case of the Sustainable Development Goals, obtaining such broad support and legitimacy has been pursued by striving for broad inclusiveness in the establishment of the goals. It has been argued that inclusiveness is key to inform deliberative processes in which different participants develop well-informed opinions and 'productive tensions' to drive reflexive reforms (Dryzek and Pickering 2017: 354). This, in turn, could foster the emergence of more flexible and adaptive architectures and facilitate the emergence and consolidation of multilevel and multi-scalar governance solutions that follow principles of institutional variety, polycentricity and analytic deliberation (that is, inclusive dialogues) (Dietz et al. 2003: 1910). In the context of sustainable development, inclusiveness has already been singled out as a key component in the gradual relaxation of strictly sovereigntist multilateralism towards what some see as more sustainable, more participatory and less state-centric formats (Eckersley 2004; Bäckstrand 2006b). Global goals can thus be successful in working towards more synergistic types of fragmentation by reducing norm conflicts, as long as the goals themselves have broad support, which in turn can be achieved by an inclusive goal-setting process.

However, even when global goals offer a common vision, normative ambiguity remains. The Sustainable Development Goals, for example, have been criticized for not providing a clear vision on sustainability (Bernstein 2017). It has been argued that vague institutional elements, such as the ambition of achieving sustainability, coincide with synergistic fragmentation, while more concrete and substantive institutional elements that are necessary for the implementation of goals coincide with more conflictive fragmentation (Fernández-Blanco, Burns and Giessen 2019). Indeed, setting goals that are as numerous and broad as the Sustainable Development Goals is bound to lead to competition for priority (Young 2017). Tension between the goals exists in the form of trade-offs (Langford 2010; Bernstein 2017), and a common global vision on the integration of the goals is lacking (Yamada 2017). Several authors have therefore highlighted the importance of systems to manage priorities (Griggs et al. 2017) and called for prioritization of goals (Spangenberg 2017). Given that goals must consider national circumstances, leaving prioritization and integration to the individual states is a logical choice. However, the adoption of integrated analytical approaches and models at the national level is lagging, posing a considerable risk for continuation of the same 'siloed' approach that has been criticized in the past (Allen, Metternicht and Wiedmann 2018), with conflicts remaining between different issue areas.

A third defining criterion of different degrees of fragmentation is overlapping actor constellations. Global goals can, again, help reduce fragmentation. Notably, the introduction of the Sustainable Development Goals in 2015 brought an unprecedented call for action from a plethora of stakeholders, including civil society, non-governmental organizations and the private sector. The involvement of such a multiplicity of actors at different scales leads to an increasingly polycentric system. Following a recent study by Jordan and colleagues (2018: 19), the effectiveness of such a polycentric system requires the presence of overarching rules or goals 'to provide a means to settle disputes and reduce the level of discord between units to a manageable level'. Especially the private sector is becoming a strong political actor in such polycentric systems, and some UN agencies see its role in achieving sustainable development as indispensable (UNCTAD 2014). The UN Global Compact, for instance, is a key network created to encourage businesses to commit themselves to the Millennium Development Goals, and now the Sustainable Development Goals. So far, almost 10,000 companies have joined the Compact, thereby committing to a set of goals to conduct business that is aligned with the Sustainable Development Goals (United Nations Global Compact

2019). The Sustainable Development Goals have even been called a 'great gift for business' for giving a clear set of long-term global priorities with which the private sector can align their strategies (Pedersen 2018). In this sense, global goals do offer guidance for a new group of actors to join the global governance system and commit to a same set of behavioural norms at the global level, perhaps steering towards more cooperative or synergistic governance fragmentation.

On the other hand, the involvement of the private sector in governing sustainable development has invoked sharp criticism as well. Large transnational companies, predominantly from Europe and the United States, have been able to represent their sectoral interests during the development of the Sustainable Development Goals (Scheyvens, Banks and Hughes 2016; Weber 2017) and have been given an active role in the form of public–private partnerships. Yet these public–private partnerships do not always lead to the desired results (Scheyvens, Banks and Hughes 2016). It has also been argued that the private sector is effectively pushing for its own corporate interests (Koehler 2015). Indeed, companies tend to engage with those goals that are most relevant to their own business interests (Abshagen et al. 2018), focusing more on 'doing no harm' than on 'doing good' (van Zanten and van Tulder 2018). Some observe also a lack of attention by business actors for those goals that are predominantly relating to environmental sustainability (Poddar, Narula and Zutshi 2019).

Taking all this together, it seems possible that goals offer an overarching set of norms, leading to more normative agreement and institutional integration, be it through soft modes of governance or orchestration. On the other hand, the involvement of particularly powerful private actors and the cherry-picking of goals could also lead to the strengthening of specific complexes around certain goals and not others. This would then result in a more modular global governance architecture, where synergistic fragmentation is present within specific complexes, but cooperative or even conflictive fragmentation is present between complexes.

Conclusions and Future Directions

Global governance through goal-setting, as an increasingly influential mechanism of global governance, poses important questions for academic research and policy analysis. For example, we need to better understand how, to what extent and with what effects global goals and their norms are embedded and integrated in existing governance arrangements at global, national and local levels. Also, what further governance reforms are needed to implement and reach the goals at various levels? The concept of orchestration in global governance constitutes an important new research area as well, focusing for example on the extent to which 'powerless' steering may have powerful effects on actors' behaviour (Chapter 11; see also Abbott et al. 2015).

Another important research question is to what extent and how the rhetoric of integration and policy coherence between the Sustainable Development Goals takes shape in governance arrangements at national and subnational levels. While the Sustainable Development Goals are meant to be indivisible and implemented coherently, unavoidable trade-offs and prioritization between goals need to be dealt with at the national and subnational levels. The question is then how the often-siloed national and subnational governance arrangements give shape to the Sustainable Development Goals, who is involved in prioritizing the goals and whether and how the rhetoric of this process of 'leaving no one behind' is being realized.

As a form of governance through goals, the Sustainable Development Goals show a level of ambition and comprehensiveness that surpasses all other forms of governance through goals. This makes them 'one of the most intriguing new global initiatives in sustainable development and environmental policy' (Biermann, Kanie and Kim 2017: 29). Governance through goals as a mechanism of global governance is not likely to disappear, nor is it likely to become less dominant with the termination of the Sustainable Development Goals by 2030. It will therefore remain of utmost importance, both for the attainment of the Sustainable Development Goals and for any future effort of global goal-setting, to continue critical examination of the various effects of global goals at the global, national and subnational levels.

References

Abbott K. W., Genschel P., Snidal, D., & Zangl, B. (eds.) (2015). *International organizations as orchestrators*. Cambridge, UK: Cambridge University Press.

Abbott, K. W., & Bernstein S. (2015). The High-level Political Forum on Sustainable Development: Orchestration by default and design. *Global Policy*, 6, 222–33.

Abshagen, M. L., Cavazzini, A., Graen, L., & Obenland, W. (2018). *Hijacking the SDGs? The private sector and the Sustainable Development Goals*. Berlin: Brot für die Welt.

Allen, C., Metternicht, G., & Wiedmann, T. (2018). Initial progress in implementing the Sustainable Development Goals (SDGs): A review of evidence from countries. *Sustainability Science*, 13 (5), 1453–67.

Bäckstrand, K. (2006a). Multi-stakeholder partnerships for sustainable development: Rethinking legitimacy, accountability and effectiveness. *European Environment*, 16 (5), 290–306.

Bäckstrand, K. (2006b). Democratizing global environmental governance? Stakeholder democracy after the World Summit on Sustainable Development. *European Journal of International Relations*, 12 (4), 467–98.

Bäckstrand, K., Campe, S., Chan, S., Mert, A., & Schäferhoff, M. (2012). Transnational public–private partnerships. In F. Biermann, & P. Pattberg (eds.), *Global environmental governance reconsidered* (pp. 123–47). Cambridge, MA: The MIT Press.

Bäckstrand, K., & Kylsäter, M. (2014). Old wine in new bottles? The legitimation and delegitimation of UN public–private partnerships for sustainable development from the Johannesburg Summit to the Rio+20 Summit. *Globalizations*, 11 (3), 331–47.

Bernstein S. (2017). The United Nations and the governance of Sustainable Development Goals. In N. Kanie, & F. Biermann (eds.), *Governing through goals: Sustainable Development Goals as governance innovation* (pp. 213–40). Cambridge, MA: The MIT Press.

Biermann, F. (2014). *Earth system governance: World politics in the Anthropocene*. Cambridge, MA: The MIT Press.

Biermann, F., Pattberg, P., van Asselt, H., & Zelli, F. (2009). The fragmentation of global governance architectures: A framework for analysis. *Global Environmental Politics*, 9 (4), 14–40.

Biermann, F., & Gupta, A. (2011). Accountability and legitimacy in earth system governance: A research framework. *Ecological Economics*, 70, 1856–64.

Biermann, F., Chan S., Pattberg, P., & Mert, A. (eds.) (2012). *Multistakeholder partnerships for sustainable development: Emergence, influence and legitimacy*. Cheltenham: Edward Elgar.

Biermann, F., Kanie, N., & Kim, R. E. (2017). Global governance by goal-setting: The novel approach of the UN Sustainable Development Goals. *Current Opinion in Environmental Sustainability*, 26–27, 26–31.

Bitzer, V., Francken, M., & Glasbergen, P. (2008). Intersectoral partnerships for a sustainable coffee chain: Really addressing sustainability or just picking (coffee) cherries? *Global Environmental Change*, 18 (2), 271–84.

Bodansky, D. (2010). The Copenhagen conference: A post-mortem. *American Journal of International Law*, 104 (2), 232–3.

Bodansky, D. (2016). The legal character of the Paris Agreement. *Review of European, Comparative & International Environmental Law*, 25 (2), 142–50.

Briant Carant, J. (2017). Unheard voices: A critical discourse analysis of the Millennium Development Goals' evolution into the Sustainable Development Goals. *Third World Quarterly*, 38 (1), 16–41.

Bridgewater, P., Kim, R. E., & Bosselmann, K. (2014). Ecological integrity: A relevant concept for international environmental law in the Anthropocene? *Yearbook of International Environmental Law*, 25, 61–78.

Chasek, P. S., Wagner, L. M, Leone F., Lebada A. M., & Risse. N. (2016). Getting to 2030: Negotiating the post-2015 sustainable development agenda. *Review of European, Comparative & International Environmental Law*, 25, 5–14.

Chasek, P. S., & Wagner, L. M. (2016). Breaking the mold: A new type of multilateral sustainable development negotiation. *International Environmental Agreements*, 16, 397–413.

Deacon, B. (2016). Assessing the SDGs from the point of view of global social governance. *Journal of International and Comparative Social Policy*, 32 (2), 116–30.

Dietz, T., Ostrom, E., & Stern, P. C. (2003). The struggle to govern the commons. *Science*, 302, 1907–12.

Dryzek, J. S., & Pickering, J. (2017). Deliberation as a catalyst for reflexive environmental governance. *Ecological Economics*, 131, 353–60.

Dryzek, J. S. (2014). Institutions for the Anthropocene: Governance in a changing earth system. *British Journal of Political Science*, 46 (4), 937–56.

Easterly, W. (2009). How the Millennium Development Goals are unfair to Africa. *World Development*, 37 (1), 26–35.

Eckersley, R. (2004). *Green state: Rethinking democracy and sovereignty*. Cambridge, MA: The MIT Press.

Elder, M., & Olsen, S. H. (2019). The design of environmental priorities in the SDGs. *Global Policy*, 10 (S1), 70–82.

Elliott, L. (2005). The United Nations' record on environmental governance: An assessment. In F. Biermann, & S. Bauer (eds.), *A world environment organization: Solution or threat for effective international environmental governance?* (pp. 27–56). Burlington: Ashgate.

Elliott, L. (2017). Cooperation on transnational environmental crime: Institutional complexity matters. *Review of European, Comparative & International Environmental Law*, 26, 107–17.

Falk, R. A. (1966). On the quasi-legislative competence of the General Assembly. *American Journal of International Law*, 60 (4), 782–91.

Feindt, P. H., & Weiland, S. (2018). Reflexive governance: Exploring the concept and assessing its critical potential for sustainable development. *Journal of Environmental Policy & Planning*, 20 (6), 661–74.

Fernández-Blanco, C. R., Burns, S. L., & Giessen, L. (2019). Mapping the fragmentation of the international forest regime complex: Institutional elements, conflicts and synergies. *International Environmental Agreements*, 19 (2), 187–205.

Franck, T. M. (1990). *The power of legitimacy among nations*. Oxford: Oxford University Press.

Fukuda-Parr, S. (2010). Reducing inequality – The missing MDG: A content review of PRSPs and bilateral donor policy statements. *IDS Bulletin*, 41 (1), 26–35.

Fukuda-Parr, S. (2014). Global goals as a policy tool: Intended and unintended consequences. *Journal of Human Development and Capabilities*, 15 (2–3), 118–31.

Fukuda-Parr, S. (2016). From the Millennium Development Goals to the Sustainable Development Goals: Shifts in purpose, concept and politics of global goal setting for development. *Gender and Development*, 24 (1), 43–52.

Fukuda-Parr, S., & McNeill, D. (2019). Knowledge and politics in setting and measuring the SDGs. *Global Policy*, 10 (S1), 5–15.

Fukuda-Parr, S., Yamin, A. E., & Greenstein, J. (2014). The power of numbers: A critical review of Millennium Development Goal targets for human development and human rights. *Journal of Human Development and Capabilities*, 15 (2–3), 105–17.

Galli, A., Durović, G., Hanscom, L., & Knežević, J. (2018). Think globally, act locally: Implementing the Sustainable Development Goals in Montenegro. *Environmental Science and Policy*, 84, 159–69.

Gasper, D., Shah, A., & Tankha, S. (2019). The framing of sustainable consumption and production in SDG 12. *Global Policy*, 10 (S1), 83–95.

Gellers, J. C. (2016). Crowdsourcing global governance: Sustainable Development Goals, civil society and the pursuit of democratic legitimacy. *International Environmental Agreements*, 16, 415–32.

Glasbergen, P., Biermann, F., & Mol, A. P. J. (eds.) (2007). *Partnerships, governance and sustainable development: Reflections on theory and practice*. Cheltenham: Edward Elgar.

Goldsmith, J. L., & Posner, E. A. (2005). *The limits of international law*. Oxford: Oxford University Press.

Griggs, D. J., Nilsson, M., Stevance, A., & McCollum, D. (2017). *A guide to SDG interactions: From science to implementation*. Paris: International Council for Science.

Gupta, J., & Vegelin, C. (2016). Sustainable Development Goals and inclusive development. *International Environmental Agreements*, 16, 433–48.

Hailu, D., & Tsukada, R. (2011). *Achieving the Millennium Development Goals: A measure of progress*. Brasilia: International Policy Centre for Inclusive Growth.

Hajer, M., Nilsson, M., Raworth, K. et al. (2015). Beyond cockpit-ism: Four insights to enhance the transformative potential of the Sustainable Development Goals. *Sustainability*, 7, 1651–60.

Hill, P. S., Ghulam, F. M., & Claudio, F. (2010). Conflict in least-developed countries: Challenging the Millennium Development Goals. *Bulletin of the World Health Organization*, 88 (8), 562.

Honniball, A., & Spijkers, O. (2014). MDGs and SDGs – Lessons learnt from global public participation in the drafting of the UN Development Goals. *Vereinte Nationen: German Review of the United Nations*, 62 (6), 251–6.

Huck, W., & Kurkin, C. (2018). The UN Sustainable Development Goals (SDGs) in the transnational multilevel system. *Heidelberg Journal of International Law*, 2, 375.

Hulme, D. (2010). Lessons from the making of the MDGs: Human development meets results-based management in an unfair world. *IDS Bulletin*, 41 (1), 15–25.

Ivanova, M. H. (2013). Reforming the institutional framework for environment and sustainable development: Rio+20's subtle but significant impact. *International Journal of Technology Management & Sustainable Development*, 12 (3), 211–31.

Jordan, A., Huitema, D., van Asselt, H., & Forster, J. (eds.) (2018). *Governing climate change: Polycentricity in action?* Cambridge, UK: Cambridge University Press.

Kabeer, N. (2010). Can the MDGs provide a pathway to social justice? The challenge of intersecting inequalities. New York: Institute of Development Studies and MDG Achievement Fund.

Kamau, M., Chasek, P., & O'Connor, D. (2018). *Transforming multilateral diplomacy: The inside story of the sustainable development goals*. New York: Routledge.

Kanie N., & Biermann F. (eds.) (2017). *Governing through goals: Sustainable Development Goals as governance innovation*. Cambridge, MA: The MIT Press.

Kanie, N., Bernstein, S., Biermann, F., & Haas, P. M. (2017). Introduction: Global governance through goal setting. In N. Kanie, & F. Biermann (eds.), *Governance through goals: Sustainable Development Goals as governance innovation*. Cambridge, MA: The MIT Press.

Keohane, R. O. (2011). Global governance and legitimacy. *Review of International Political Economy*, 18 (1), 99–109.

Kim, R. E., & Bosselmann, K. (2015). Operationalizing sustainable development: Ecological integrity as a Grundnorm of international law. *Review of European, Comparative & International Environmental Law*, 24, 194–208.

Kim, R. E., & Bosselmann, K. (2013). International environmental law in the Anthropocene: Towards a purposive system of multilateral environmental agreements. *Transnational Environmental Law*, 2, 285–309.

Kim, R. E. (2016). The nexus between international law and the Sustainable Development Goals. *Review of European, Comparative & International Environmental Law*, 25 (1), 15–26.

Koehler, G. (2015). Seven decades of 'development', and now what? *Journal of International Development*, 27 (6), 733–51.

Langford, M., & Winkler, I. (2014). Muddying the water? Assessing target-based approaches in development cooperation for water and sanitation. *Journal of Human Development and Capabilities*, 15 (2–3), 247–60.

Langford, M. (2010). A poverty of rights: Six ways to fix the MDGs. *IDS Bulletin*, 41 (1), 83–91.

Le Blanc, D. (2015). Towards integration at last? The Sustainable Development Goals as a network of targets. *Sustainable Development*, 23, 176–87.

Lodefalk, M., & Whalley, J. (2002). Reviewing proposals for a world environmental organisation. *The World Economy*, 25, 601–17.

MacFeely, S. (2019). The big (data) bang: Opportunities and challenges for compiling SDG indicators. *Global Policy*, 10 (S1), 121–33.

Merry, S. E. (2019). The Sustainable Development Goals confront the infrastructure of measurement. *Global Policy*, 10 (S1), 146–8.

Meuleman, L., & Niestroy, I. (2015). Common but differentiated governance: A metagovernance approach to make the SDGs work. *Sustainability*, 7, 12295–321.

Patel, Z., Greyling, S., Simon, D., Arfvidsson, H., Moodley, N., Primo, N., & Wright, C. (2017). Local responses to global sustainability agendas: Learning from experimenting with the urban Sustainable Development Goal in Cape Town. *Sustainability Science*, 12 (5), 785–97.

Pattberg, P., & Widerberg, O. (2016). Transnational multistakeholder partnerships for sustainable development: Conditions for success. *AMBIO*, 45 (1), 42–51.

Pedersen, C. S. (2018). The UN sustainable development goals (SDGs) are a great gift to business! *Procedia CIRP*, 69, 21–4.

Persson, Å., Weitz, N., Nilsson, M. (2016). Follow-up and review of the Sustainable Development Goals: Alignment vs. internalization. *Review of European, Comparative & International Environmental Law*, 25, 59–68.

Poddar, A., Narula, S. A., & Zutshi, A. (2019). A study of corporate social responsibility practices of the top Bombay Stock Exchange 500 companies in India and their alignment with the Sustainable Development Goals. *Corporate Social Responsibility and Environmental Management*, 1–22.

Raustiala, K. (2005). Form and substance in international agreements. *American Journal of International Law*, 99 (3), 581–92.

Sachs, J. D. (2012). From Millennium Development Goals to Sustainable Development Goals. *The Lancet*, 379, 2206–11.

Sarwar, M. B., & Nicolai, S. (2018). *What do analysis of Voluntary National Reviews for Sustainable Development Goals tell us about 'leaving no one behind'?* London: Overseas Development Institute.

Scheyvens, R., Banks, G., & Hughes, E. (2016). The private sector and the SDGs: The need to move beyond 'business as usual'. *Sustainable Development*, 24 (6), 371–82.

Solberg, E. (2015). From MDGs to SDGs – The political value of common global goals. *Harvard International Review*, 37 (1), 58–61.

Spangenberg, J. H. (2017). Hot air or comprehensive progress? A critical assessment of the SDGs. *Sustainable Development*, 25 (4), 311–21.

Stevens, C., & Kanie, N. (2016). The transformative potential of the Sustainable Development Goals (SDGs). *International Environmental Agreements*, 16, 393–6.

UNCTAD (2014). *Investing in the SDGs: An action plan*. UNCTAD: World Investment Report 2014.

Underdal, A., & Kim, R. E. (2017). The Sustainable Development Goals and multilateral agreements. In N. Kanie, & F. Biermann (eds.), *Governing through goals: Sustainable Development Goals as governance innovation* (pp. 241–58). Cambridge, MA: The MIT Press.

United Nations General Assembly (2012). *The future we want*. UN Doc. A/RES/66/288.

United Nations Global Compact (2019). *United Nations Global Compact*. Available at: www.unglobalcompact.org. Accessed: 17 June 2019.

Van der Hoeven, R. (2014). Full employment target: What lessons for a Post-2015 Development Agenda? *Journal of Human Development and Capabilities*, 15 (2–3), 161–75.

Van Zanten, J. A., & van Tulder, R. (2018). Multinational enterprises and the Sustainable Development Goals: An institutional approach to corporate engagement. *Journal of International Business Policy*, 1 (3–4), 208–33.

Victor, D. G., Raustiala, K., & Skolnikoff, E. B. (eds.) (1998). *The implementation and effectiveness of international environmental commitments: Theory and practice.* Cambridge, MA: The MIT Press.

Visseren-Hamakers, I. J., & Verkooijen, P. (2013). The practice of interaction management: Enhancing synergies among multilateral REDD+ Institutions. *Forest and Nature Governance World Forests,* 14, 133–49.

Visseren-Hamakers, I. J., Arts, B., & Glasbergen, P. (2011). Interaction management by partnerships: The case of biodiversity and climate change. *Global Environmental Politics,* 11 (4), 89–107.

Voß, J.-P., & Kemp, R. (2006). Sustainability and reflexive governance: Introduction. In J. Voß, D. Bauknecht, & R. Kemp (eds.), *Reflexive governance for sustainable development* (pp. 3–30). Cheltenham: Edward Elgar.

Weber, H. (2017). Politics of 'leaving no one behind': Contesting the 2030 Sustainable Development Goals agenda. *Globalizations,* 14 (3), 399–414.

Yamada, T. (2017). Corporate water stewardship: Lessons for goal-based hybrid governance. In N. Kanie & F. Biermann (eds.), *Governing through goals: Sustainable Development Goals as governance innovation* (pp. 187–209). Cambridge, MA: The MIT Press.

Young, O. R. (2017). Conceptualization: Goal setting as a strategy for earth system governance. In N. Kanie, & F. Biermann (eds.), *Governance through goals: Sustainable Development Goals as governance innovation* (pp. 31–52). Cambridge, MA: The MIT Press.

Zelli, F., Biermann, F., Pattberg, P., & van Asselt, H. (2010). The consequences of a fragmented climate governance architecture: A policy appraisal. In F. Biermann, P. Pattberg, & F. Zelli (eds.), *Global climate governance beyond 2012. Architecture, agency and adaptation* (pp. 25–34). Cambridge, UK: Cambridge University Press.

13

Hierarchization

RAKHYUN E. KIM, HARRO VAN ASSELT, LOUIS J. KOTZÉ,
MARJANNEKE J. VIJGE AND FRANK BIERMANN

Unlike government-dominated top-down decision-making in a domestic context, world politics works largely through interstate bargaining between sovereign states. There are few instances of intended or planned hierarchy where actors or institutions are vertically stacked above one another. Hierarchy is actively and deliberately avoided. For example, it has been repeatedly emphasized that a new international agreement 'should not undermine existing relevant legal instruments and frameworks and relevant global, regional and sectoral bodies' (Scanlon 2018), and that no rights and obligations of the parties under any other international agreement shall be affected by it (Axelrod 2011). What we observe as a consequence are regime complexes, or 'non-hierarchical but loosely coupled systems of institutions' (Keohane and Victor 2011: 8). The advent of new types of actors, norms and institutions in world politics over the past two decades has caused further entanglement of global institutions (Biermann and Pattberg 2008). In fact, the very notion of global governance is premised on the assumption that various actors and institutions exist side by side and that hierarchy among them is absent or difficult to discern (Dingwerth and Pattberg 2006).

A key consequence of such a non-hierarchical architecture of global governance has been a degree of what is often perceived as disorder. For example, in the absence of a supreme court in global governance, international courts and tribunals decide on cases without necessarily following precedents. Similarly, international organizations (Alvarez 2005) and conferences of the parties to multilateral environmental agreements adopt decisions that may point in different directions (Wiersema 2009). It is often left undetermined which rule or norm prevails in the case of inconsistencies. An obvious concern here is the likely decrease in the effectiveness of global governance architectures; but there are also important political consequences that go beyond concerns of effectiveness. For instance, powerful states could take advantage of 'chaotic' situations to exert their power and gain more leverage over global governance outcomes at the expense of the less powerful (Paulus 2005; Benvenisti and Downs 2007). The general view is that the lack of hierarchical relations between international norms, actors and global governance institutions has created conditions of complexity, uncertainty and unpredictability, with important implications for global justice, stability, continuity and order.

To address these concerns, the hierarchization of the global institutional architecture is often proposed in both academic writing and in policy proposals. Proponents of hierarchization seek to improve stability, predictability and durability of governance processes and outcomes by ordering relationships between various actors, norms and institutions in the same way as hierarchization in domestic jurisdictions is seen to foster order. An early example of the deliberate pursuit of hierarchy within the international order came at the end of the Second World War. Governments were then determined 'to save succeeding generations from the scourge of war' (UN Charter, preamble) and established the United Nations with a mandate to maintain international peace and security. The UN Security Council was created with far-reaching powers and the UN Charter given primacy over all other international agreements (Article 103; see Liivoja 2008). These reforms after 1945 can be seen as having had some positive impact on establishing order through one central mechanism: the hierarchization of global governance.

Hierarchy is lacking, however, in many other domains, and earth system governance is a prime example. Here, the governance architecture and the laws that underpin it have been characterized by their weak constitutional nature (Bodansky 2009; Kotzé 2016), their lack of an overarching goal (Kim and Bosselmann 2013) and the absence of a central governing authority (Biermann 2000; Vijge 2013). Policy entrepreneurs continue to formulate proposals that centre on improving hierarchization. These include proposals for establishing a sustainability *Grundnorm* in global governance (Kim and Bosselmann 2013); a global environmental constitution (Kotzé 2012; Bosselmann 2015); a law of the atmosphere (Najam 2000; Sand and Wiener 2016); an international environmental court (Murphy 1999; Pedersen 2012; Lehmen 2015); a world environment organization (Palmer 1992; Biermann 2000); and most recently a Global Pact for the Environment (Aguila and Viñuales 2019). If realized, these reform measures have the potential to radically transform the architecture of earth system governance.

In this chapter, we review recent research on hierarchization in earth system governance and the political and legal processes that establish, maintain and legitimize it. We begin by conceptualizing both hierarchy and hierarchization. We present three mutually non-exclusive forms of hierarchization – systematization, centralization and prioritization – all involving different actors and rationales, mechanisms and strategies, while achieving different purposes with varying governance outcomes. We illustrate our argument with empirical examples including the proposed Global Pact for the Environment, the proposal to establish a world environment organization and the Sustainable Development Goals. We conclude with an assessment of the benefits and drawbacks of hierarchization as an approach to some of the challenges inherent in earth system governance, and offer suggestions for future research.

Conceptualization

We define *hierarchy* in global governance as a vertically nested structure in which actors and institutions at a lower rank are bound or otherwise compelled to obey, respond to, or contribute to higher-order norms and objectives. The place in a hierarchy corresponds to relative status or authority. A prominent example is the hierarchy of norms in international law, and specifically the debate revolving around constitutionalizing international law (Kotzé 2015), where erga omnes obligations (applicable to all) or even peremptory norms such as jus cogens (binding on all) are generally considered as superior to other types of norms (Shelton 2006). In this chapter, however, we go beyond such a legalistic conception and use the concept of hierarchy rather broadly. Reflecting the past decade's research on global governance, we conceptualize hierarchy also to include other recently emerging forms of nested structures between new types of actors as well as norms and institutions. Our conceptualization includes, for example, the practice of global goal-setting that identifies a certain set of non-binding, yet influential, global priorities among many possible priorities. We do not equate hierarchy with government, although we note that a vision of a single global constitutional polity with a global legislature, executive and judiciary is arguably the most extreme example of hierarchy in the international system.

Generally speaking, a hierarchy can emerge in a system through intended and unintended processes, and it can emerge top-down by central decisions or bottom-up by informal association. A bottom-up hierarchy may emerge in a growing network through preferential attachment, wherein system elements become differentiated hierarchically according to varying degrees of power or influence. For example, Google would assume a top position in the hierarchy of the Internet, but it does not stand formally 'above' others. It is just another website but with many links that make it a central hub.

However, the type of hierarchy most observers discuss in earth system governance is more formal with defined hierarchical relationships among elements and processes in a governance architecture. Here, *hierarchization* is a deliberate process driven by a normative and governance aim to transform a horizontal global governance architecture into one with a clearer hierarchy that resembles more accurately the governance systems of domestic states. This could be done through creating new institutions above or below an existing institution or rearranging existing institutions hierarchically, or by creating superior and inferior norms. Hierarchization in earth system governance is therefore broadly about institutionalizing a hierarchy among norms, institutions, actors and governance priorities. It is different from the clustering or coordination of treaties, for example through the Biodiversity Liaison Group, which does not create a stable and persistent hierarchy.

Once hierarchization is seen as a planned process, we need to ask about the purpose behind any quest for more hierarchy in global governance. We derive at least three rationales, or desired outcomes, of hierarchization from the literature.

(1) First, hierarchization is sought to address complex coordination problems. According to organizational studies, hierarchy supports a division of labour, which in turn leads to enhanced coordination and cooperation as well as reduced conflict (Halevy, Chou and Galinsky 2011). Similarly, in earth system governance, hierarchization has been suggested to fill the coordination gaps of fragmented or differentiated institutions (Zelli and van Asselt 2013; Zürn and Faude 2013) and improve policy integration (Biermann, Davies and van der Grijp 2009). In line with polycentric governance theory, coordination without hierarchy is also possible (Ostrom 2010; Jordan et al. 2018), and self-organized coordination seems in fact to be the dominant mode of coordination among multilateral environmental agreements (Kim and Bosselmann 2013). And yet, in the absence of hierarchical steering of individual institutions from above, there is uncertainty over how they will collectively behave. Hierarchization could reduce such uncertainty by providing clear direction to coordination.

(2) Second, hierarchization may improve institutional fit, or the fit between institutional arrangements and the defining features of the problems they address (Young 2002; Galaz et al. 2008). A horizontal governance architecture is not well aligned with, for example, the 'hierarchy of [planetary] boundaries', where climate change and biosphere integrity serve as 'core planetary boundaries through which the other boundaries operate' (Steffen et al. 2015: 8). More fundamentally, the finite biophysical carrying capacity of the earth has not been formally recognized and reflected in the global governance architecture (Kim and Bosselmann 2015). This is so despite a scientific consensus that 'sustainability must be conceptualized as a hierarchy of considerations, with the biophysical limits of the Earth setting the ultimate boundaries within which social and economic goals must be achieved' (Fischer et al. 2007: 621; see also Costanza et al. 2015; Kim and Bosselmann 2015). Hierarchization could establish an order within the architecture of earth system governance that matches better with earth system science.

(3) Third, and with specific reference to the juridical dimensions of global governance, hierarchization may address normative conflicts that arise between

different levels and actors of global governance (domestic, subregional, regional and international); between different treaty regimes that often focus on specific issues such as biodiversity conservation and trade in endangered species, but that have a common goal of environmental protection; and between the different types of international law norms such as international conventions, custom, general principles of law and judicial decisions (Vidmar 2012).

Hierarchization is often not a predictable, linear process. Hierarchy is more often than not a result of a power struggle between various actors advocating and promoting different objectives, processes, norms and institutions. Some powerful actors may have a vested interest in the status quo and resist attempts at creating a hierarchy in the architecture of earth system governance. The resistance by some powerful industrialized countries against creating a stronger top-down climate treaty and accompanying institutional enforcement regime is a case in point. On the other hand, other powerful actors may also propose hierarchization. France and Germany, for example, have actively pushed for an upgrade of the United Nations Environment Programme into a fully fledged international organization, and France has played the lead role in driving and promoting the initiative for a Global Pact for the Environment.

Because of the political nature of hierarchization, it is important to critically scrutinize the process and identify whose goals, norms or institutions have come out on top of the hierarchy, and why. For example, while the 17 Sustainable Development Goals cover a broad range of issues, many major issues have not been included, such as ozone depletion and population growth. Why, then, were these specific 17 goals and 169 targets identified as global priorities, and not others? Are the needs and interests of all stakeholders fairly represented in this set of global goals? Relatedly, to what extent do concerns about lack of democracy and representation arise in global governance? There may also be questions about the possible imposition of norms from some countries on others. This concern relates to the literature on the fragmentation of international law and governance where institutional diversity is seen as a necessary reflection of societal diversity (Koskenniemi and Leino 2002; Fischer-Lescano and Teubner 2004; see also Chapter 8).

Research Findings

Hierarchization can take many different forms. We identify at least three mutually non-exclusive categories of hierarchization in global governance, namely, systematization, centralization, and prioritization.

(1) *Systematization*. Arranging actors and institutions according to an organized system could lead to a nested hierarchy. Although not all systems are hierarchically

organized, a key component of a system is its overarching goal or purpose that trumps other auxiliary objectives of individual system elements. Systematization could therefore involve establishing and defining an ultimate objective of a governance architecture and make it binding on all international regimes and institutions therein (Kim and Bosselmann 2013). Without such systematization of relationships, an architecture risks being a random collection of actors and institutions without a common goal. Here, secondary rules about how global governance processes work (e.g., rules of procedure), as opposed to primary rules of conduct, play a critical role (Bodansky 2006). In a nested hierarchy, the relationships between norms at the same level are still not explicitly defined, but higher-order norms may serve as adjudicators to reconcile any potential conflicts between norms at a lower level (Vidmar 2012).

(2) *Centralization.* Hierarchy can also be created through centralization of a governance architecture. A key example here is creating a new supranational authority with more centralized decision-making, norm creation, dispute resolution and enforcement authority. Such a hierarchical intergovernmental organization could be equipped with majority decision-making (Biermann 2014). Existing examples include the United Nations Security Council and the World Trade Organization, which have increased the degree of centralization, hence hierarchical order, in a decentralized institutional landscape. The proposal for a Deliberative Global Citizens' Assembly is expected to bring about a similar effect (Dryzek, Bächtiger and Milewicz 2011). The notion of centralization should, however, not be mistaken for efforts aimed at creating a monocentric architecture with a single authority. There are typically multiple governing authorities in a global governance architecture, which together lead to a polycentric system with 'multiple and overlapping cores of control hierarchies' (Duit et al. 2010: 366). But centralization would aim to introduce a hierarchy in the architecture by promoting one decision-making authority over other authorities.

(3) *Prioritization.* Prioritization is another form of hierarchization. It refers to ordering of issues or interests according to their relative (perceived) importance. For example, some argue that environmental concerns need to be given 'principled priority' over others (Lafferty and Hovden 2003). In the legal domain, prioritization is evident in proposals to elevate ecological care as the most important fundamental objective of the state and of its constitution (Steinberg 1998). Prioritization could also include the practice of global goal-setting through which certain issues are identified as global priorities (Biermann, Kanie and Kim 2017). Examples include the Paris Agreement's two-degrees Celsius target (Morseletto, Biermann and Pattberg 2017) as well as the Sustainable Development Goals (Kanie and Biermann 2017). Some commentators have gone further and called for ranking of global priorities by recognizing, for example, poverty reduction and planetary

stability as 'twin priorities' (Griggs et al. 2013) or 'a prosperous, high quality of life that is equitably shared and sustainable' as an overarching goal for the Sustainable Development Goals (Costanza et al. 2015: 13). As global priorities, these goals and targets are expected to have some degree of hierarchical steering effect on earth system governance (Chapter 12).

We now discuss these three categories of hierarchization with examples specifically focusing on developments over the last decade.

Systematization

The idea of hierarchization through systematization goes to the heart of a long-standing debate in public international law: to what extent is there a normative hierarchy in which some norms are superior and others inferior, and how can such norms be identified? The background to this question derives from concerns about conflicts between horizontal norms (e.g., between norms in a trade agreement and an environmental treaty). In case of such a conflict, how do we know which norm prevails? The context also derives from attempts to raise the relative normative status of some environmental norms above others, specifically with a view to creating higher-order ecological norms that would be binding on states regardless of their consent (Kotzé 2015).

Normative hierarchy refers to 'the relationship between and ordering of legal norms according to their superiority in terms of their objectives, importance of their content as well as the universal acceptance of their superiority' (Kotzé and Muzangaza 2018: 282–3). In domestic legal systems, the determination of such a hierarchy would usually lead to the main source of authority: the constitution. In international law, however, where it is far more difficult to discern a constitutional structure, rules, sources and procedures emanating from different sectors are deemed equivalent (Broude and Shany 2011). While no formal hierarchy between international legal norms thus exists, 'something like an informal hierarchy' may still be present (International Law Commission 2006: para. 327). This informal international normative hierarchy presumes the existence of an international value system, in which some norms are of such fundamental importance or necessity that they are considered to be superior (De Wet and Vidmar 2012).

In this hierarchy, three types of norms are usually identified (Shelton 2006). First, there are peremptory norms from which no derogation is possible, known as jus cogens. Such norms are erga omnes, that is, owed to the entire international community. There is some agreement that norms such as the prohibition of genocide and the prohibition of torture could be regarded jus cogens norms. Yet, there is still no consensus on the full list of norms that qualify as jus cogens, nor is there agreement on how to identify such norms in the first place, notwithstanding ongoing efforts (e.g., International Law Commission 2019). This uncertainty extends to international environmental norms. Yet, some have proposed to consider including some fundamental norms in earth system governance as jus cogens, for instance the wrongful emission of ozone-depleting substances (Biermann 2014).

The second type of norms consists of norms contained in treaties that claim superiority by subjugating other norms. While such 'conflict (or savings) clauses' are found in many agreements, only one treaty claims superiority over any other treaty: the United Nations Charter, which in its Article 103 provides that the Charter shall prevail in the event of a conflict with any other international agreement. In earth system governance, conflict clauses are more limited in scope, usually ceding priority to the rights and obligations arising from existing treaties (Wolfrum and Matz 2003). There are exceptions, however, for instance with Article 311.3 of the United Nations Convention on the Law of the Sea claiming priority over existing and future agreements, making it akin to a peremptory norm.

A third type of norms that suggests the presence of a normative hierarchy is what is called soft law. Though the notion of soft law remains contested (Klabbers 1996), soft law is usually considered to include instruments that are non-legally binding or that may be non-enforceable (Boyle 1999). Soft law instruments have been a hallmark of international environmental governance, with declarations, action plans, guidelines and recommendations playing a key role in the progressive development of international environmental law (Friedrich 2013). While the influence of soft law norms should therefore not be underestimated (see also Chapter 12), given their non-legally binding nature, they can be considered inferior to other norms.

Any systematization that leads to a normative hierarchy raises an all-important question: what is the basis for the systematic ordering and the resulting hierarchy? Rephrased in Hartian terms, what is the underlying 'rule of recognition' (Hart 1994: 94); or in Kelsenian terms, what is the *Grundnorm* (Kelsen 1960)? As discussed below, these questions have also arisen – and preliminary answers to them have been put forward – in the context of earth system governance.

The discussion on the existence of superior and inferior norms has found resonance in the emerging debate on global environmental constitutionalism (Bosselmann 2015; Kotzé 2016). The promise of global environmental constitutionalism is that elevating norms to protect the environment to global 'constitutional' norms might result in more effective legal protection. Throughout the years, states and non-state actors have made efforts to try and compile a set of international environmental norms that could resemble an international environmental constitution. Some examples are the 1972 Stockholm Declaration on the Human Environment, the 1982 World Charter for Nature, and the ongoing Earth Charter initiative (Kotzé 2019a). However, it is widely accepted that international environmental law still lacks a strong constitutional order (Bodansky 2009).

Against this background, it is worth highlighting the latest effort to elevate a set of principles to a higher level: the proposed Global Pact for the Environment. Driven by France, the proposed Pact is the result of an exercise involving legal experts from all over the globe to formulate a set of legally binding international environmental principles (Aguila and Viñuales 2019). The proposed Pact includes a variety of well-established principles, such as the prevention, precaution and polluter-pays principles. In addition, the Pact also puts forward several emerging principles (e.g., the principles of resilience and non-regression), and suggests the inclusion in a legally binding instrument of principles that had hitherto only been included in soft law instruments (e.g., the right to an ecologically sound environment) (Knox 2019).

The proposed Pact's objectives seem to be three-pronged, namely (1) to be the first globally binding framework instrument of the entire body of international environmental law; (2) to entrench all major principles of international environmental law in one document; and (3) to develop progressively the law to provide a globally recognized right to live in an ecologically sound environment, with associated procedural environmental rights. As reflected in these objectives, the Global Pact shows the ambition and potential to strengthen the constitutional order in international environmental law.

Following the Pact proposal, a United Nations General Assembly resolution called for the discussion of 'possible gaps in international environmental law', which may ultimately lead to 'the convening of an intergovernmental conference to adopt an international instrument' (United Nations General Assembly 2018: para. 2). The ensuing intergovernmental discussions – in which the idea for a Global Pact met with significant resistance, meaning that its adoption is unlikely to take place before 2022 (Earth Negotiations Bulletin 2019) – have highlighted some concerns that will be informative for any attempt to create a normative hierarchy through systematization (Kotzé 2019b).

Among the concerns expressed about these developments, we note two that are particularly relevant to our discussion. First, in line with the discussion above, the inclusion and exclusion of certain norms in a Global Pact will likely be contested. Some will view the inclusion of some norms as unwarranted, for instance because the legal status of the norm may be in dispute. Others will lament the exclusion of other norms (Kotzé and French 2018; Kotzé 2019a). Second, while some principles may be widely accepted, their formulation may differ depending on the issue at hand. Any attempt at harmonizing and codifying these principles may both be resisted and have the unintended side-effect of creating uncertainty in the subfields of international environmental governance (Biniaz 2017).

But global environmental constitutionalism, and the hierarchy in international environmental law it may bring about, do not need to come about only through the adoption of a binding global environmental constitution. Other, perhaps even more viable and realistically achievable, alternatives exist. One such alternative is the formal adoption of a global right to a healthy environment (possibly outside the Global Pact process described above) (Knox 2019). To date, the United Nations General Assembly, which is considered the final arbiter on the formal creation and inclusion of international human rights into the body of international law, has not yet seen its way open to proclaiming a binding international right to a healthy environment (Alston 1984). Having the social and morally justified objective of environmental protection recognized as a right internationally, could be a critically important means to establishing regulatory priorities in earth system governance through hierarchization (Kotzé 2018); although admittedly, the two concerns raised immediately above will likely also be an issue in this instance.

Centralization

We now discuss attempts to partially centralize global governance architectures as a strategy for achieving more effective outcomes.

Our focus is, first, on the role of centralization in global networks of organizations. Here, authority and, eventually, organizational hierarchy can arise from an organization's ability to steer other intergovernmental organizations, to create or influence norms through drafting and coordinating multilateral agreements, to resolve disputes among member states or to enforce member states' compliance with certain rules. In a situation of such high organizational hierarchy, even a single intergovernmental organization can steer or dominate a governance architecture. The World Trade Organization at the centre of international trade governance is a case in point (Young 2008; Charnovitz 2012).

In contrast to trade governance, however, the architecture of earth system governance is characterized by a largely non-hierarchical coexistence of multiple intergovernmental organizations with overlapping mandates. Within this architecture, no single organization has the authority to steer other intergovernmental organizations or multilateral agreements (Biermann 2014). In the 1960s, environmental concerns led to first calls for the establishment of authoritative intergovernmental institutions for the environment, which would effectively increase the hierarchization of global governance in this area. In 1972, however, governments could only agree on establishing a United Nations Environment Programme, instead of an autonomous United Nations specialized agency that would rank higher in the United Nations hierarchy (Ivanova 2007; Linnér and Selin 2013; Vijge 2013). Though the United Nations Environment Programme has the mandate to coordinate environmental activities at the global level, it has long lacked legal authority and resources to effectively execute this mandate (Zelli and van Asselt 2013; Biermann 2014). Debates around the reform of the earth system governance architecture continued after 1972 and led to the establishment of a few weak institutions to coordinate environmental activities, such as the now-defunct Commission on Sustainable Development in 1992 and the Environmental Management Group in 1999 (Meyer-Ohlendorf and Knigge 2007). The establishment of additional non-hierarchical 'centres' of decision-making within a continuing decentralized or diffused architecture might have contributed to, rather than reduced, the horizontal nature of the architecture of earth system governance (Andresen 2007; Hoare and Tarasofsky 2007; Vijge 2013).

This has led to calls for a strengthened decentralized network of organizations (see, e.g., Najam 2003; Oberthür and Gehring 2004; Haas 2007; Meadowcroft 2007) as well as for enhanced coordination across multilateral environmental agreements (Von Moltke 2005; Selin 2010). Proposals that go a step further towards a more substantial reform of the architecture of earth system governance typically aim to centralize authority within one overarching intergovernmental organization. We discuss two categories of proposals that would constitute different degrees of centralization (for categorizations, see Biermann 2000; Lodefalk and Whalley 2002; Biermann and Bauer 2005; Biermann, Davies and van der Grijp 2009).

Some of the most prominent reform proposals entail an upgrade of the United Nations Environment Programme from a subsidiary body to a specialized agency, often referred to as a United Nations Environment Organization. Such an upgrade would increase the organization's authority by providing it with its own general assembly, universal membership, a broadened mandate, centralized legislative authority, an independent budget and an overall increase in political weight within the United Nations system (Biermann 2000; Biermann and Bauer 2005). However, this organizational upgrade would centralize authority within earth system governance only to a certain extent, since an upgrade of the United Nations Environment Programme does not entail a transfer of power from or the integration of existing intergovernmental organizations. This means that a United Nations Environment Organization would not necessarily have more authority than existing intergovernmental organizations of the United Nations. with environmental mandates, such as the World Health Organization or the Food and Agriculture Organization.

A category of reform proposals that would entail a much higher degree of centralization revolves around the establishment of a supranational agency that can steer other intergovernmental organizations in relation to their environmental or sustainable development-related mandates and activities. Such an organization would be situated high in the United Nations hierarchy, comparable for instance to the United Nations Security Council. It could arise from merging and subsuming existing United Nations agencies, possibly including the United Nations Environment Programme, and incorporate the currently independent secretaries of multilateral environmental agreements. Such an organization might also have enforcement powers over member states in regulating compliance with environmental standards, thereby following the model of, and being a counterweight to, the World Trade Organization (discussed in Biermann 2000; Lodefalk and Whalley 2002; Biermann and Bauer 2005; Kanie 2007).

Some proposals in the wider domain of sustainable development referred to a Sustainable Development Board or Council that could merge organizations such as the Commission on Sustainable Development and have authority over a wide range of organizations and institutions. Since such an entity would cut across various domains of global governance, it could potentially have much more authority than an organization within the environmental domain alone. Though the above proposals for a centralization of authority received some traction among member states, United Nations representatives and even high-level representatives from the World Trade Organization, these proposals ultimately never got very far.

Much attention has been paid instead to the recent replacement of the Commission on Sustainable Development with the High-level Political Forum, which spearheaded the formation of the Sustainable Development Goals. This Forum is mandated to enhance coordination of the sustainable development agenda within the United Nations system; but it has limited authority and resources. This has given rise to scholarly debates around orchestration, a governance strategy that stands in stark contrast with centralization, since it involves 'soft' modes of power exerted through intermediary organizations that guide and support actions (Chapter 11). Yet, the effectiveness of the Forum as an orchestrator is yet to be evaluated.

In sum, none of the far-reaching reform proposals of the last 40 years have been implemented, even though some progress in that direction has been made. In explaining why proposals for centralization of the earth system governance architecture have never been realized, it is important to note that many actors within this architecture have an interest in maintaining the status quo. Some powerful member states, such as China and the United States, are sceptical of, and often even opposed to, centralization proposals (Najam 2005). Many member states are wary of an organization that could become as powerful as the World Trade Organization, which could enforce compliance with environmental regulations and hence encroach on their sovereignty (Young 2008). This can explain why, as Ivanova and Roy (2007: 50) argue, member states 'deliberately create weak and underfunded international organizations with overlapping and even conflicting mandates'. Also, intergovernmental organizations and secretariats of multilateral environmental agreements have an interest in maintaining their authority and mandates and are therefore often not prepared to defer to a hierarchical organization for their (overlapping) environmental activities (Vijge 2013; see also Biermann 2001; Charnovitz 2012).

Even though a strong centralization of earth system governance through authoritative supranational institutions has not taken place, the latest reform efforts did result in enhanced levels of authority for the United Nations Environment Programme. The United Nations Conference on Sustainable Development in 2012 (known also as the Rio+20 conference) resulted in the United Nations Environment Programme adopting some key features of a specialized agency while maintaining the status of a subsidiary body. The United Nations Environment Programme acquired an increased budget and mandate, and, most importantly, universal membership with the creation of its United Nations Environment Assembly, thereby enhancing its formal authority among member states and intergovernmental organizations (Ivanova 2013). The effects of the United Nations Environment Programme's increased authority have not yet been thoroughly studied in the literature, but this reform seems to have muted, temporarily at least, the debates around the further centralization of the earth system governance architecture.

Prioritization

We now turn to prioritization as a strategy for further hierarchization within architectures of earth system governance. We focus here on global goal-setting, that is, the agreement on internationally agreed non-legally binding policy objectives that are time-bound, measurable and aspirational in nature, as discussed in more detail by Vijge and colleagues in Chapter 12 of this volume. Yet, while the focus of Vijge and colleagues is on global goal-setting as a governance mechanism, we are rather interested here in hierarchies among goals and between goals and non-goals, and how this affects the structure and effectiveness of governance architectures.

Generally speaking, global goals are selected through a goal-setting process in which stakeholders consider multiple issues of global significance and ultimately agree on a relatively small number of goals. These global goals broadly reflect *priorities* of the international community. For example, Young (2017: 33) explains that the 'whole point of goal setting is to single out a small number (sometimes just one) of concerns and to accord them priority in the allocation of scarce resources, including staff time and political capital', among competing objectives. Goal-setting therefore establishes a hierarchy of priorities, which can be instrumental as a means of galvanizing attention and mobilizing resources to make a sustained push to achieve measurable results within a fixed time frame (Young 2017).

Goal hierarchy in global governance then refers to the hierarchy of priorities between issues covered by a certain set of global goals as opposed to those that did not become part of the goal set. Although goal hierarchy is a relatively recent concept in global governance research, such goal hierarchies are not new. Multiple goal hierarchies have existed in global governance, some of which date back to the 1960s (Fukuda-Parr 2014), including the seven 'major goals for the survival, protection and development of children' adopted in 1991. Other examples of goal hierarchies are 'global environmental goals' that are derived from multilateral environmental agreements and the decisions of their governing bodies, some of which date back to the 1800s (Mitchell 2003).[1] Some of these goals satisfy our conceptualization of global goals, including the Paris Agreement's two-degrees global temperature objective (Morseletto, Biermann and Pattberg 2017).

Goal hierarchy in global governance has features that are distinct from other types of hierarchies discussed earlier in this chapter. For example, goal hierarchy is relatively flexible because global goals reflect priorities at the time of negotiation and have a predetermined expiry date. Furthermore, goal hierarchy is different from normative hierarchy as global goals are not 'superior' to other non-goals in a normative sense, but policy problems they address are rather considered more urgent than others. For example, the protection of the ozone layer is not selected among the targets of the Sustainable Development Goals, while it was part of the Millennium Development Goals. This development does not reflect or affect the status of the norm to protect the ozone layer in international law. However, the omission indicates that the issue is no longer a priority that requires urgent attention, since the ozone layer is predicted to recover by mid-century.

Goal-setting now constitutes an established global governance strategy (Biermann, Kanie and Kim 2017; Kanie and Bierman 2017). Analysts have highlighted that global goals may have the potential to transform our societies (Hajer et al. 2015; Stevens and Kanie 2016). However, we know little about when and how global priorities affect change. The literature so far suggests at least two key roles performed by global goal-setting, namely mobilization and orchestration. First, global goals help mobilize financial and other resources (Sachs 2015), especially from non-traditional financial sources such as business, venture capital and sovereign wealth funds (Mawdsley 2018), by offering a strong signal of the direction of policy in the medium to long term. Second, global goals serve as tools for orchestrating or aligning sectoral international institutions and organizations towards achieving the common objective of sustainable development (Kim 2016; Underdal and Kim 2017; Stevens 2018). Similarly, as higher-order priorities,

---

[1] The United Nations Information Portal on Multilateral Environmental Agreements (InforMEA) lists 289 such global environmental goals.

global goals may help resolve conflicts or manage trade-offs between lower-ranked objectives, such as priorities identified at the target level in the Sustainable Development Goals (Kim 2016).

Not all global goals, however, exert the same level of steering effects. There are many possible explanations for such variation, among which is the degree of goal hierarchy or prioritization. The strength of prioritization depends partially on the attributes of global goals in question, namely their content and intensity on what needs to be done and to what degree (Latham and Locke 1991). Following this logic, Underdal and Kim (2017) argue that the effectiveness of global goals in enhancing the overall performance of global governance depends on (a) the degree of agreement on a small and manageable set of goals; (b) the degree to which goals provide clear guidance for both agents and principals; and (c) the degree to which goals enhance the willingness and ability of agents to work effectively together to achieve the goals set for them. Global goals meet these conditions to varying extents, and this variation accounts for varying degrees of goal hierarchy they establish. Not all of the Sustainable Development Goals are deemed to fully satisfy the conditions, and as a result, some of the goals may provide 'scant guidance for prioritizing scarce resources' (Underdal and Kim 2017: 242).

We now turn to hierarchization among goals. While global goals establish a hierarchy of priorities in global governance, we observe that in almost all cases, global goals are not hierarchically differentiated among themselves. The lack of hierarchy within a goal framework is not necessarily a result of all goals being equally important, but rather because of the political nature of the goal-setting process. Young (2017), for example, observes that when setting priorities involves a relatively large number of self-interested actors, there is a danger that the group will end up with too many goals or individual goals will be incompatible or even contradictory. The Sustainable Development Goals is a case in point, where priority setting has led to 'competition for priority attention and conflict over the allocation of scarce resources' (Young 2017: 46). This non-hierarchical nature of the goal framework itself poses a challenge of managing trade-offs between competing priorities (Nilsson, Griggs and Visbeck 2016; Stafford-Smith et al. 2016). As a potential solution to this challenge, some that creating a hierarchy within a goal set by adopting a single (or sometimes twin) overarching priority of priorities is necessary to address trade-offs between goals (Griggs et al. 2013; Costanza et al. 2015; Kim 2016).

Conclusions and Future Directions

When compared to the other reform strategies discussed elsewhere in this book, hierarchization is the most far-reaching, and inevitably the most controversial, policy response to structural complexities of global governance. Yet, the proliferation of various 'building blocks' of the global governance architecture such as international institutions and norms (?) has continued to fuel a long-standing debate on structural transformation through creating a formal hierarchy.

This reform debate has been particularly heated in certain domains that suffer from the lack of hierarchy, including earth system governance. The architecture of earth system governance has often been compared with that of global trade governance or global ocean governance, which are arguably more hierarchical. As we have attempted to show, however, there are some recent signs of hierarchization in earth system governance. These include the upgrading of the Governing Council of the United Nations Environment Programme into the United Nations Environment Assembly in 2012, the inclusion of key environmental goals as part of the Sustainable Development Goals framework in 2015 and the possibility of adopting a Global Pact for the Environment.

The three key types of hierarchy – systematization, centralization and prioritization – perform different functions and have varying effects on the performance of global environmental governance. For example, higher-order norms may serve as adjudicators to reconcile potential conflicts between norms at a lower level. A central organization could improve coordination between over 1,000 international treaties and organizations. Priority goals may mobilize and steer action in certain directions.

The hierarchies are complementary. In earth system governance, for example, any serious attempt at strengthening organizational hierarchy has faced severe resistance. Evidently, proposals for a fully fledged international organization have not so far garnered sufficient political support. Yet, the possibility of a stronger normative hierarchy through global environmental constitutionalism is looming on the horizon. At the same time, global environmental goals such as the Paris Agreement's two-degrees objective are being increasingly used as a way of prioritization. Based on our analysis of and experience with other more hierarchical global governance systems, it is reasonable to expect that a stronger hierarchy in earth system governance will lead to mostly positive outcomes.

But there are also drawbacks of hierarchization as an approach to some of the challenges inherent in global governance. A key drawback is its rigidity; it is difficult to create a hierarchy and to subsequently modify it when necessary. Actors with vested interests in the current non-hierarchical global governance architecture would resist any serious attempt at disturbing the status quo for a new or different hierarchical order. Furthermore, hierarchy is relatively inflexible and it sets a path that is not easily amenable to change. For example, the United Nations Charter has largely set the course of the United Nations system and it has never been amended, even though the world has changed drastically since 1945.

This could potentially be a problem for addressing environmental problems that are in constant flux, and require a more flexible approach.

Despite these contestations, hierarchization in our view will remain an important policy response to the increasing level of complexity in global governance. More research is warranted to understand causes and effects of hierarchy and hierarchization in earth system governance. Some of the key research questions remaining to be examined are as follows. What triggers certain forms of hierarchization? What explains variations in the level of hierarchy across different global governance architectures? What are the key consequences on the functioning of international institutions as well as global governance as a whole? In this chapter, we sought to contribute towards addressing these questions by offering a clearer conceptualization of hierarchy and hierarchization in global governance. Yet, more empirical and theoretical research in this area is still urgently needed.

References

Aguila, Y., & Viñuales, J. (2019). A Global Pact for the Environment: Conceptual foundations. *Review of European, Comparative & International Environmental Law*, 29, 3–12.

Alston, P. (1984). Conjuring up new human rights: A proposal for quality control. *American Journal of International Law*, 78, 607–21.

Alvarez, J. E. (2005). *International organizations as law-makers*. Oxford: Oxford University Press.

Andresen, S. (2007). Key actors in UN environmental governance: Influence, reform and leadership. *International Environmental Agreements*, 7 (4), 457–68.

Axelrod, M. (2011). Savings clauses and the 'chilling effect': Regime interplay as constraints on international governance. In S. Oberthür, & O. S. Stokke (eds.), *Managing institutional complexity: Regime interplay and global environmental change* (pp. 87–114). Cambridge, MA: The MIT Press.

Benvenisti, E., & Downs, G. W. (2007). The empire's new clothes: Political economy and the fragmentation of international law. *Stanford Law Review*, 60, 595–631.

Biermann, F. (2000). The case for a world environment organization. *Environment*, 42 (9), 22–32.

Biermann, F. (2001). The emerging debate on the need for a world environment organization: A commentary. *Global Environmental Politics*, 1, 45–55.

Biermann, F. (2014). *Earth system governance: World politics in the Anthropocene*. Cambridge, MA: The MIT Press.

Biermann, F., Davies, O., & van der Grijp, N. (2009). Environmental policy integration and the architecture of global environmental governance. *International Environmental Agreements*, 9 (4), 351.

Biermann, F., Kanie, N., & Kim, R. E. (2017). Global governance by goal-setting: The novel approach of the UN Sustainable Development Goals. *Current Opinion in Environmental Sustainability*, 26–27, 26–31.

Biermann, F., & Pattberg, P. (2008). Global environmental governance: Taking stock, moving forward. *Annual Review of Environment and Resources*, 33, 277–94.

Biniaz, S. (2017). *10 questions to ask about the proposed 'Global Pact for the Environment'*. New York: Sabin Centre for Climate Change Law.

Bodansky, D. (2006). Does one need to be an international lawyer to be an international environmental lawyer? *American Society of International Law Proceedings*, 100, 303–7.

Bodansky, D. (2009). Is there an international environmental constitution? *Indiana Journal of Global Legal Studies*, 16, 565–84.

Bosselmann, K. (2015). Global environmental constitutionalism: Mapping the terrain. *Widener Law Review*, 21, 171–85.

Boyle, A. E. (1999). Some reflections on the relationship of treaties and soft law. *International and Comparative Law Quarterly*, 48, 901–13.

Broude, T., & Shany, Y. (eds.) (2011). *Multi-sourced equivalent norms in international law*. Oxford: Hart Publishing.

Charnovitz, S. (2012). Organizing for the green economy: What an international green economy organization could add. *Journal of Environment and Development*, 21 (1), 44–7.

Costanza, R., McGlade, J., Lovins, H., & Kubiszewski, I. (2015). An overarching goal for the UN Sustainable Development Goals. *Solutions*, 5, 13–16.

De Wet, E., & Vidmar, J. (2012). Introduction. In E. De Wet, & J. Vidmar (eds.), *Hierarchy in international law: The place of human rights* (pp. 1–13). Oxford: Oxford University Press.

Dingwerth, K., & Pattberg, P. (2006). Global governance as a perspective on world politics. *Global Governance*, 12, 185–203.

Dryzek, J. S., Bächtiger, A., & Milewicz, K. (2011). Toward a deliberative global citizens' assembly. *Global Policy*, 2, 33–42.

Duit, A., Galaz, V., Eckerberg, K., & Ebbesson, J. (2010). Governance, complexity, and resilience. *Global Environmental Change*, 20, 363–8.

Earth Negotiations Bulletin (2019). Summary of the third substantive session of the Ad Hoc Open-ended Working Group towards a Global Pact for the Environment: 20–22 May 2019. *Earth Negotiations Bulletin*, 35, 1–12.

Fischer, J., Manning, A. D., Steffen, W. et al. (2007). Mind the sustainability gap. *Trends in Ecology and Evolution*, 22, 621–4.

Fischer-Lescano, A., & Teubner, G. (2004). Regime collisions: The vain search for legal unity in the fragmentation of global law. *Michigan Journal of International Law*, 25, 999–1073.

Friedrich, J. (2013). *International environmental 'soft law'*. Berlin: Springer.

Fukuda-Parr, S. (2014). Global Goals as a policy tool: Intended and unintended consequences. *Journal of Human Development and Capabilities*, 15, 118–31.

Galaz, V., Olsson, P., Hahn, T., Folke, C., & Svedin, U. (2008). The problem of fit between governance systems and environmental regimes. In O. R. Young, L. A. King, & H. Schroeder (eds.), *Institutions and environmental change: Principal findings, applications and research frontiers* (pp. 147–86). Cambridge, MA: The MIT Press.

Griggs, D., Stafford-Smith, M., Gaffney, O. et al. (2013). Sustainable development goals for people and planet. *Nature*, 495, 305–7.

Haas, P. M. (2007). Turning up the heat on global environmental governance. *The Forum*, 5.

Hajer, M., Nilsson, M., Raworth, K. et al. (2015). Beyond cockpit-ism: Four insights to enhance the transformative potential of the Sustainable Development Goals. *Sustainability*, 7, 1651–60.

Halevy, N., Y Chou, E., Galinsky, A. D. (2011). A functional model of hierarchy. *Organizational Psychology Review*, 1, 32–52.

Hart, H. L. A. (1994). *The concept of law*. Oxford: Oxford University Press.

Hoare, A., & Tarasofsky, R. (2007). *International environmental governance*. London: Chatham House.

International Law Commission (2006). *Fragmentation of international law: Difficulties arising from the diversification and expansion of international law*. UN Doc. A/CN.4/L.682.

International Law Commission (2019). *Peremptory norms of general international law (jus cogens)*. UN Doc. A/CN.4/L.936.

Ivanova, M. H. (2007). Designing the United Nations Environment Programme: A story of compromise and confrontation. *International Environmental Agreements*, 7 (4), 337–61.

Ivanova, M. H. (2013). Reforming the institutional framework for environment and sustainable development: Rio+20's subtle but significant impact. *International Journal of Technology Management & Sustainable Development*, 12 (3), 211–31.

Ivanova, M. H., & Roy, J. (2007). The architecture of global environmental governance: Pros and cons of multiplicity. In L. Swart, & E. Perry (eds.), *Global environmental governance: Perspectives on the current debate* (pp. 48–66). New York: Center for UN Reform Education.

Jordan, A., Huitema, D., van Asselt, H., & Forster, J. (eds.) (2018). *Governing climate change: Polycentricity in action*? Cambridge, UK: Cambridge University Press.

Kanie, N. (2007). Governance with multilateral environmental agreements: A healthy or ill-equipped fragmentation? In L. Swart, & E. Perry (eds.), *Global environmental governance: Perspectives on the current debate* (pp. 67–86). New York: Center for UN Reform Education.

Kanie, N., & Biermann, F. (eds.) (2017). *Governing through goals: Sustainable Development Goals as governance innovation*. Cambridge, MA: The MIT Press.

Kelsen, H. (1960). *Reine Rechtslehre*. Vienna: Deuticke.

Keohane, R. O., & Victor, D. G. (2011). The regime complex for climate change. *Perspectives on Politics*, 9, 7–23.

Kim, R. E. (2016). The nexus between international law and the Sustainable Development Goals. *Review of European, Comparative & International Environmental Law*, 25, 15–26.

Kim, R. E., & Bosselmann, K. (2013). International environmental law in the Anthropocene: Towards a purposive system of multilateral environmental agreements. *Transnational Environmental Law*, 2, 285–309.

Kim, R. E., & Bosselmann, K. (2015). Operationalizing sustainable development: Ecological integrity as a Grundnorm of international law. *Review of European, Comparative & International Environmental Law*, 24, 194–208.

Klabbers, J. (1996). The redundancy of soft law. *Nordic Journal of International Law*, 65, 167–82.

Knox, J. H. (2019). The Global Pact for the Environment: At the crossroads of human rights and the environment. *Review of European, Comparative & International Environmental Law*, 29, 40–7.

Koskenniemi, M., & Leino, P. (2002). Fragmentation of international law? Postmodern anxieties. *Leiden Journal of International Law*, 15, 553–79.

Kotzé, L. J. (2012). Arguing global environmental constitutionalism. *Transnational Environmental Law*, 1, 199–233.

Kotzé, L. J. (2015). Constitutional conversations in the Anthropocene: In search of environmental jus cogens norms. *Netherlands Yearbook of International Law*, 46, 241–71.

Kotzé, L. J. (2016). *Global environmental constitutionalism in the Anthropocene*. Oxford: Hart Publishing.

Kotzé, L. J. (2018). In search of a right to a healthy environment in international law. In J. Knox, & R. Pejan (eds.), *The human right to a healthy environment* (pp. 136–54). Cambridge, UK: Cambridge University Press.

Kotzé, L. J. (2019a). A global environmental constitution for the Anthropocene? *Transnational Environmental Law*, 8, 11–33.

Kotzé, L. J. (2019b). International environmental law's lack of normative ambition: An Opportunity for the Global Pact for the Environment? *Journal of European Environmental and Planning Law*, 16, 213–36.

Kotzé, L. J., & French, D. (2018). A critique of the Global Pact for the Environment: A stillborn initiative or the foundation for Lex Anthropocenae? *International Environmental Agreements*, 18, 811–38.

Kotzé, L. J, & Muzangaza, W. (2018). Constitutional international environmental law for the Anthropocene? *Review of European, Comparative & International Environmental Law*, 27, 278–292.

Lafferty, W., & Hovden, E. (2003). Environmental policy integration: Towards an analytical framework. *Environmental Politics*, 12, 1–22.

Latham, G. P., & Locke, E. A. (1991). Self-regulation through goal setting. *Organizational Behavior and Human Decision Processes*, 50, 212–47.

Lehmen, A. (2015). The case for the creation of an international environmental court: Non-state actors and international environmental dispute resolution. *Colorado Natural Resources, Energy, and Environmental Law Review*, 26, 179–217.

Liivoja, R. (2008). The scope of the supremacy clause of the United Nations Charter. *International and Comparative Law Quarterly*, 57, 583–612.

Linnér, B. O., & Selin, H. (2013). The United Nations Conference on Sustainable Development: Forty years in the making. *Environment and Planning C: Government and Policy*, 31 (6), 971–87.

Lodefalk, M., & Whalley J. (2002). Reviewing proposals for a world environmental organisation. *The World Economy*, 25, 601–17.

Mawdsley, E. (2018). 'From billions to trillions': Financing the SDGs in a world 'beyond aid'. *Dialogues in Human Geography*, 8, 191–5.

Meadowcroft, J. (2007). Who is in charge here? Governance for sustainable development in a complex world. *Journal of Environmental Policy & Planning*, 9 (3–4), 299–314.

Meyer-Ohlendorf, N., & Knigge, M. (2007). A United Nations Environment Organization. In L. Swart, & E. Perry (eds.), *Global environmental governance: Perspectives on the current debate* (pp. 124–41). New York: Center for UN Reform Education.

Mitchell, R. B. (2003). International environmental agreements: A survey of their features, formation, and effects. *Annual Review of Environment and Resources*, 28, 429–461.

Morseletto, P., Biermann, F., & Pattberg, P. (2017). Governing by targets: Reductio ad unum and evolution of the two-degree climate target. *International Environmental Agreements*, 17, 655–76.

Murphy, S. D. (1999). Does the world need a new international environmental court? *George Washington Journal of International Law and Economics*, 32, 333–49.

Najam, A. (2003). The case against a new international environmental organization. *Global Governance*, 9, 367–84.

Najam, A. (2005). Neither necessary, nor sufficient: Why organizational tinkering won't improve environmental governance. In F. Biermann, & S. Bauer (eds.), *A world environment organization: Solution or threat for effective international environmental governance?* (pp. 235–56). Burlington: Ashgate.

Nilsson, M., Griggs, D., & Visbeck, M. (2016). Map the interactions between Sustainable Development Goals. *Nature*, 534, 320–2.

Oberthür, S., & Gehring, T. (2004). Reforming international environmental governance: An institutionalist critique of the proposal for a world environment organisation. *International Environmental Agreements*, 4, 359–81.

Ostrom, E. (2010). Polycentric systems for coping with collective action and global environmental change. *Global Environmental Change*, 20, 550–7.

Palmer, G. (1992). New ways to make international environmental law. *American Journal of International Law*, 86, 259–83.

Paulus, A. (2005). Jus cogens in a time of hegemony and fragmentation: An attempt at a reappraisal. *Nordic Journal of International Law*, 74, 297–334.

Pedersen, O. W. (2012). An international environmental court and international legalism. *Journal of Environmental Law*, 24, 547–58.

Sachs, J. D. (2015). Goal-based development and the SDGs: Implications for development finance. *Oxford Review of Economic Policy*, 31, 268–78.

Sand, P. H., & Wiener, J. B. (2016). Towards a new international law of the atmosphere. *Tulsa Law Review*, 7, 195–223.

Scanlon, Z. (2018). The art of 'not undermining': Possibilities within existing architecture to improve environmental protections in areas beyond national jurisdiction. *ICES Journal of Marine Science*, 75, 405–16.

Selin, H. (2010). *Global governance of hazardous chemicals: Challenges of multilevel management*. Cambridge, MA: The MIT Press.

Shelton, D. (2006). Normative hierarchy in international law. *American Journal of International Law*, 100, 291–323.

Stafford-Smith, M., Griggs, D., Gaffney, O. et al. (2016). Integration: The key to implementing the Sustainable Development Goals. *Sustainability Science*, 13, 1–9.

Steffen, W., Richardson, K., Rockström, J. et al. (2015). Planetary boundaries: Guiding human development on a changing planet. *Science*, 347, 1259855.

Steinberg, R. (1998). *Der ökologische Verfassungsstaat*. Berlin: Suhrkamp Verlag.

Stevens, C. (2018). Scales of integration for sustainable development governance. *International Journal of Sustainable Development and World Ecology*, 25, 1–8.

Stevens, C., & Kanie, N. (2016). The transformative potential of the Sustainable Development Goals (SDGs). *International Environmental Agreements*, 16, 393–6.

Underdal, A., & Kim, R. E. (2017). The Sustainable Development Goals and multilateral agreements. In N. Kanie, & F. Biermann (eds.), *Governing through goals: Sustainable Development Goals as governance innovation* (pp. 241–58). Cambridge, MA: The MIT Press.

United Nations General Assembly (2018). *Towards a global pact for the environment*. UN Doc. A/RES/72/277.

Vidmar, J. (2012). Norm conflicts and hierarchy in international law: Towards a vertical international legal system? In E. De Wet, & J. Vidmar (eds.), *Hierarchy in international law: The place of human rights* (pp. 13–41). Oxford: Oxford University Press.

Vijge, M. J. (2013). The promise of new institutionalism: Explaining the absence of a World or United Nations Environment Organisation. *International Environmental Agreements*, 13 (2), 153–76.

Von Moltke, K. (2005). Clustering international environmental agreements as an alternative to a world environment organization. In F. Biermann, & S. Bauer (eds.), *A world environment organization: Solution or threat for effective international environmental governance?* New York: Routledge.

Wiersema, A. (2009). The new international law-makers? Conferences of the parties to multilateral environmental agreements. *Michigan Journal of International Law*, 31, 231–87.

Wolfrum, R., & Matz, N. (2003). *Conflicts in international environmental law*. Berlin: Springer.

Young, O. R. (2002). *The institutional dimensions of environmental change: Fit, interplay, and scale*. Cambridge, MA: The MIT Press.

Young, O. R. (2008). The architecture of global environmental governance: Bringing science to bear on policy. *Global Environmental Politics*, 8 (1), 14–32.

Young, O. R. (2017). Conceptualization: Goal setting as a strategy for earth system governance In N. Kanie, & F. Biermann (eds.), *Governing through goals: Sustainable Development Goals as governance innovation* (pp. 31–52). Cambridge, MA: The MIT Press.

Zelli, F., & van Asselt, H. (2013). The institutional fragmentation of global environmental governance: Causes, consequences, and responses. *Global Environmental Politics*, 13 (3), 1–13.

Zürn, M., & Faude, B. (2013). On fragmentation, differentiation, and coordination. *Global Environmental Politics*, 13, 119–30.

Part IV

Future Directions

14

Taking Stock and Moving Forward

FRANK BIERMANN, RAKHYUN E. KIM, KENNETH W. ABBOTT,
JAMES HOLLWAY, RONALD B. MITCHELL AND MICHELLE SCOBIE

Given the scope of the research programme that we present here, it is hardly possible to draw one overarching conclusion. There is no one-liner, no elevator pitch, no single finding when it comes to the complexities of the broader architectures of earth system governance. A few overriding conclusions, however, stand out, and we present them in the next section of this chapter. Following that, we identify four promising new research trends. Finally, we sketch a set of transformative policy proposals regarding the architecture of earth system governance.

Crosscutting Contributions

Four crosscutting contributions are supported by all chapters in this book, though to varying degrees.

First, this book brings much-needed conceptual clarity to a profuse but meandering debate. Concepts such as 'interplay', 'complexes', 'integration', 'interlinkages' and 'fragmentation' pervade the burgeoning literature in this field, with little agreement, so far, on how specific terms relate to others. With the collective insights of 42 experts collaborating in the 14 chapters in this book, we can better define, compare and relate the various conceptualizations in which the architecture debate is awash. While we cannot claim that our book resolves all conceptual contestation, the organization of the volume, the depth of the individual chapters and the collective effort of all our contributors will clearly help strengthen conceptual clarity in this field. The glossary presented at the end of the volume further advances this aim.

Second, this book shows that architectures matter. Whereas individual institutions and distinct regimes were at the core of the debate from the 1970s through to the 1990s, when pioneering scholars such as Oran Young, Arild Underdal, Ronald Mitchell, Steinar Andresen, Sebastian Oberthür and others advanced our understanding of environmental 'regime effectiveness', the new wave of research since the 2000s has shown that institutions do not operate in a void. Instead, they are enmeshed in complex structures of which they are merely a part. All contributions to this volume emphasize the importance of such structures, which we conceptualize as the architectures of earth system governance. The book provides substantial evidence that it matters at the micro level how institutions interact with others; that it matters at the meso level how institutions are entangled with others in larger regime complexes; and that it matters at the macro level how institutions are affected by broader architectures that are more or less fragmented or polycentric.

Third, the book offers numerous insights on the policy interventions that can address problems of conflictive interlinkages, complexity and fragmentation. Given the increasing structural complexity of the architecture of earth system governance, it is not surprising that highly diverse policy responses are available, and in fact used, to deal with these challenges. These responses range from interplay management to policy integration, orchestration, global goal-setting and, as the most radical intervention, the hierarchization of norms, institutions and priorities. Traditionally, most responses were targeted at the level of dyads, that is, interactions between merely two institutions or organizations. More recently, however, the range of responses has expanded to incorporate attempts at overall coordination and structural reform of the entire architecture of earth system governance. Even so, the 'old' modes have received renewed attention in recent years as well.

Fourth, the book is firmly located beyond the statist perspective of the 1980s and 1990s, which was informed – at least in the North American and European international relations discourse – by the meta-debates between neoliberal institutionalism and neorealism. Most chapters in this book go well beyond this statist perspective and include in their analysis the increasing role of non-state actors in earth system governance as norm-creators, global orchestrators, builders of transnational institutions and so on. The conceptual turn during the late 1990s from intergovernmental relations to governance is fully reflected throughout this volume. Undoubtedly, the governance architectures that we are discussing comprise both public and private actors. They include states as well as actors that seek to represent civil society, science organizations and major corporations and their global associations and networks. This does not imply, however, that the role of governments is necessarily shrinking. While non-state actors assume novel roles and add to the complexity of global steering processes, governments – and, in particular, the governments of the larger countries – remain powerful actors within the complexities of novel governance architectures.

New Research Trends

In addition to these crosscutting themes and findings, numerous new lines of research have become central over the last decade, many of them flowing from the work presented in this book. We identify three promising trends and research directions as most relevant and interesting.

Complexity and Polycentricity

To start with, many scholars have turned in recent years to studying large sets of international institutions under new headings, especially complexity and polycentricity.

Research on complexity has two dimensions: the governance of complexity and the complexity of governance. In the former use of the term, research is not about analyzing institutional complexity as such, but about finding ways to govern complex systems (Duit and Galaz 2008; Underdal 2010; Le Prestre 2017; Young 2017). Much of this research is motivated by systems or complexity thinking. In particular, researchers have applied the Conant-Ashby theorem (Ruhl 2008; Duit et al. 2010; Kim and Mackey 2014), which suggests that for governance to be effective, the complexity of governance needs to match the complexity of the system being governed (Conant and Ashby 1970).

In the latter use of the term, complexity refers as a variable to a *quality* of a governance architecture. In this sense, a governance architecture can be more or less complex, and hence potentially more or less effective, fair, adaptive or whatever performance criterion one wants to use. In this meaning, complexity has been central in several recent studies. Conceptually, however, this use of 'complexity' has not added much to the more widely used term 'fragmentation'. Architectures of global governance can be described as more or less fragmented, or more or less complex, with little analytical difference. One more practical consequence lies in the possibility of interdisciplinary exchange and cooperation. The notion of a fragmented governance system, for one, is more widely used in the legal literature and in the policy world; here, analysts can link their studies more easily to United Nations processes or legal reform debates. Also, fragmentation is easily conceptualized as a process, which allows for comparative research over time, examining processes of governance fragmentation and defragmentation. The notion of complexity, in contrast, opens avenues for collaboration and mutual inspiration with the field of complexity studies, as well as with those science disciplines that are deeply involved with the study of complexity, such as theoretical physics, biology, ecology, system analysis and information sciences. This link or bridge to the natural sciences allows for the employment of sophisticated tools and methods, such as the modelling of networks and agent-based modelling. Like ecological systems, for example, scholars might be able to claim that complexity within governance architectures emerges through self-organization of their elementary units, similar to the evolution of ecosystems.

Broadly speaking, governance scholars have operationalized institutional complexity with two variables: diversity and multiplicity. The logic is that a system with more diverse or a greater number of institutions is likely to be more complex. For example, Zelli, Möller and van Asselt (2017: 670) understand institutional complexity as 'a diversity of international institutions that legally or functionally overlap in addressing a given issue area of global governance'. This simultaneous use of 'complexity' and 'diversity' might invite criticisms of tautological reasoning unless compelling conceptual differences between complexity and diversity can be shown. Van Asselt and Zelli (2014) consider the number of institutions as a useful measure of institutional complexity; in future research this might need to be strengthened by additional information on their interlinkages and relative weight and power.

In general, analysts in this line of studies observe an increasing level of institutional complexity at the global level, and attribute it to 'the rise of private and hybrid authority manifested by collaborative governance arrangements' (Widerberg 2016: 84), in addition to more traditional international public institutions (also Green 2013; Green and Auld 2016; Hickmann 2017; Zelli, Möller and van Asselt 2017). These scholars argue that the emergence of private authorities and their private certification schemes does not bring together disconnected institutions but rather creates an extra layer of complexity (Gulbrandsen 2009; Kalfagianni and Pattberg 2013; van der Ven, Rothacker and Cashore 2018). As noted above, a similar observation could be expressed in terms of increased fragmentation of a global governance architecture.

This suggests again that both terms are largely synonymous descriptions of qualities of governance architectures, yet both bring different opportunities for cross-disciplinary exchange and cooperation. It is conceivable, of course, that the two concepts are today synonymous largely because the complexity research programme has not advanced very far yet, inasmuch as it still views complexity in the same terms as described by fragmentation. If the complexity research programme could move to actually analyzing governance systems as complex systems, the two approaches might become more different over time.

The second conception of complexity just discussed is also similar to polycentricity. Polycentricity too describes a structural quality of an architecture and is hence in line with the notion of fragmentation (Chapter 8). Polycentricity owes much to the work of Elinor Ostrom and as such often has a positive normative connotation (Aligica and Tarko 2011). For many scholars, polycentricity is not only seen as a *description* of a governance architecture but also as a *virtue*. Polycentric governance systems are under certain circumstances assumed to be, following Ostrom's work, more effective than centralized monocentric ones. Recently, the notion of polycentricity has been systematically employed to study global climate governance, and here with a more nuanced normative assessment than this concept normally carries (Jordan et al. 2015, 2018).

Regarding the notion of regime complexes, however, we see a clear difference between this term, on the one hand, and institutional complexity on the other, despite linguistic similarity. Complexity is a quality of an architecture of institutions, and as such is comparable to both fragmentation and polycentricity. A regime complex, instead, describes a unit at the meso level of governance (Orsini, Morin and Young 2013), that is, an assemblage of international regimes, other institutions and actors, which can all be described as one 'complex' (see in more detail Chapter 7). Like governance architectures at the macro level, regime complexes at the meso level can be described by qualities such as their levels of fragmentation or polycentricity.

Recent research has sought to identify and explain the effects of institutional complexity and polycentricity on, for example, the performance of agents operating within given policy domains. But the perception that researchers bring to the analysis often predefines the approach taken as well as its conclusions. On the one hand, complexity is often used interchangeably with disorder, chaos and uncertainty; something that needs to be reduced or managed. Moreover, the term is associated with multiplicity, which implies for some an increased probability of 'inconsistent international legal commitments' (Axelrod 2014: 987). Legal indeterminacy, normative ambiguity and regulatory uncertainty, institutional complexity or fragmentation, are all supposed to make international cooperation more difficult, for example, in attempting to meet the environmental, economic and legal challenges of transnational environmental crime (Elliott 2017). In short, like fragmentation, complexity is often associated with negative effects (e.g., Drezner 2009).

On the other hand, complexity is also studied with a more optimistic approach – again, not different from the literature on governance fragmentation. For example, complexity is seen as a necessary ingredient for adaptability, flexibility and resilience of a governance architecture. Studies have identified some degree of order or (organized) complexity in the structure of systems of international institutions that were previously imagined to be chaotic. For example, a distinct order has emerged out of the seemingly chaotic and complex institutional landscape of global carbon standards (Green 2013). In a similar vein, Hickmann (2017) argues that the spread of sub- and non-state climate initiatives does not lead to a loss of state authority, but rather enhances the centrality of state-based forms of governance, because transnational actors often use the climate regime as a point of reference and build upon the norms and rules stipulated in international agreements.

These findings have important implications for policy responses to institutional complexity. Scholars have suggested various policy strategies that are more or less explicitly captured in concepts such as harnessing (Axelrod and Cohen 1999; Ruhl, Katz and Bommarito 2017), taming (Barabási 2005), embracing (Hirsch et al. 2011) or managing institutional complexity (Oberthür and Stokke 2011; Lubell 2013; Pickering, Betzold and Skovgaard 2017). Despite subtle differences and slightly varying connotations, what is common to these concepts is the view that 'the complexity of an interconnected society and its governance require a complexity-informed approach' (Teisman and Gerrits 2014: 17). In other words, we need to delve deeper into the study of institutional complexity in order to provide evidence-based policy advice to navigate our societies through mounting global challenges and promote the transformation of governance structures.

Evolutionary Dynamics

Despite the connotation of immovability that the concept of architecture carries, such macro-level structures are not static. Architectures of global governance are fluid and dynamic, continuously changing in response to pressures and governance processes. Scholars have so far devoted only limited analytical attention to these dynamics due to the prevailing focus on elementary institutions and their interactions. Over the past decade, however, a new research focus has emerged, seeking to unravel how governance architectures evolve over time. This research strand describes and analyzes architectural change through innovative empirical and methodological approaches. It also suggests possible explanations for observed changes from diverse theoretical perspectives.

A *longitudinal perspective* on governance architectures has proven useful for a fuller understanding of actor configurations, as well as the processes that produce distinct structural patterns. Integration (Chapter 9), orchestration (Chapter 11), fragmentation (Chapter 8) and hierarchization (Chapter 13) are all processes that can reshape an architecture. But these processes cannot be properly understood through snapshots representing particular moments in time. A longitudinal analysis is thus necessary to identify changes, for example in the degree of fragmentation over time (Kim 2013; Greenhill and Lupu 2017). One could also develop a metric to measure the degree of fragmentation, but it will be difficult to interpret the result without placing that number into a comparative perspective as well.

Over the past decade, analysts have made a number of efforts to theorize the evolutionary dynamics of governance architectures. We introduce three research strands here, by no means an exhaustive list. First, scholars have used the concept of *punctuated equilibrium* to study dynamics in global energy governance (Colgan, Keohane and Van de Graaf 2012). This strand has produced insights into when and under what conditions innovations occur that generate abrupt structural changes, and is particularly well suited to explaining transformative change in governance architectures (Kettl 2015). Second, an interesting theoretical development is the application of *organizational ecology*. Here, scholars have tried to understand institutional change in global governance by focusing on 'populations' of institutions as the unit of analysis. This has enabled analysis of the influence of institutional environments, especially their organizational density and resource availability, on organizational behaviour and viability (Abbott, Green and Keohane 2016). Organizational ecology has been useful in explaining, for example, why international organizations fail to spread more evenly within a governance architecture when they expand and proliferate (Morin 2018). A third theoretical lens on the evolution of governance architectures is the concept of *complex adaptive systems*. Here, scholars emphasize endogenous processes of selection akin to natural selection as a key explanatory variable and apply this empirically to environmental as well as non-environmental institutional systems (Kim and Mackey 2014; Pauwelyn 2014; Morin, Pauwelyn and Hollway 2017).

These three theoretical approaches are derived from the natural sciences and were originally designed to explain how ecosystems evolve. The models thus tend to overlook the role of human agency in shaping global governance. It is assumed that individual agents are homogeneous and react to structural constraints in a similar and predictable manner. But social systems are different. Architects are heterogeneous. They possess varying levels of power, authority and legitimacy, seek to further particular interests and exercise agency (Schroeder 2010; Bouteligier 2011; Dellas, Pattberg and Betsill 2011; Newell, Pattberg and Schroeder 2012; Mukhtarov and Gerlak 2013). And while architects have the capacity to shape the governance architectures within which they operate, existing frameworks and dominant norms simultaneously shape their roles and capacities (Dellas, Pattberg and Betsill 2011). Therefore, the precise mechanisms by which architectures evolve, through the myriad decisions of relatively autonomous architects, remain obscure and require further study (e.g., Hollway 2015; Rabitz 2017).

Furthermore, given the density of the architectures of earth system governance and their long history and evolution, more research is needed on the factors that can explain not only the creation of international institutions, but also their 'deaths', and the associated impacts on the global governance architecture. While new international institutions are regularly created, a considerable number of institutions become inactive or dormant. In some instances they even cease to exist or 'die away' (Jinnah 2011; Eilstrup-Sangiovanni 2018). As more and more international institutions complete their life cycles, we expect to observe a considerable change in the structure and dynamics of earth system governance.

In sum, a richer understanding of the evolutionary dynamics of governance architectures requires an integrative framework that brings together three bodies of literature: ideally, it should combine insights from theories of institutional change (e.g., Hall 2010; Young 2010a, 2010b; Marcoux 2011; Hall 2015, 2016); institutional interactions and inter-organizational relations (e.g., Young 2002; R. Biermann 2008; Gehring and Oberthür 2009; R. Biermann and Koops 2017); and macro-level structural change (e.g., Kim 2013; Morin, Pauwelyn and Hollway 2017).

Transformative Reform

Along with analyzing adaptive or evolutionary change, earth system governance researchers are also investigating how to *deliberately* transform governance architectures that are no longer fit-for-purpose in addressing problems of earth system transformation (O'Brien 2012). This research builds on previous work on how to better design (Young 2011) or adapt (Boyd and Folke 2012; Ruhl 2012; Tomozeiu and Joss 2014) individual international regimes and institutions. The same key concepts apply, such as fit, interplay and scale (Young 2002), but these concepts are applied at much larger scales. For example, one might ask whether the entire architecture of global biodiversity governance is doing more harm than good. If so, and if this is found to result from excessive fragmentation, one would then ask how to remedy it (Jóhannsdóttir, Cresswell and Bridgewater 2010). Essentially, this research programme aims to transform the structures of global governance architectures so as to make them more impactful.

This research programme is not entirely new; it dates back at least to the 1990s, with debates about the need for a world environment organization (e.g., Biermann and Bauer 2005). Another early example is the proposal for a law of the atmosphere, which would have had profound implications for the governance of climate change, stratospheric ozone depletion and air pollution. These debates resurface from time to time in different forms. For example, the idea of a more centralized steering mechanism in earth system governance found support in proposals for a United Nations Sustainable Development Council (Biermann et al. 2012; Kanie et al. 2012; Biermann 2014) and a reformed United Nations Trusteeship Council (Biermann 2014; Kim and Bosselmann 2015; see also Chapter 13). Some proposals have eventually been adopted. For example, the upgrading of the Governing Council of the United Nations Environment Programme to the United Nations Environment Assembly with universal membership was a significant political outcome of the 2012 United Nations Conference on Sustainable Development in

Rio de Janeiro. Recently, the International Law Commission has considered the case for a law of the atmosphere (Sand and Wiener 2016; Sand 2017). And international discussions are underway on a Global Pact for the Environment (Kotzé and French 2018), intended to become a framework agreement of international environmental law.

A new trend in this debate is the turn to more integrative and systems analysis in addressing the question of *how* to transform. A serious answer to this question requires a thorough understanding of the structure and dynamics of the architecture in question. One approach perceives an architecture as a system of international institutions and seeks to work towards transformation by targeting key leverage points in the system (Meadows 2008). Leverage points are places in a system where actors can intervene and create radical shifts in its structure or function with relatively little effort. For example, systems analysis has advanced to the point where we can pinpoint missing links between key institutions. Researchers do not simply suggest that we need more policy and institutional coherence, but now identify which exact 'dots' need to be connected (van Asselt and Zelli 2013) or which linkages need to be strengthened for maximum transformative effects (Jinnah 2011; Abbott 2014; Betsill et al. 2015).

Overall, years of analysis back up an emerging scientific consensus on the need for major transformations in the architecture of earth system governance and a 'constitutional moment' in world politics that could possibly even include amendments to the Charter of the United Nations (Biermann et al. 2012; Kanie et al. 2012). Similar suggestions have been made for international environmental law (Kotzé 2012, 2016), based on the conclusion that that body of law lacks a clear goal, which is at least in part responsible for its ineffectiveness (Bodansky 2009; Kim and Bosselmann 2013). While these proposals would intervene at the level of the governance architecture, other analyses suggest even more fundamental paradigm shifts. For instance, some scholars make a case for going beyond environmental rights (Boyd 2012; Gellers 2015) to recognizing the rights of nature, as seen in recent examples in certain national jurisdictions (Boyd 2017). A proposal for transforming the architecture of earth system governance by introducing sustainability as a fundamental norm on par with equality, freedom and justice is another example (Bosselmann 2017; see also Chapter 13).

In this line of investigation, we observe an emerging trend of linking the literature on governance architectures with the literature on transformative governance. The transformative governance approach is intended to respond to, manage and trigger regime shifts in coupled socio-ecological systems, at multiple scales (Chaffin et al. 2016). Therefore, reforming a governance architecture would have to consider not only transformation *in* global governance, but also what sorts of complex impacts on the earth system the reform would aim to achieve. For example, scholars have suggested structural reforms that would create conditions for the effective governance of interacting planetary boundaries (Galaz et al. 2012a, 2012b).

Conceptually differentiating between governance *for* and *of* transformation can be useful in this respect (Patterson et al. 2017). Governance *for* transformation is governance that creates the necessary institutional conditions for transformation in socio-technical-ecological systems to emerge. This is an indirect, bottom-up way of transformation by, for example, identifying key barriers to change and removing them. Governance *of* transformation refers to top-down modes of governance that actively trigger transformation and steer its trajectory. The two concepts are not mutually exclusive. For example, a governance measure might actively trigger a major reform, which in turn creates necessary conditions for socio-technical-ecological transformation. According to one study, for example, the emergence of a modular governance architecture has created enabling conditions for transnational standard-setters to govern sustainability transitions (Manning and Reinecke 2016). Another prominent example could be the Sustainable Development Goals, which may have created conditions for societal transformation, yet at the same time may also exert top-down steering effects that need to be assessed in-depth (Stevens and Kanie 2016; see also Chapter 12).

Research Strategies

In addition to the substantive trends discussed above, we see an emerging trend of broadening the scope of research on earth system governance architectures beyond the confines of traditional international relations research. Yet we still see a need to expand research strategies further to a much broader ambit.

One notable trend is expanding research towards more collaborative programmes based on interdisciplinarity. Often, researchers work on a project together, each with their own disciplinary backgrounds. However, integrating these perspectives better, within and beyond Western knowledge practices, carries the potential for further innovation, even though transcending disciplines may also mean losing clear standards by which research on global governance architectures can be judged.

In addition, transdisciplinary approaches that involve non-academic stakeholders in research have become more prominent. Societal actors are important, for example, to highlight problems that should shape research agendas. While not all science must be immediately 'policy-relevant', if architecture science does hold some emancipatory potential, then this should be directed at the problems that concern society the most, whether this be the fragmentation of legal rules or the democratic deficit of global governance. Societal actors can also contribute to the 'co-production' of knowledge (Jentoft, McCay and Wilson 1998; Jasanoff 2004) and to identifying solutions that reduce inequality, increase capacity, foster legitimacy and improve effectiveness.

Overall, we see a number of areas for the further development of research strategies in this field.

First, we need more collaboration and dialogue among researchers who study different elements and processes of earth system governance architectures. Knowledge of particular aspects of an architecture often remains isolated in bubbles. The separate strands of research on, for example, international institutions and their interlinkages at the micro level, regime complexes at the meso level and governance architectures at the macro level should more frequently be brought together. New holistic insights may emerge through *research on the cross-level dynamics* within an architecture.

Second, holistic insights may also arise from more empirical analysis, which we see as imperative to advance theory. In particular, many scholars have highlighted the importance of more *systematic comparative approaches*. Here, comparison is not limited to comparing institutions and their dyadic relationships, but includes comparison of larger structures such as regime complexes and entire governance architectures. Comparative analysis will help explain why certain structures are, for example, more or less fragmented, and what that means for a range of characteristics of interest, including their performance. This again requires a stronger emphasis on global interdependencies. The architectures of earth system governance have become more complex over the years, increasing the interdependence of international institutions. If actions in one institution affect the performance of another, the effectiveness of individual institutions cannot be explained without accounting for their interdependency. Future research needs to pay more attention *to complex interdependencies* when assessing and optimizing the performance of both individual institutions and governance architectures.

In this context, it remains important to better integrate research on intergovernmental institutions with research on non-state actors and their transnational networks, focusing on how each of them shapes and is shaped by the architectures of earth system governance. The past decade has witnessed an explosion of private codes, schemes and partnerships, and more research is needed on the impacts of these relatively new institutions. Furthermore, concerns over their legitimacy have been documented in several chapters of this book; this must be examined systematically in relation to equity and justice concerns.

Third, these approaches require expanding the *methodological toolbox* of governance scholars, including greater reliance on combinations of methods. For example, social network analysis has become popular in studying earth system governance architectures (Kim 2013; Hollway and Koskinen 2016; Widerberg 2016). More traditional methods might also need adjustments, and better integration, in mixed-methods approaches that investigate larger sets of actors and institutions. Traditional statistical inference, such as regression analysis, typically assumes a sample of independent observations or treats any potential dependence as nuisance. But the premise of the study of governance architectures is that the actors and institutions constituting the architecture are interdependent. This should make statistical network models, which not only account for observational dependencies but actually highlight them, particularly attractive. A range of statistical network models are now available for different purposes (Block, Stadtfeld and Snijders 2016; Block et al. 2018), including tie-based (Butts 2008; Lusher, Koskinen and Robins 2013) and actor-oriented models (Snijders, van de Bunt and Steglich 2010; Stadtfeld, Hollway and Block 2017).

More fundamentally, research on the architectures of earth system governance can contribute to society by highlighting the global structures that constrain governance and identifying actors and points of agency through which these structures can be changed. Such research not only can inform judicious governance choices, but also holds emancipatory potential. In the Anthropocene, as we move rapidly towards a warmer world, requiring societies to adapt, research on global governance architectures should consider the potential for structural institutional reform.

Finally, research on governance architectures can also make durable societal contributions through education. Highlighting global governance constraints and agency to students or the public can raise awareness, strengthen knowledge and promote thinking critically and creatively. Active learning may help, but most existing simulations and case-study exercises were designed for the study of a single negotiation or problem, not the long-run evolution of entire architectures. Global governance architectures – macro-level webs of principles, institutions and practices – can appear remote, clinical, complicated and slow to change. In competition with the latest and loudest news, emotionally manipulative 'big lies' and scandals of a 'post-truth' era (Peters 2017), architectures may lose attention. The didactic challenge will be to find ways to render complex global governance architectures more comprehensible.

Reforming Governance Architectures

While most chapters in this volume discuss some policy reforms, they generally focus on detailed analysis of existing architectures and their elements; broader governance transformations have not been central. In this section, therefore, we explore such larger transformations. We rely here on two assessments that large groups of leading scholars associated with the Earth System Governance Project compiled in 2012 and 2017, with a view to sketching far-reaching transformations of the architectures of global governance (Biermann et al. 2012; Earth System Governance Project 2017). The groups concluded in these reports that incrementalism – the main approach so far – will not suffice to stimulate societal change at the level and speed needed. The challenges of the Anthropocene require instead novel institutional strategies that are bolder in scope, swifter in implementation and more adaptive in character. The group offered an ambitious roadmap for institutional change and the fundamental reform of governance architectures, combining reform proposals that could be achieved within just a decade with others that are more far-reaching.

First, the author team from the Earth System Governance Project argued for a global constitutive agreement that would draw on key principles enshrined in the outcomes of the major conferences in Stockholm (1972) and Rio de Janeiro (1992), as well as in human rights and other treaty regimes, merging them into a constitutional framework that would fill the normative gaps left by the 1945 United Nations Charter (Chapter 13). Such a contract could be a stand-alone document, or could become an integral part of the United Nations system. The constitutive agreement could resolve current normative conflicts between economic, social and environmental institutions and help mainstream social and environmental standards into economic institutions, hence reducing the adverse impacts of economic globalization and global governance fragmentation (Chapter 8). Treaty norms could be clustered more systematically around the interlinkages among our planet's socio-ecological systems. Concretely, the proposed architecture would nest international institutions under a limited number of global umbrella treaties. Similar nesting processes already take place, for instance with respect to the international regulation of hazardous wastes and biodiversity, where multiple treaty secretariats coordinate decision-making, monitoring and enforcement.

As this book is written, governments are negotiating a Global Pact for the Environment. This is clearly a step in the right direction – even though it is too early to tell whether negotiation of the Pact will succeed, what form it will take and what its effects will be. The versions of the Pact currently under discussion, however, would most likely not have the full integrative impact that the earth system governance research community originally called for (see Kotzé and French 2018; Kotzé 2019 and our discussion in Chapter 13).

Second, the group proposed to strengthen the integration of sustainable development organs and agencies through a new World Sustainable Development Council within the United Nations (Biermann et al. 2012; Biermann 2014; Earth System Governance Project 2017). This council would be built on the ethics of planetary stewardship and given an earth trusteeship mandate, reflecting the deep-rooted idea that states should act as trustees of the earth (Kim and Bosselmann 2015). In addition to states, a chamber of the council could be devoted to civil society, including non-governmental organizations and scientific communities. As trustee of the earth, the council could ensure that the entire spectrum of human rights is respected, and that countries share rights, responsibilities and risks in accordance with the precautionary principle and the principle of common but differentiated responsibilities and respective capabilities.

To be effective, this council could rely not only on traditional modes of geographical representation, but should give special prominence to the 20 largest economies in North and South. These states could hold at least 50 per cent of votes in the council, with the other 50 per cent reserved for representatives of smaller countries (Biermann et al. 2012). A strong role for the largest economies would allow the World Sustainable Development Council to have a meaningful influence in areas such as economic and trade governance. The 20 largest economies in North and South – broadly represented today by the Group of 20 – represent about two-thirds of the world's population and around 90 per cent of global gross national product, justifying their role.

At the 2012 United Nations Conference on Sustainable Development, governments agreed to establish the High-level Political Forum on Sustainable Development to fulfil a similar role. Yet its structure is very different, and several scholars doubt its future significance, although, again, it is too early to tell whether the High-level Political Forum will succeed as an orchestrator of global governance (see our discussions in, Chapters 11, 12 and 13).

In addition, the original United Nations Charter and institutional system made no provision for the protection of planetary ecological systems. The only adjustment has been the creation of the United Nations Environment Programme in 1972, which has not – despite all efforts by its dedicated staff and funders – managed to live up to the long-term challenges of the Anthropocene. Effective earth system governance requires a powerful organ that focuses on planetary ecological concerns. The governance architecture that researchers from the Earth System Governance Project proposed would hence include a strong world environment organization, initially along the lines of the World Health Organization or the International Labour Organization (Biermann et al. 2012; Earth System Governance Project 2017). This approach has been supported by the African Union, European Union and many state governments. To be sure, the United Nations Environment Programme has been strengthened in recent years, making it a more effective and independent agency, for instance by establishing the United Nations Environment Assembly with universal membership. This body may perform some of the functions suggested for a world environment organization (Chapter 3). Still, further strengthening of the 'environmental pillar' of the global governance system seems essential.

Third, the group of earth system governance scholars argued for more general reforms of international institutions as well. The dynamic nature of Anthropocene challenges requires that international institutions be more adaptive in responding to change, both in socio-ecological problems and in our knowledge of their nature and causes. International fisheries treaties face quite different challenges than they did 20, let alone 50, years ago: greater threats from marine pollution, warming sea temperatures and ocean acidification; increased demand for fish, size of fleets and efficiency of technologies; and improved knowledge of fish stocks and population biodynamics. Managing dynamic problems like these effectively requires dynamic and responsive institutions.

Numerous short-term actions could be taken, for example, introducing procedures that ensure that new scientific information is quickly taken up, or that systematically collect and review information about a treaty's impact. Such measures, however, would lead only to incremental improvements. While the search for incremental change is important, it is not enough. The earth system governance researchers therefore recommended transformative changes in international decision-making, making it more democratic, more adaptive, and more rapidly implemented, with fewer veto points. More precisely, the group advocated stronger reliance on qualified majority voting. Political systems that build on such a majority-based rule arrive more quickly at more far-reaching decisions. Earth system transformation is too urgent to be left to the veto power of individual countries, as is the case under the consensus decision rules common in many treaties. However, it is also evident that qualified majority voting with binding effect is rare in international politics, so this approach must be further developed (Biermann 2014; Kemp 2014).

Fourth, stronger global institutions raise important questions of legitimacy and accountability. Governance through United Nations-type institutions tends to give a large role to international and domestic bureaucracies, at the expense of national parliaments and direct citizen involvement. Accountability can be strengthened by granting stakeholders better access to decision-making, through special rights enshrined in agreements, or through stronger participation in councils that govern resources and commissions that hear complaints. Greater transparency would empower citizens and consumers to hold governments and businesses accountable, providing incentives for better governance (Gupta and Mason 2014). The inclusive negotiations around the Sustainable Development Goals in 2012–2015 are a valuable example (Chapter 12). Stronger consultative rights in intergovernmental institutions for civil society representatives, parliamentarians and citizens could also be a step forward. However, this would require the development of transparent and effective accountability mechanisms for civil society representatives vis-à-vis their constituencies, as well as mechanisms to account for imbalances in the strength of civil society across countries from North to South, and for power and resource differentials across segments of civil society.

The author team from the Earth System Governance Project (2017) hence proposed a polycentric, pluriform system of global accountability that would complement the United Nations General Assembly with additional assemblies representing the parliaments, civil society organizations and citizens of the world. These might include a Global Parliamentary Assembly, modelled on regional parliamentary assemblies in South America, Africa and Europe; a Global Assembly of Civil Society, modelled on regional bodies such as the European Social and Economic Committee, which integrates unions, employers and other societal representatives in regional decision-making; and a Global Citizens Assembly, which would bring together individual citizens, who could even be selected through random drawings (Dryzek, Bächtiger and Milewicz 2011). In addition, the group argues for a High Commissioner of Future Generations, who would have speaking rights in international institutions and could provide guidance as to their long-term impacts (Pearce 2012). Finally, the proposed governance architecture would open other global institutions for meetings and assemblies of representatives of cities and federal states. This polycentric, pluriform system of global deliberation and decision-making would allow for a much more representative, deliberative and hence accountable and legitimate system of global cooperation.

Fifth, global cooperation goes beyond intergovernmental agreements. Vast networks of non-state initiatives led by industry, activists and scientists, multisectoral partnerships and cities has sprung up, giving new strength and enthusiasm to global cooperation (Chapter 4). Such governance 'beyond the state' can help avoid capture by powerful interests. Yet to be effective, non-governmental initiatives require the involvement of multiple stakeholders, appropriate national regulatory frameworks and accountability mechanisms, along with strong consumer demand – all of which are not always present. Transnational labelling schemes cover a sizable share of global markets only for a handful of goods; so far they hardly offer a solution to sustainability problems such as forest conservation and poverty eradication (International Institute for Sustainable Development 2014). Thus, the proposed governance architecture would provide for novel cooperative frameworks that promote successful non-state-driven, transnational governance: regulations that create incentives for firms to seek certification; better-focused procurement policies; stronger legitimation; and better monitoring of sustainability effects. International organizations could play a powerful role in catalyzing and steering novel forms of private and public–private governance (Abbott and Snidal 2010).

Sixth, the group argued that special attention must be paid to the poorest billion of humankind, who will suffer most from earth system transitions and global economic changes. Policies are rarely made *by* poor and marginalized people – they are made only *for* poor people, by others who believe they understand or represent poor people's preferences and aspirations. Global policy processes that affect poor and marginalized people must thus as far as possible enable those people's participation in preparation, implementation, monitoring and adaptation. While the traditional dichotomy of 'North' and 'South' may be less relevant today, extremely high consumption levels in industrialized countries and in some parts of emerging economies require special and urgent action (Lebel, Lorek and Daniel 2010), while many poorer societies and marginalized groups lack the capacities to take forceful action in mitigating and adapting to global environmental change. Overall, strong financial and technological support for poorer countries remains an inevitable part of an effective earth system governance architecture.

Conclusions

'The mother art is architecture. Without an architecture of our own we have no soul of our own civilization.' This statement – attributed to Frank Lloyd Wright – also applies to architectures of earth system governance. While individual actions count, individual institutions matter, and each negotiation or political process has value – in the end, it is governance architectures that determine political outcomes. Architectures comprising diverse institutions that interact in numerous ways create structural power that affects our societies and all humankind. Global governance architectures may sometimes create inequalities, but they can also be a source of transformation. They shape the structure of global trade; the freedom of maritime transport; the flows of global communication; the stocks of global fisheries; and last but not least, the global politics surrounding the ongoing climate crisis, the destruction of our planet's biological diversity, and myriad other issues of earth system transformation.

Yet overall, the social sciences still lack sufficient knowledge about the emergence, dynamics and impacts of global governance architectures. This book has sought to address that gap. We have sought to increase conceptual clarity; synthesize a decade of intense research; and chart directions for future research. While the volume surely has not provided conclusive answers to all the problems identified, it has made one point clear: global governance architectures are of utmost importance. The 'architecture lens' offers a bird's-eye view on the global governance landscape that is highly valuable in explaining outcomes of world politics. As a result, the governance architecture research programme will continue to flourish.

References

Abbott, K. W. (2014). Strengthening the transnational regime complex for climate change. *Transnational Environmental Law*, 3, 57–88.

Abbott, K. W., Green, J. F., & Keohane, R. O. (2016). Organizational ecology and institutional change in global governance. *International Organization*, 70, 247–77.

Abbott, K. W., & Snidal, D. (2010). International regulation without international government: Improving IO performance through orchestration. *Review of International Organizations*, 5 (3), 315–44.

Aligica, P. D., & Tarko, V. (2011). Polycentricity: From Polanyi to Ostrom, and beyond. *Governance*, 25 (2), 237–62.

Axelrod, M. (2014). Clash of the treaties: Responding to institutional interplay in European Community–Chile Swordfish negotiations. *European Journal of International Relations*, 20, 987–1013.

Axelrod, R., & Cohen, M. D. (1999). *Harnessing complexity: Organizational implications of a scientific frontier*. New York: The Free Press.

Barabási, A.-L. (2005). Taming complexity. *Nature Physics*, 1, 68–70.

Betsill, M. M., Dubash, N. K., Paterson, M., van Asselt, H., Vihma, A., & Winkler, H. (2015). Building productive links between the UNFCCC and the broader global climate governance landscape. *Global Environmental Politics*, 15, 1–10.

Biermann, F. (2000). The case for a world environment organization. *Environment*, 42, 22–32.

Biermann, F. (2014). *Earth system governance. World politics in the Anthropocene*: Cambridge, MA: The MIT Press.

Biermann, F., & Bauer, S. (eds.) (2005). *A world environment organization: Solution or threat for effective international environmental governance?* Aldershot: Ashgate.

Biermann, F., Abbott, K. W., Andresen, S. et al. (2012). Navigating the Anthropocene: Improving earth system governance. *Science*, 335, 1306–7.

Biermann, R. (2008). Towards a theory of inter-organizational networking: The Euro-Atlantic security institutions interacting. *Review of International Organizations*, 3, 151–177.

Biermann, R., & Koops, J. A. (eds.) (2017). *Palgrave handbook of inter-organizational relations in world politics*. London: Palgrave Macmillan.

Block, P., Stadtfeld, C., & Snijders, T. AB. (2016). Forms of dependence: Comparing SAOMs and ERGMs from basic principles. *Sociological Methods and Research*, 48 (1), 202–39.

Block, P., Koskinen, J., Hollway, J., Steglich, C., & Stadtfeld, C. (2018). Change we can believe in: Comparing longitudinal network models on consistency, interpretability and predictive power. *Social Networks*, 52, 180–91.

Bodansky, D. (2009). Is there an international environmental constitution? *Indiana Journal of Global Legal Studies*, 16, 565–84.

Bosselmann, K. (2017). *The principle of sustainability: Transforming law and governance*. London: Routledge.

Bouteligier, S. (2011). Exploring the agency of global environmental consultancy firms in earth system governance. *International Environmental Agreements*, 11, 43–61.

Boyd, D. R. (2012). The constitutional right to a healthy environment. *Environment*, 54, 3–15.

Boyd, D. R. (2017). *The rights of nature: A legal revolution that could save the world*. Toronto: ECW Press.

Boyd, E., & Folke, C. (eds.) (2012). *Adapting institutions: Governance, complexity and social-ecological resilience*. Cambridge, UK: Cambridge University Press.

Butts, C. T. (2008). A relational event framework for social action. *Sociological Methodology*, 38 (1), 155–200.

Chaffin, B. C., Garmestani, A. S., Gunderson, L. H. et al. (2016). Transformative environmental governance. *Annual Review of Environment and Resources*, 41, 399–423.

Colgan, J. D., Keohane, R. O., & Van de Graaf, T. (2012). Punctuated equilibrium in the energy regime complex. *Review of International Organizations*, 7, 117–43.

Conant, R. C., & Ashby, W. R. (1970). Every good regulator of a system must be a model of that system. *International Journal of Systems Science*, 1, 89–97.

Dellas, E., Pattberg, P., & Betsill, M. M. (2011). Agency in earth system governance: Refining a research agenda. *International Environmental Agreements*, 11, 85–98.

Drezner, D. W. (2009). The power and peril of international regime complexity. *Perspectives on Politics*, 7, 65–70.

Dryzek, J. S., Bächtiger, A., & Milewicz, K. (2011). Toward a deliberative global citizens' assembly. *Global Policy*, 2, 33–42.

Duit, A., & Galaz, V. (2008). Governance and complexity: Emerging issues for governance theory. *Governance: An International Journal of Policy, Administration, and Institutions* 21 (3), 311–35.

Duit, A., Galaz, V., Eckerberg, K., & Ebbesson, J. (2010). Governance, complexity and resilience. *Global Environmental Change*, 20 (3), 363–8.

Earth System Governance Project (2017). *Submission to the Global Challenges Foundation*. Stockholm: Global Challenges Foundation.

Eilstrup-Sangiovanni, M. (2018). Death of international organizations: The organizational ecology of intergovernmental organizations, 1815–2015. *Review of International Organizations*, 70, 1–32.

Elliott, L. (2017). Cooperation on transnational environmental crime: Institutional complexity matters. *Review of European, Comparative & International Environmental Law*, 26, 107–17.

Galaz, V., Biermann, F., Crona, B. et al. (2012a). 'Planetary boundaries' – exploring the challenges for global environmental governance. *Current Opinion in Environmental Sustainability*, 4 (1), 80–7.

Galaz, V., Crona, B., Österblom, H., Olsson, P., & Folke, C. (2012b). Polycentric systems and interacting planetary boundaries: Emerging governance of climate change–ocean acidification–marine biodiversity. *Ecological Economics*, 81, 21–32.

Gehring, T., & Oberthür, S. (2009). The causal mechanisms of interaction between international institutions. *European Journal of International Relations*, 15, 125–56.

Gellers, J. C. (2015). Explaining the emergence of constitutional environmental rights: A global quantitative analysis. *Journal of Human Rights and the Environment*, 6, 75–97.

Green, J. F. (2013). Order out of chaos: Public and private rules for managing carbon. *Global Environmental Politics*, 13, 1–25.

Green, J. F., & Auld, G. (2016). Unbundling the regime complex: The effects of private authority. *Transnational Environmental Law*, 6, 259–84.

Greenhill, B., & Lupu, Y. (2017). Clubs of clubs: Fragmentation in the network of intergovernmental organizations. *International Studies Quarterly*, 61, 181–95.

Gulbrandsen, L. H. (2009). The emergence and effectiveness of the Marine Stewardship Council. *Marine Policy*, 33, 654–60.

Gupta, A., & Mason, M. (eds.) (2014). *Transparency in global environmental governance: Critical perspectives*. Cambridge, MA: The MIT Press.

Hall, N. (2015). Money or mandate? Why international organizations engage with the climate change regime. *Global Environmental Politics*, 15, 79–97.

Hall, N. (2016). *Displacement, development and climate change: International organizations moving beyond their mandates*. London: Routledge.

Hall, P. A. (2010). Historical institutionalism in rationalist and sociological perspective. In J. Mahoney, & K. Thelen (eds.), *Explaining institutional change: Ambiguity, agency, and power* (pp. 204–24). Cambridge, UK: Cambridge University Press.

Hickmann, T. (2017). The reconfiguration of authority in global climate governance. *International Studies Review*, 19 (3), 430–51.

Hirsch, P. D., Adams, W. M., Brosius, J. P., Zia, A., Bariola, N., & Dammert, J. L. (2011). Acknowledging conservation trade-offs and embracing complexity. *Conservation Biology*, 25, 259–64.

Hollway, J. (2015). The evolution of global fisheries governance, 1960–2010. University of Oxford.

Hollway, J., & Koskinen, J. (2016). Multilevel embeddedness: The case of the global fisheries governance complex. *Social Networks*, 44, 281–94.

Jasanoff, S. (2004). *States of knowledge: The co-production of science and the social order*. London: Routledge.

Jentoft, S., McCay, B. J., & Wilson, D. C. (1998). Social theory and fisheries co-management. *Marine Policy*, 22 (4–5), 423–36.

Jinnah, S. (2011). Climate change bandwagoning: The impacts of strategic linkages on regime design, maintenance, and death. *Global Environmental Politics*, 11, 1–9.

Jóhannsdóttir, A., Cresswell, I., & Bridgewater, P. (2010). The current framework for international governance of biodiversity: Is it doing more harm than good? *Review of European, Comparative and International Environmental Law*, 19, 139–49.

Jordan, A. J., Huitema, D., Hildén, M. et al. (2015). Emergence of polycentric climate governance and its future prospects. *Nature Climate Change*, 5, 977–82.

Jordan, A., Huitema, D., Schoenefeld, J., van Asselt, H., & Forster, J. (2018). Governing Climate Change polycentrically: Setting the scene. In A. Jordan, D. Huitema, H. van Asselt, & J. Forster (eds.), *Governing climate change: Polycentricity in action?* (pp. 3–26). Cambridge, UK: Cambridge University Press.

Kalfagianni, A., & Pattberg, P. (2013). Fishing in muddy waters exploring the conditions for effective governance of fisheries and aquaculture. *Marine Policy*, 38, 124–32.

Kanie, N., Betsill, M. M., Zondervan, R., Biermann, F., & Young, O. R. (2012). A charter moment: Restructuring governance for sustainability. *Public Administration and Development*, 32, 292–304.

Kemp, L. (2014). Framework for the future? Exploring the possibility of majority voting in the climate negotiations. *International Environmental Agreements*, 16, 757–79.

Kettl, D. F. (2015). *The transformation of governance: Public administration for the twenty-first century*. Baltimore, MD: Johns Hopkins University Press.

Kim, R. E. (2013). The emergent network structure of the multilateral environmental agreement system. *Global Environmental Change*, 23 (5), 980–91.

Kim, R. E., & Bosselmann, K. (2013). International environmental law in the Anthropocene: Towards a purposive system of multilateral environmental agreements. *Transnational Environmental Law*, 2, 285–309.

Kim, R. E., & Bosselmann, K. (2015). Operationalizing sustainable development: Ecological integrity as a *Grundnorm* of international law. *Review of European, Comparative & International Environmental Law*, 24, 194–208.

Kim, R. E., & Mackey, B. (2014). International environmental law as a complex adaptive system. *International Environmental Agreements*, 14, 5–24.

Kotzé, L. J. (2012). Arguing global environmental constitutionalism. *Transnational Environmental Law*, 1, 199–233.

Kotzé, L. J. (2016). *Global environmental constitutionalism in the Anthropocene*. Portland: Bloomsbury Publishing.

Kotzé, L. J., & French, D. (2018). A critique of the global pact for the environment: A stillborn initiative or the foundation for *Lex Anthropocenae*? *International Environmental Agreements*, 18, 811–38.

Kotzé, L. J. (2019). International environmental law's lack of normative ambition: An opportunity for the Global Pact and its Gap Report? *Journal of European Environmental and Planning Law*, in press.

Lebel, L., Lorek, S. & Daniel, R. (eds.) (2010). *Sustainable production consumption systems: Knowledge, engagement and practice*. Dordrecht: Springer Dordrecht.

Le Prestre, P. (2017). *Global ecopolitics revisited: Toward a complex governance of global environmental problems*. London: Routledge.

Lubell, M. (2013). Governing institutional complexity: The ecology of games framework. *Policy Studies Journal*, 41, 537–59.

Lusher, D., Koskinen, J., & Robins, G. (2013). *Exponential random graph models for social networks: Theory, methods, and applications*. Cambridge, UK: Cambridge University Press.

Manning, S., & Reinecke, J. (2016). A modular governance architecture in-the-making: How transnational standard-setters govern sustainability transitions. *Research Policy*, 45, 618–33.

Marcoux, C. (2011). Understanding institutional change in international environmental regimes. *Global Environmental Politics*, 11, 145–51.

Meadows, D. H. (2008). *Thinking in systems: A primer*. White River Junction: Chelsea Green.

Morin, J. F. (2018). Concentration despite competition: The organizational ecology of technical assistance providers. *Review of International Organizations*, 70, 1–33.

Morin, J. F., Pauwelyn, J., & Hollway, J. (2017). The trade regime as a complex adaptive system: Exploration and exploitation of environmental norms in trade agreements. *Journal of International Economic Law*, 20, 365–90.

Mukhtarov, F., & Gerlak, A. K. (2013). River basin organizations in the global water discourse: An exploration of agency and strategy. *Global Governance*, 19, 307–26.

Najam, A. (2000). Future directions: The case for a 'Law of the Atmosphere'. *Atmospheric Environment*, 34, 4047–9.

Newell, P., Pattberg, P., & Schroeder, H. (2012). Multiactor governance and the environment. *Annual Review of Environment and Resources*, 37, 365–87.

Nilsson, M., & Persson, Å. (2012). Can earth system interactions be governed? Governance functions for linking climate change mitigation with land use, freshwater and biodiversity protection. *Ecological Economics*, 81, 10–20.

O'Brien, K. (2012). Global environmental change II: From adaptation to deliberate transformation. *Progress in Human Geography*, 36, 667–76.

Oberthür, S. (2009). Interplay management: enhancing environmental policy integration among international institutions. *International Environmental Agreements*, 9 (4), 371–91.

Oberthür, S., & Stokke, O. S. (eds.) (2011). *Managing institutional complexity: Regime interplay and global environmental change*. Cambridge, MA: The MIT Press.

Orsini, A., Morin, J. F., & Young, O. R. (2013). Regime complexes: A buzz, a boom or a boost for global governance? *Global Governance*, 19 (1), 27–39.

Patterson, J., Schulz, K., Vervoort, J. et al. (2017). Exploring the governance and politics of transformations towards sustainability. *Environmental Innovation and Societal Transitions*, 24, 1–16.

Pauwelyn, J. (2014). At the edge of chaos? Foreign investment law as a complex adaptive system, how it emerged and how it can be reformed. *ICSID Review*, 29 (2), 263–88.

Pearce, C. (2012). Ombudspersons for future generations: A proposal for Rio+20. *UNEP: Perspectives*, 6, 1–12.

Peters, M. A. (2017). Education in a post-truth world. *Educational Philosophy and Theory*, 49 (6), 563–6.

Pickering, J., Betzold, C., & Skovgaard, J. (2017). Managing fragmentation and complexity in the emerging system of international climate finance. *International Environmental Agreements*, 17, 1–16.

Rabitz, F. (2017). *The global governance of genetic resources: Institutional change and structural constraints*. London: Routledge.

Ruhl, J. B. (2008). Law's complexity: A primer. *Georgia State University Law Review*, 24, 885–911.

Ruhl, J. B. (2012). Panarchy and the law. *Ecology and Society*, 17.

Ruhl, J. B., Katz, D. M., & Bommarito, M. J. (2017). Harnessing legal complexity. *Science*, 355, 1377–8.

Sand, P. H. (2017). The discourse on 'protection of the atmosphere' in the International Law Commission. *Review of European, Comparative & International Environmental Law*, 26 201–9.

Sand, P. H., & Wiener, J. B. (2016). Towards a new international law of the atmosphere. *Göttingen Journal of International Law*, 7, 195–223.

Schroeder, H. (2010). Agency in international climate negotiations: The case of indigenous peoples and avoided deforestation. *International Environmental Agreements*, 10, 317–32.

Snijders, T. A. B, van de Bunt, G. G., & Steglich, C. E. G. (2010). Introduction to stochastic actor-based models for network dynamics. *Social Networks*, 32 (1), 44–60.

Stadtfeld, C., Hollway, J., & Block, P. (2017). Dynamic network actor models: Investigating coordination ties through time. *Sociological Methodology*, 47 (1), 1–40.

Stevens, C., & Kanie, N. (2016). The transformative potential of the Sustainable Development Goals (SDGs). *International Environmental Agreements*, 16, 393–6.

Teisman, G., & Gerrits, L. (2014). The emergence of complexity in the art and science of governance. *Complexity, Governance and Networks*, 1 (1), 17–28.

Tomozeiu, D., & Joss, S. (2014). Adapting adaptation: The English eco-town initiative as governance process. *Ecology and Society*, 19.

Underdal, A. (2010). Complexity and challenges of long-term environmental governance. *Global Environmental Change*, 20, 386–93.

Van Asselt, H., & Zelli, F. (2014). Connect the dots: Managing the fragmentation of global climate governance. *Environmental Economics and Policy Studies*, 16, 137–55.

Van der Ven, H., Rothacker, C., & Cashore, B. (2018). Do eco-labels prevent deforestation? Lessons from non-state market driven governance in the soy, palm oil, and cocoa sectors. *Global Environmental Change*, 52, 141–51.

Widerberg, O. (2016). Mapping institutional complexity in the Anthropocene: A network approach. In P. Pattberg, & F. Zelli (eds.), *Environmental politics and governance in the Anthropocene: Institutions and legitimacy in a complex world* (pp. 81–102). London: Routledge.

Young, O. R. (2002). *The institutional dimensions of environmental change: Fit, interplay, and scale*. Cambridge, MA: The MIT Press.

Young, O. R. (2010a). *Institutional dynamics: Emergent patterns in international environmental governance*. Cambridge, MA: The MIT Press.

Young, O. R. (2010b). Institutional dynamics: Resilience, vulnerability and adaptation in environmental and resource regimes. *Global Environmental Change*, 20, 378–85.

Young, O. R. (2011). Effectiveness of international environmental regimes: Existing knowledge, cutting-edge themes, and research strategies. *Proceedings of the National Academy of Sciences of the United States of America*, 108, 19853–60.

Young, O. R. (2017). *Governing complex systems: Social capital for the Anthropocene*. Cambridge, MA: The MIT Press.

Zelli, F., Möller, I., & van Asselt, H. (2017). Institutional complexity and private authority in global climate governance: The cases of climate engineering, REDD+ and short-lived climate pollutants. *Environmental Politics*, 26, 669–93.

Glossary

This glossary presents definitions of key terms used in research on architectures of earth system governance. The glossary draws in part on the synthesis of the Institutional Dimensions of Global Environmental Change project (Young, King and Schroeder 2008) and on the synthesis of the Global Governance Project (Biermann and Pattberg 2012), hence following a trajectory in conceptual development. However, most terms, to the extent that they build on earlier glossaries, have been adjusted in light of the current state of debate.

Accountability. A relationship whereby those who govern are subject to control and held accountable by their constituencies, for instance through obligations to report on activities or the possibility of being rewarded or punished.

Architecture. The overarching macro-level system of public and private institutions, principles, norms, regulations, decision-making procedures and organizations that are valid or active in a given area of global governance. Usually, a governance architecture will consist of regime complexes, regimes and international organizations.

Earth system governance. The interrelated system of formal and informal rules, rule-making mechanisms and actor networks at all levels of human society (from local to global) that are set up to steer societies towards preventing, mitigating and adapting to environmental change and earth system transformations, within the normative context of sustainable development.

Fragmentation. A non-binary, dynamic quality of a global governance architecture that is marked by multiple and often overlapping institutions and actor constellations active in the same area, and the resulting normative conflicts.

Global goals. Internationally agreed, non-legally binding policy objectives that are time-bound, measurable and aspirational in nature, such as the Sustainable Development Goals. This definition excludes legally binding international rules and norms such as those established through multilateral agreements.

Governance. The process of steering or guiding societies towards collective outcomes that are deemed socially desirable and away from those that are socially undesirable.

Hierarchization. A deliberate process to transform a horizontal global governance architecture into one with a clearer hierarchy that resembles more accurately the governance systems of domestic states, for instance through creating new institutions above or below an existing institution or rearranging existing institutions hierarchically, or by creating superior and inferior norms. Hierarchization in earth system governance is broadly about institutionalizing a hierarchy among norms, institutions, actors and governance priorities.

Hierarchy. In global governance, a vertically nested structure whereby actors and institutions in a lower rank are bound or otherwise compelled to obey, respond to or contribute to higher-order norms and objectives. The place in a hierarchy corresponds to relative status or authority.

Horizontal fragmentation. The segmentation of governance into parallel, often overlapping, rule-making systems maintained by different groups of actors at the same level of decision-making.

Horizontal institutional interlinkages. Connections between policy processes, rules, norms and principles of two or more institutions at the same level of decision-making, for instance between international regimes.

Institutional complexity. Used broadly interchangeably with governance *fragmentation*.

Institutional conflict. A situation in which two or more international institutions with (partially) contradictory rules or rule-related behaviour interact and functionally overlap in the international arena, negatively affecting the effectiveness of at least one of the institutions.

Institutional interplay. A situation in which the operation of one institutional arrangement affects the outputs, outcomes and impacts of another or others.

Institutions. Clusters of rights, rules and decision-making procedures that give rise to a social practice, assign roles to participants and guide interactions among occupants of these roles. At the international level, institutions can be fully fledged international organizations, specialized bodies and programmes, treaty secretariats or norms guiding actor behaviour.

International bureaucracies. Hierarchically organized groups of civil servants that act to pursue a policy in the international area, within the mandate of an international organization or regime and within the decisions of the assembly of its member states.

International organizations. Institutional arrangements that combine a normative framework, a group of member states and a bureaucracy as administrative core.

International regimes. A set of implicit or explicit principles, norms, rules and decision-making procedures generated by states, around which the expectations of actors converge in a specific issue area.

Multilevel governance. A type of governance that is fragmented vertically between different layers of rule-making and rule implementation, ranging from supranational to international, national and subnational layers of authority.

Non-governmental organizations. Institutional actors that are not part of, and not predominantly funded by, governments. They usually represent social, cultural, legal and environmental advocacy groups with primarily non-commercial goals.

Orchestration. A mode of governance in which one actor (the orchestrator) enlists one or more intermediary actors (intermediaries) to govern a third set of actors (targets) in line with the orchestrator's goals. It is a strategy of *indirect* governance, which contrasts with both mandatory regulation and direct collaborative approaches.

Policy convergence. The process in which institutional frameworks and regulatory approaches in a policy area become more and more similar.

Policy divergence. The process in which policymaking in different countries or constituencies (increasingly) differs according to local or national needs and preferences.

Policy integration. In environmental terms, the incorporation of environmental concerns and objectives into non-environmental policy areas, such as energy, transport and agriculture, as opposed to pursuing such objectives through purely environmental policy practices.

Polycentricity. The non-binary dynamic quality of a governance architecture that is marked by the simultaneous presence of different loci of authority, power and decision-making. Can largely be used interchangeably with governance *fragmentation*.

Regime. See *international regime*.

Regime complex. A set of partially overlapping and non-hierarchical institutions governing a particular issue area.

Sustainable development. Broadly defined as development that meets the needs of the present without compromising the ability of future generations to meet their own needs.

Transnational institutions. Sets of norms, rules and decision-making procedures that are made and implemented across borders predominantly through the activities of non-state actors. They include private certification schemes, reporting initiatives and accounting standards.

Treaty secretariats. Bureaucracies within international treaties, conventions or protocols, such as the secretariats of the United Nations Framework Convention on Climate Change or the Convention on Biological Diversity.

Vertical fragmentation. The segmentation of governance into different layers and clusters of rule-making and rule implementing between supranational, international, national and subnational layers of authority.

Vertical institutional interlinkages. Connections between policy processes, rules, norms and principles of two or more institutions at different levels of policymaking.

Index

Accountability, 87, 246, 259, 313, 314
Adaptation, 239
   Adaptability, 145, 303
Agency, 59, 208–10, 215–16, 284, 312
Agent-based modelling, 47, 301
Agreement, 62, 139, 148
   Bilateral, 168
   Intergovernmental, 1, 6, 108, 257, 275, 303–4, 315
   Multilateral, 60, 111, 166, 168, 284
   Paris, 42, 62, 85, 105, 258
   Trade, 18, 142, 222
Allocation, 20, 44, 287
Anthropocene, 17, 70, 310, 312
Architectural voids, 20–1
Architecture, 4–7, 310–15
Areas beyond national jurisdiction, 98–9, 100, 102, 111–13
Autonomy, 59
   Bureaucratic, 66–8

Boundary Organization, 15, 146
Brokers, 15, 66
Bureaucracies, 7, 58
   International, 7, 68–71

Centralization, 146, 280, 284–7
Coherence, 145, 186, 194, 210–11, 227
Complex adaptive system, 150, 305
Complexity, 301–4
   Institutional, 220, 302, 303, 323
Conflict, 125
   Institutional, 125, 323
Constitution, 158
   Global, 276, 277
Coordination, 216–19
   Cross-issue, 61
Cyberspace, 107–11

Division of labour, 13, 99, 170

Earth system, 61, 103
   -governance, 18, 61
   -science, 278
Earth System Governance Project, 1, 10
   Science and Implementation Plan, 10, 236, 247
Effectiveness, 81–5
Emergence, 1, 76
Environmental policy integration, 19, 189
Equity, 3
European Union, 14, 122, 186, 189–90
Evolution, 142–4, 304–6

Fragmentation
   Conflictive, 161, 267, 268
   Cooperative, 161, 268
   Synergistic, 161, 267, 268

Global goals, 79, 255, 287–9, 322
Goal-setting, 254, 258, 288–9
Governance
   Earth system. *See* Earth system governance
   Global, 4, 7–10
   Multi-level, 13–15, 323
   Regional, 15
   Through goals, 255–6, 269
   Transnational, 61–2, 76, 80–1, 85, 89–90
Government, 7–10, 212

Hierarchization, 10, 277–8, 322
Hierarchy, 277, 322
   Goal, 288
   Normative, 281
   Organizational, 284
High-level Political Forum (HLPF), 237, 260, 312

Inclusiveness, 260–3
Institutions, 5
   Intergovernmental, 37–8
   Private, 214–15
   Transnational, 75, 324

Integration
　　Policy. *See* Environmental policy integration
Interaction
　　Institutional, 18, 123
　　Vertical, 13
Interlinkages
　　Institutional, 120
International Environmental Agreements Database, 24
International environmental law, 16, 283, 307
International law, 224, 257, 281–2
International organizations, 64, 85, 166, 213–14
　　Intergovernmental organizations, 60, 243, 285
　　International non-governmental organizations, 22
Inter-organizational relations, 130, 306
Interplay management, 207–8

Justice, 3, 69, 172, 315

Legitimacy, 11, 86–9, 215
Level, 24
　　Macro, 4, 10, 158, 310
　　Meso, 138, 303
　　Micro, 10, 300

Methodology, 21–4, 229
Multilateral environmental agreements, 123, 149, 222, 285, 288

Nesting, 144, 311
　　Institutional, 160, 162
Non-environmental policy, 18, 183, 202
Non-governmental organizations, 41, 77–8, 323
Non-regime or non-governance. *See* Architectural void
Norms, 15–18
　　Crosscutting, 15–16
　　Overarching, 15–16

Orchestration, 233, 234–6, 241–5, 323
　　Orchestrator, 9, 233, 243–4
Organizational ecology, 127, 145, 150, 305
Overlap
　　Institutional, 126, 127, 146, 147

Paris Agreement. *See* Agreement, Paris
Performance, 10–13
Planetary boundaries, 20, 278
Policy
　　Concern, 211
　　Domain, 18–20
　　Integration. *See* Environmental policy integration
Polycentricity, 84, 302–3, 323
　　Polycentric order, 16
　　Polycentric system, 171, 267

Power, 38, 44, 70, 90
　　Shift, 167
　　Veto, 313
Principles
　　Common but differentiated responsibilities, 3, 312
　　Polluter-pays, 283
　　Precautionary, 15, 227
　　Prevention, 283
　　Resilience and non-regression, 283
Prioritization, 184, 188, 224–7, 280, 287–9
Priority, 20, 184, 281, 289
Public–private partnerships, 78–9, 237, 268
Punctuated equilibrium, 143, 305

Reflexivity, 247, 261
Reform, 6, 9, 306
　　Global governance, 276, 285–7, 310–15
　　United Nations, 6, 287
Regime complex, 10, 138–41
Regimes
　　International, 46, 121, 216, 217

Scale, 14, 164
Secretariats
　　Treaty, 7, 58, 61, 324
Self-organization, 169, 302
Social network analysis, 11, 127, 148–9, 150
Spill-over, 138
　　Negative, 140, 147, 166
　　Positive, 166
Sustainable development, 18, 79, 237–8, 288
　　Goals. *See* Global goals
Systematization, 279–80, 281–4

Trade agreement. *See* Agreement, Trade
Transboundary, 39, 46, 69, 75, 119
Transformation, 2, 10, 306–8
Transnational governance. *See* Governance, Transnational
Transnational network, 5, 80, 242
Transparency, 49, 87, 89, 313
Treaty, 42, 48, 166, 311

United Nations, 60, 213, 284
United Nations Environment Assembly (UNEA), 287, 306, 312
United Nations Environment Programme (UNEP), 166, 186, 217, 284, 312
United Nations General Assembly (UNGA), 9, 16, 257, 284
United Nations Security Council, 1, 280, 286

Voluntary certification programme, 76–8

World environment organization, 6, 186, 312